D0918760

Analysing Survival Data
from Clinical Trials
and Observational Studies

Analysing Survival Data from Clinical Trials and Observational Studies

Ettore Marubini and Maria Grazia Valsecchi

Institute of Medical Statistics and Biometry
University of Milan
Italy

JOHN WILEY & SONS

Chichester · New York · Brisbane · Toronto · Singapore

Other Wiley Editorial Offices

John Wiley & Sons, Inc., 605 Third Avenue,
New York, NY 10158-0012, USA

Jacaranda Wiley Ltd, 33 Park Road, Milton,
Queensland 4064, Australia

John wiley & Sons (Canada) Ltd, 22 Worcester Road,
Rexdale, Ontario M9W 1L1, Canada

John Wiley & Sons (SEA) Pte Ltd, 37 Jalan Pemimpin #05-04,
Block B, Union Industrial Building, Singapore 2057

Library of Congress Cataloging-in-Publication Data

Marubini, Ettore.
   Analysing survival data from clinical trials and observational
studies / Ettore Marubini and Maria Grazia Valsecchi.
      p.   cm.—(Statistics in practice)
   Includes bibliographical references and index.
   ISBN 0 471 93987 0
   1. Clinical trials—Statistical methods.   2. Survival analysis
(Biometry)   I. Valsecchi, Maria Grazia.   II. Title.   III. Series:
Statistics in practice (Chichester, England)
   R853.C55M37   1994   94-30632
   610'.72—dc20   CIP

British Library Cataloguing in Publication Data

A catalogue record for this book is available from the British Library

ISBN 0 471 93987 0

Typeset in 10/12pt Photina by Thomson Press (India) Ltd, New Delhi
Printed and bound in Great Britain by Biddles Ltd, Guildford and King's Lynn

To Mariola (E. M.)
To Adriano
and to my parents (M. G. V.)

Contents

4   Non-parametric methods for the comparison of survival curves   91

5   Distribution functions for failure time T 143

9 The study of prognostic factors and the assessment of treatment effect   **295**

10 Competing risks   **331**

11 Meta-analysis   **365**

Preface

In 1950 Berkson and Gage published their paper "Calculation of survival rates for cancer"[†]; this may sensibly be considered the starting point of the present wide application of methods of survival analysis in the biomedical field. In the last 25 years, methodology in the analysis of failure time has developed rapidly and has tended to find prompt application in different branches of medicine, particularly in oncology.

The present spread of personal computers and the availability of pertinent software of simple use exposes researchers with biological or medical backgrounds to the risk of uncritical use of the statistical techniques. On the other hand there are a number of biostatisticians cooperating with clinicians in planning clinical studies and in analysing their findings who are not specialists in survival theory. This book is intended for applied researchers involved in the analysis of survival data collected in clinical trials and observational studies. It aims to be a practical guide for medical research professionals with limited statistical knowledge, while also providing a fairly rigorous outline of the underlying statistical and scientific principles which will be useful to biostatisticians.

The book reflects our teaching experience of survival analysis methods in *ad hoc* courses sponsored by the Italian Region of the Biometric Society; the audience included both physicians with some experience in applying statistics, and biostatisticians without experience in the analysis of survival data. Chapters 1 to 4, 6 and 9 deal with basic methods of survival analysis which allow for the estimation and comparison of survival curves, with the Cox regression model, which is nowadays the most widely used method of regression analysis, and with the study of prognostic factors. We have tended in these chapters to keep the mathematical level to a minimum, and to place in sections identified by an asterisk (*) the topics requiring more advanced statistical background. In addition the examples are worked out in detail to enable the reader to check the computation procedures step by step. The remaining chapters need some knowledge of calculus and matrix algebra. The notation "log" indicates logarithms in base e throughout the book.

Chapter 1 serves as an introduction in which the main features of failure time data are discussed and some data sets which will be utilized in the following chapters are presented.

[†]*Proceedings of the Staff Meetings of the Mayo clinic*, **25**, 270–286, 1950.

Chapter 2 is not concerned with time failure in particular, but deals with general principles and some controversial issues in clinical trials, thus applying to any end-point. It aims at focusing attention on traps and pitfalls one can meet in planning controlled clinical trials or in interpreting reports on trial results.

Chapter 3 is devoted to the estimation of a survival curve. Both the product limit method and the life table approach are introduced together with the estimation of their standard errors and the construction of confidence limits and confidence bands.

Chapter 4 faces the problem of the comparison of survival curves by non parametric methods. The Mantel–Haenzel (M–H) procedure, in its original form for the two-groups comparison, is presented first, together with the estimator of the constant of proportionality of hazards and its confidence interval. Secondly, the extension of the M–H test to include strata is considered, and thirdly the Mantel–Byar procedure, making allowance for time-dependent covariates, is commented upon. Finally, the method for comparing more than two groups is outlined.

Chapters 5, 6, 7 and 8 are concerned with the use of regression models to study the effect of covariates (risk or prognostic factors) on survivorship. In particular, Chapter 5 serves as an introduction to the following chapters by showing the relationship between density, survival and hazard functions; furthermore, by resorting to the exponential distribution, it shows how to model the hazard function by means of covariates. Maximum likelihood estimators and their variances are discussed along with the likelihood ratio, the Wald and the Rao score tests. The relationship among these tests is shown by studying the curvature of the likelihood function. The Cox regression model is introduced in Chapter 6 both in its original form and by considering extensions allowing for stratification and time-dependent covariates. Validity assumptions to be fulfilled for the proper use of the Cox model are debated in Chapter 7. This includes a discussion of graphical methods and statistical tests for verifying the proportional hazards assumption, and the definition and use of diagnostics to assess goodness-of-fit. Chapter 8 extends the subject matter of Chapter 5, not only by considering distributions alternative to the exponential one in the class of proportional hazards models but also by presenting the class of accelerated failure time models and showing the relationship of these latter with the proportional hazards and the proportional odds models.

Chapter 9 can be divided into two parts. The first deals with problems related to the use of prognostic factors both for adjustment of treatment effect and for definition of risk groups. The second part aims at giving to clinical researchers a set of measures suitable for evaluating treatment effectiveness from a practical viewpoint along with suggestions for computing approximate confidence intervals.

In the last decade in the oncological branch of medicine, as well as in other branches, there has been an increasing trend towards analysing failure times of different, non-independent events. Chapter 10 discusses some statistical tools for obtaining unbiased curves of cumulative incidence in the presence of competing risks, for comparing two such curves and for applying regression analysis in order to study the role of prognostic factors in each type of event.

The final chapter deals with statistical methods suitable for performing meta-analyses. Owing to the impetus given to this area mainly by Tom Chalmers and Richard Peto, we thought it advisable to present the statistical procedures for handling both binary data and failure times. The fixed effects and random effects models have been presented by focusing more on concepts and examples than on the theoretical aspects. This is because we are convinced that clinical researchers should be familiar with the statistical assumptions of the models in order to be able to critically review results of meta-analyses, which are becoming increasingly adopted as a tool for the evaluation of treatment effectiveness.

Most of the methods presented in this book are implemented in sections devoted to survival analysis which are included in well-known statistical packages such as BMDP (1990), SAS (1991) and SPSS (1991). Their use in clinical research is quite common and requires basic knowledge of the package language and features. Researchers who are oriented more to epidemiological than to clinical applications may find it useful to refer to EGRET (1993) and EPICURE (1993). A more advanced knowledge of statistics and programming is required for the use of GLIM (1993) for survival data analysis. Software specifically developed to allow for non-basic types of analysis will be referred to through the text, when we are aware that it is available on request.

We wish to thank Prof. U. Veronesi, Head of the National Cancer Institute (INT) of Milan, Prof. C. Mangioni, Chairman of the Italian Group of Gynaecologic Oncology (GICOG), and Prof. G. Paolucci, Chairman of the Italian Group of Paediatric Oncology (FONOP), who kindly made available the databases for the examples on breast cancer, ovarian cancer and childhood leukemia. We are grateful to M. Gail, S. Green, J. Peto and P. Sasieni for helpful discussions and suggestions on various topics. We wish to acknowledge the valuable contributions of A. Manzari, D. Silvestri, P. Boracchi, M. Mezzetti and I. Cortinovis in checking the examples and reading the draft manuscript, and of D. Labadini for typing the manuscript. Finally, we are grateful to M. Healy and to other colleagues and collaborators for their support and encouragement to embark on this project.

Milano, April 1994   E. M.

                                                        M. G. V.

Series Preface

Statistics in Practice is an important international series of texts which provide direct coverage of statistical concepts, methods and worked case studies in specific fields of investigation and study.

With sound motivation and many worked practical examples, the books show in down-to-earth terms how to select and use a specific range of statistical techniques in a particular practical field within each title's special topic area.

The books meet the need for statistical support required by professionals and research workers across a wide range of employment fields and research environments. The series covers a variety of subject areas: in medicine and pharmaceutics (e.g. in laboratory testing or clinical trials analysis); in social and administrative fields (e.g. for sample surveys or data analysis); in industry, finance and commerce (e.g. for design or forecasting); in the public services (e.g. in forensic science); in the earth and environmental sciences, and so on.

But the books in the series have an even wider relevance than this. Increasingly, statistical departments in universities and colleges are realising the need to provide at least a proportion of their course-work in highly specific areas of study to equip their graduates for the work environment. For example, it is common for courses to be given on statistics applied to medicine, industry, social and administrative affairs and the books in this series provide support for such courses.

It is our aim to present judiciously chosen and well-written workbooks to meet everyday practical needs. Feedback of views from readers will be most valuable to monitor the success of this aim.

Vic Barnett
Series Editor
1994

1

The Scope of Survival Analysis

1.1 INTRODUCTION

In January 1662 the book *Natural and Political Observations upon the Bill of Mortality* was published in London by John Graunt. Thanks to this publication, John Graunt was granted admission to the Royal Society on 26 February of the same year. The book reported on registrations of births and deaths collected over a few decades in the registers of London parishes. It suggested a new scientific perspective: for the first time, death was regarded as an event which deserved collective study. Registered deaths were classified into homogeneous groups according to age, period, sex and cause of death.

Some years later, using data from the population of Breslau (now Wrocław in Poland) for the period 1687–1691, Edmund Halley devised the first life table. The table was very similar to those still used in demographic and actuarial studies.

The present state of the art in survival analysis is the product of a long process, which has undergone a great impulse in the last 60 years. Researchers from the biological, medical and epidemiological fields, as well as from industry, have stimulated this process with their scientific problems. Greenwood's article providing the formula for the standard error of the survival probability dates back to 1926. The Second World War strongly stimulated research on reliability and in particular on the "lifetime" of military equipment. Interest in this type of research continued after the war, especially in the electronics industry. Knowledge of techniques and materials used in the production processes led researchers in this field to formulate hypotheses on the density functions of the studied events. Consequently, particular emphasis was put on the formulation of parametric models (e.g. exponential, log-normal, Weibull), on the estimation of their parameters, and on the evaluation of their performance. The last 30 years have seen widespread application of methods of survival analysis to clinical data. In the clinical context, survival time is used to indicate not only "time to death" but time to any event, generally defined as failure (i.e. disease progression, metastasis or toxic event). The difficulty in finding empirical and theoretical evidence to support a particular family of failure time distributions stimulated the development of non-parametric methods. The seminal paper by Cox in 1972 on the proportional hazards regression model, whilst representing the natural evolution of standard non-parametric methods, disclosed further research on robust and efficient methods for the analysis of failure time data.

Survival analysis in the biomedical field deals with the following problems:

(1) estimation of failure time distributions;
(2) comparison of survival of different groups of subjects and estimation of treatment effect;
(3) prognostic evaluation of different variables, jointly or singularly considered, such as biochemical, histological and clinical characteristics.

1.2   CHARACTERISTICS OF SURVIVAL DATA AND PROBLEMS IN SURVIVAL ANALYSIS

The characteristics of survival data will be described through some examples. In particular, examples which come from our own experience in medical research and which will be used later through the text are accompanied by a table and a concise description of design, aims and data.

1.2.1   Characteristics of survival data

Survival data arise when the aim is to study the time elapsed from some particular starting point to the occurrence of an event. The starting point of observation is usually a medical intervention such as first diagnosis of a given disease, a surgical intervention or the beginning of a treatment in a clinical study. In epidemiological studies, the starting point may be birth or the beginning of exposure to some risk factor. The terminal event may be death or a prespecified event of interest. Survival analysis is useful whenever the researcher is interested not only in the frequency of occurrence of a certain type of event, but also in the time process underlying such occurrence.

The GISSI-2 study (1990) is a multicentre randomized trial planned to compare streptokinase (SK) and alteplase (rt-PA) in the treatment of evolving acute myocardial infarction (AMI) (figure 1.1). The trial's 2×2 factorial design also addresses the question of the role of subcutaneous heparin (HEP) after thrombolysis.

In the protocol of this trial, recruiting patients in Italian coronary care units, in order to pursue the primary goal of the study, a combined end-point was adopted: mortality and extensive left ventricular damage observed in the in-hospital period. This is an example of an end-point for which time appears to be relatively unimportant, and therefore the authors properly analysed the

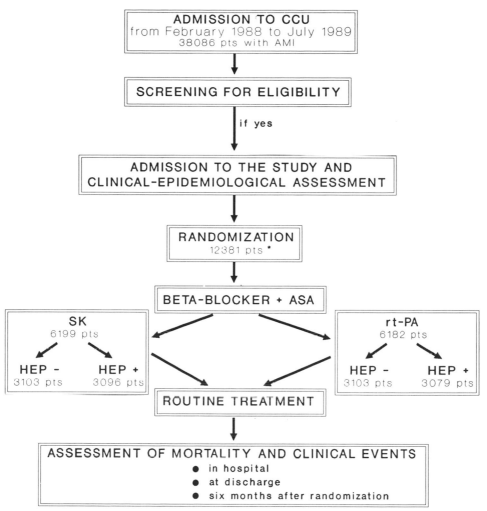

Figure 1.1   GISSI-2 international protocol for treatment of acute myocardial infarction (AMI): design and numbers of accrued patients.

results in terms of proportion of patients having the event of interest in the four treatment groups. In May 1988 a decision was taken to investigate mortality alone in a larger study population. To achieve this aim the International Study Group (1990) was formed and from October 1988 patients were recruited in other countries according to the same factorial design. The primary aim of the international trial was to compare the effects of SK and rt-PA both on the in-hospital and on the six-month mortality. Implicitly, death attributable to a specific cause, myocardial infarction, is the end-point in this analysis and time elapsed between randomization and death becomes the main response variable.

The time until the occurrence of an event will be represented by the random variable (r.v.) T. Usually, even if the event is not death, the r.v. T is called the "survival time"; a broader term which is often used for T is "failure time". More generally, T is just a non-negative variable. For example, in a reliability experiment conducted in a wind tunnel on some mechanical aircraft components, the experimenters studied the distribution of failures as the velocity of the wind increased from a starting point. The r.v. T was defined as the velocity at which a failure occurred and methods for survival analysis were appropriately used to study its distribution (Halperin, 1952).

1.2.2   Censoring

A distinctive characteristic of survival data is that the event of interest may not be observed on every experimental unit. This feature is known as censoring. Censoring can arise because of time limits and other restrictions depending on the nature of the experiment.

Let us consider again the experiment on N pieces of aircraft equipment. To carry over the experiment until all pieces fail could be too costly, and uninteresting after a certain velocity is reached. Therefore, the experiment could be modified in two ways: first, the experimenter terminates the observation when some prespecified velocity T^* is reached in the wind tunnel. This induces the so-called "type I censoring", in which the velocity at break-up will be known exactly only for pieces which break before T^*; and second, the experimenter terminates the observation when a number r of failures is reached, with $r < N$. This determines the so-called "type II censoring". In this design, the number of failures is not a random variable.

Both types of censoring designs, or generalizations of them, may be usefully adopted, provided that N, r and T^* are adequately chosen and appropriate methods are adopted for the analysis (Lawless, 1982).

In clinical and epidemiological studies, censoring is mainly caused by a time restriction, and is therefore of type I (Lagakos, 1979). For example, in clinical trials on chronic diseases, ethical, scientific and economic reasons suggest that the study continue until a prespecified time point (cut-off date). Then, the time to the event of interest is known precisely only on those subjects who present the event before that time point. For the remaining subjects, it is only known that the time to the event is greater than the observation time. This is referred to as "administrative censoring" and the incomplete data are called "right censored"; the subjects providing them are called *withdrawn alive* from the study. However, in medical and epidemiological studies, the censoring process is usually more complicated than that generated in a laboratory setting. Causes of censoring other than the predetermined duration of the study are often in effect and seldom can be precisely determined. Most commonly, enrolled subjects may be unwilling or unable, for some reasons, to continue participating in the study and providing follow-up information. These subjects are called *lost to follow-up* or *drop-outs*. As those withdrawn alive, they give incomplete data, which are also "right censored".

Unlike the laboratory setting, where all pieces can be put on test at the same moment, clinical and epidemiological studies have a recruitment period. Entries in the study are scattered during this period and consequently, for all subjects withdrawn alive, censored time is not necessarily the same at the cut-off date, when the study is closed or the analysis is performed. We shall distinguish between calendar time and time of observation since entry in a study, i.e. follow-up time, respectively represented in figure 1.2(a) and (b) for the same four individuals. The calendar time scheme reflects study planning, while the follow-up time scheme is oriented towards data analysis.

In figure 1.2(a), for each subject, the date of entry, during the enrolment period of one year from $1/1/87$ to $1/1/88$, corresponds to the dot. The study is closed after one year from the termination of the enrolment, at the cut-off date of $31/12/88$. A minimum planned follow-up time of one year is foreseen for all subjects entered, while subjects have potential total follow-up times (the sum of the continuous and dotted lines) which differ because of the scattered entry.

In figure 1.2(b) each individual's entry is at the origin of the time axis, $t = 0$, that represents the starting point of the observation. The observation ends when the subject dies (D) or drops out (L) or when the study terminates (W). Thus the actual observation time may be less than the minimum follow-up guaranteed by design if a patient is lost, as for patient number 2.

Survival data are best represented by a pair of variables (T, δ), where T is time since entry into the study and δ is an indicator of failure, assuming values of $\delta = 1$ if the event of interest is observed and $\delta = 0$ if the time is censored. Formally, suppose U is the true survival time we ultimately wish to investigate, but we cannot always observe, and V is the potential censoring time. Then, the observed time is:

$$T = \min(U, V) \quad \text{and} \quad \delta = 1 \quad \text{if} \quad U \leqslant V \quad \text{or} \quad \delta = 0 \quad \text{if} \quad U > V \qquad (1.1)$$

that is, $T = U$ only when the observation is uncensored. A sample of N individuals gives the set of observations $\{(t_i, \delta_i), i = 1, 2, \ldots, N)\}$.

While in clinical trials the beginning of observation commonly coincides with the beginning of "exposure" to treatment, this may not be the case in epidemiological studies. For instance, Doll *et al.* (1970) studied a cohort of workers at a nickel refinery with the aim of evaluating the association between exposure to nickel and risk of lung and nasal sinus cancer. The cohort was identified using the refinery pay-sheets and included men who had been employed in the years

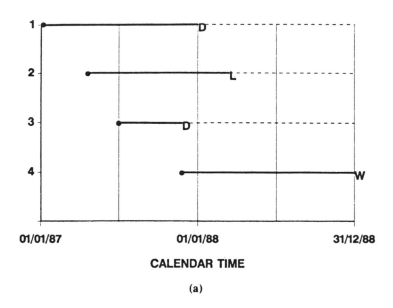

CALENDAR TIME

(a)

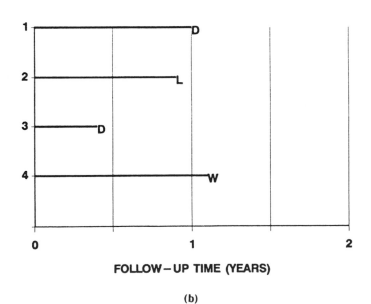

FOLLOW–UP TIME (YEARS)

(b)

Figure 1.2 Life experience in terms of (a) calendar time and (b) follow-up time of the same four individuals. Entry date is represented by the '●'; D = dead, L = lost to follow-up, W = withdrawn alive.

between 1934 and 1949, with at least five years of employment. The 968 subjects in this cohort were followed up until 31 December 1981: by that time, 788 deaths had occurred, for which the date and cause were recorded from death certificates, 18 subjects were lost to follow-up and the remaining subjects were withdrawn alive at the cut-off date above. In this type of data, the entry into the study does not coincide with the beginning of exposure to nickel, which dates back to the time of first employment in the refinery. These data are said to be "left truncated" since observation begins some time after the time origin which is of relevance for the aim of the study. Instead of analysing survival from study entry, it is advisable to account for left truncation, that is for the time elapsed from first exposure to study entry. An analogous problem could arise in studying the natural history of a disease from a hospital database in which patients' entry is at the first contact with the hospital and not necessarily at first diagnosis of the disease.

1.2.3   Problems in survival analysis

Methods for survival analysis assume that in a sample of N individuals, observations $\{(t_i, \delta_i), \ i = 1, 2, \ldots, N\}$ are independent. Furthermore, another assumption is typically made in methods for censored survival data, that the random variables U and V are independent. In practical terms, this means that the survival experience of censored individuals can be accurately estimated by using the data on the uncensored patients available at the time points following censoring. Conversely, censoring carries no prognostic information about subsequent survival experience; in other words, those who are censored are no more or less likely to die than the rest of the individuals in the sample. The assumption of independence is justified with administrative censoring, when V is the time due for withdrawal. In this context, the occurrence of censoring is only due to the scattered entry and to time limitation in the study.

In other situations, however, it is not as evident that censoring is non-informative. This is the case, for example, in clinical trials in which some patients may drop out of the study (i.e. be lost to follow-up) because of the occurrence of critical events related to drug toxicity or to the worsening of clinical conditions. For situations in which censoring is informative, censored survival models can be developed; however, these latter make parametric assumptions on the shape of the distribution of (U, V) that are often untestable. In practice, the presence of informative censoring should be prevented with careful study design and conduction. When the number of patients lost to follow-up is small, very little bias is likely to result from applying methods based on non-informative censoring.

A particular type of informative censoring, which is worth mentioning here and which will be considered later in the book, is that related to competing risks. This type of censoring is encountered when it is desired to estimate the distribution of time to failure from a particular cause, and failure times from other causes are considered as censored observations for the cause of interest. As an example, consider the data coming from a randomized clinical trial comparing the Halsted mastectomy with the quadrantectomy plus axillary dissection and radiotherapy in early breast cancer patients (Veronesi *et al.*, 1981). The trial randomized 701 patients who were diagnosed as having

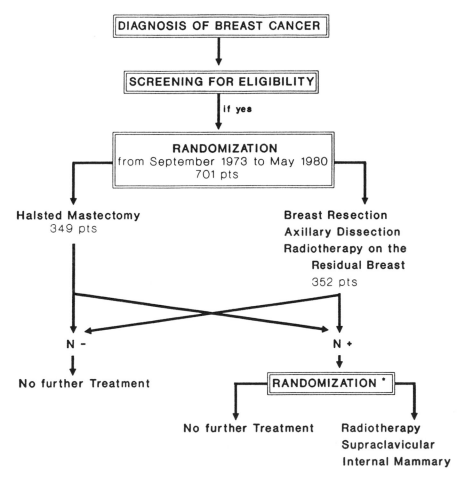

(*) From January 1975 this randomization was quitted: all patients with positive nodes were treated with 12 cycles of polychemotherapy (cyclophosphamide - methotrexate - fluoracil).

Figure 1.3   Randomized trial on radical (Halsted mastectomy) vs conservative surgery in early breast cancer, conducted at the Istituto Nazionale per lo Studio e la Cura dei Tumori (INT) of Milan: design and numbers of accrued patients.

a breast cancer less than 2 cm in diameter without palpable lympho nodes, at the Istituto Nazionale per lo Studio e la Cura dei Tumori (INT) in Milan (figure 1.3).

The researcher is interested here in studying time free from different types of competing events such as distant disease (metastases), ipsilateral breast tumor, controlateral tumor and death.

Another aspect faced by methods of survival analysis is the need to account for heterogeneity in the data. Most often, the individuals being studied differ with regard to many covariates which could in principle be related to survival. Heterogeneity is present even if extreme features of individuals are avoided by fixing some eligibility criteria for entry into the study. Classical theory of linear regression and least-squares estimation does not apply to censored data, and parametric methods have been specifically extended to accommodate censoring. By using these methods, the estimate of treatment effect may be adjusted for

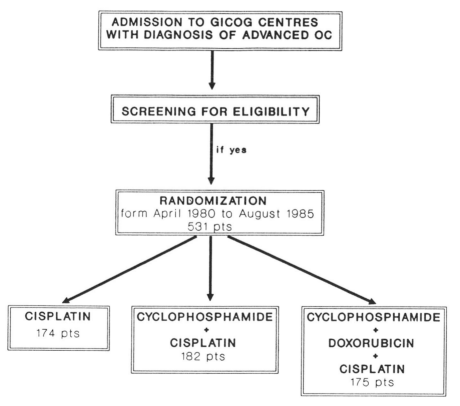

Figure 1.4   A randomized trial on treatment of advanced ovarian cancer (OC) conducted by GICOG centres in Italy: design and numbers of accrued patients.

patient and disease characteristics, usually called covariates. Furthermore, the assessment of the prognostic role of such characteristics is possible.

In a randomized trial on advanced ovarian cancer patients, the primary objective was to compare the effect on survival of three different treatments: cisplatin, cyclophosphamide/cisplatin and cyclophosphamide/doxorubicin/cisplatin. A secondary objective was that of comparing the effect of the three treatments on progression-free survival. Residual tumor size after first surgery, and Karnofsky index, i.e. an index grading the patient's general clinical condition, were found to be relevant predictors of survival (GICOG, 1987). Figure 1.4 gives design and numbers of patients accrued in the trial.

A specific feature of survival analysis is that covariates may be time-dependent, that is their value may vary during the period of observation. Age of an individual and calendar period of observation clearly have a time-dependent nature. This must be accounted for in long-term studies, such as cohort studies, when it is desirable to stratify the mortality experience into age groups and chronological subperiods, of five years' length, for example. Clinical studies do not commonly need to consider these variables, since they cover a relatively short period and aim at estimating treatment effect on survival rather than the mortality process itself. However, other time-dependent variables may come into play. Consider, for example, the study on breast cancer outlined in figure 1.3 and suppose we wish to study whether the risk of developing distant disease varies in relationship to the occurrence of an ipsilateral breast tumor, i.e. local disease. Then, while we are observing time from surgery to metastasis, we account for the moment in which the status of the patient varies from free of disease to affected by ipsilateral tumor. In other situations, treatment itself may best be represented by a variable changing value during the observation time. In a national study on treatment of childhood acute lymphoblastic leukemia, conducted by AIEOP (Associazione Italiana Ematologia Oncologia Pediatrica) transplantation is included in the therapeutic programme as an option for subjects presenting a recurrence. To study the effect of transplant on survival, as compared to standard treatment, we consider a subject as belonging to the control group (chemotherapy alone) until he finds a donor and is transplanted.

2

Randomized Clinical Trials: General Principles and Some Controversial Issues

2.1 INTRODUCTION

Chronic diseases including cancer are, nowadays, the main concern of medicine in developed countries. Chronic disease is a complex event characterized by several features. Among these are a multifactorial aetiology, nearly nonspecific and often endogeneous, and a long-term course with continuous interactions between the host biological system and the surrounding environment. This makes it difficult to trace the "natural history" of chronic diseases and to develop appropriate therapeutic strategies.

Let us now focus our attention on the problem of evaluating the effectiveness of a treatment, whether pharmacological, radiological or surgical; as regards the strength of evidence concerning efficacy of treatment, Green and Byar (1984) suggested a hierarchy of different types of study in current use in medicine as reported in table 2.1 Each of these types of study has been and is still useful in developing medical knowledge, but we need to consider which type of study should inspire the greatest confidence in assessing treatment effectiveness.

If we think of the table as an eight-level pyramid and proceed from the top to the bottom, the base on which we can build our conclusion becomes broader and broader.

Regardless of the amount of information collected, the types of study mentioned in the first five levels share a common characteristic: they imply that the investigator is a careful observer of biological and clinical events capable of

Table 2.1 Hierarchy of strength of evidence concerning effectiveness of treatment (from Green and Byar, 1984, modified).

1. Anedoctal case reports
2. Case series without controls
3. Series with literature controls
4. Analysis using computer databases
5. "Case-control" observational studies
6. Series based on historical control groups
7. Single randomized controlled clinical trials
8. Meta-analysis of randomized controlled clinical trials

picking up statistical associations and identifying possible cause–effect relationships.

From the sixth level downwards the pyramid refers to clinical trials. By this we mean a rigorously designed and executed scientific experiment which generates clinical data with the aim of evaluating one or more treatments on a patient population. Unlike the types of study mentioned in the previous five levels, the trial implies that the clinical investigator has control of the process by which a treatment is assigned to a patient.

The sixth level concerns historically controlled trials and we consider here the best of this kind of study, in which all consecutive patients referred to a given institution and meeting proper criteria specified by the study protocol receive certain treatment for a preselected period of time. In the following period all similarly selected patients receive the next treatment of interest. It is sensible to think that these studies should be less subject to bias than the observational studies previously mentioned if uniform diagnostic criteria and consistent evaluation of outcomes are utilized along the two time periods (Gehan and Freireich, 1974). In fact the choice between the two treatments depends only on the day on which the patient is admitted to the hospital rather than on his or her prognostic factors. Nonetheless the investigator must be very cautious in interpreting the results obtained and aware that even if no source of bias can be identified, bias may exist without any apparent explanation.

An example of the risks of a practice involving use of data from one trial as historical controls for the next one in breast cancer was published by Micciolo *et al.* (1985). The authors resorted to data collected in three consecutive randomized trials performed at the Istituto Nazionale per lo Studio e la Cura dei Tumori of Milan. The sequence of the trials with treatments, number of patients and period is reported in table 2.2.

The results of the final evaluation of the relative effectiveness of the treat—ments under study were investigated using both randomized and historical controls.

Table 2.2   Sequence of the trials with treatments, number of patients and period (reproduced by permission of the publisher from Micciolo *et al.*, *Controlled Clinical Trials*, **6**, 259–70, 1985. © 1985 by Elsevier Science Inc.).

Trial	Control group	Treatment group	Period
SURG[a]	Halsted radical, 200	Extended radical, 181	1964–68
AD1[b]	Surgery, 179	Surgery + CMF 12/1, 207	1973–75
AD2[c]	Surgery + CMF 12/2, 211	Surgery + CMF 6, 216	1975–78

[a]SURG trial refers to the subset of 381 patients, among 716, with histologically positive axillary lymph nodes.
[b]AD1 trial aimed at evaluating the effect, after surgery, of adjuvant chemotherapy with CMF (cyclophosphamide, methotrexate, fluouracil) given for 12 cycles (CMF 12/1) in patients with positive axillary lymph nodes.
[c]AD2 trial aimed at reducing the duration of adjuvant treatment without decreasing the therapeutic effect; thus the patients were randomized to receive CMF either for 12 cycles (CMF 12/2) or for 6 cycles (CMF 6).
In all trials the outcomes were: disease free survival (from radical mastectomy to any first failure evidence) and overall survival time.

Although the eligibility criteria did not change from trial to trial, the results obtained by means of both univariate and multivariate analyses showed that the baseline features of the breast cancer patients accrued in the three consecutive trials covering a period of some 15 years changed considerably, perhaps because of the effect of several different factors which could not be precisely identified. Furthermore, the results obtained by using historical control data, after adjustment for the differential composition in the baseline characteristics, enabled the authors to make evident a biased conclusion in favour of the new treatment tested, which, in a historically controlled study, is always the last one. Clinical knowledge progressively accumulated during the trials appeared to imply subtle improvements in the care of patients, tending to influence their survival and to bias the comparisons in the historically controlled studies. The authors concluded that with the improvements expected nowadays from treatments available for breast cancer, "...no alternative to the use of randomized concurrent controls can be accepted."

The writings of the French physician P.C.A. Louis (1787–1872) appear to have had an important role in beginning the modern clinical trial. Influenced by Laplace's contributions to probability theory, in 1836 Louis published the results of his researches on the effect of bloodletting in treatment of pneumonia. He performed a clinical trial in which three groups of patients were formed by deliberately delaying the onset of bloodletting. The comparison of the populations of survivors in these groups enabled Louis to demonstrate that bloodletting was not beneficial, clashing with the current opinion of his colleagues. This trial may be sensibly considered a landmark in the development of medicine through a scientific approach. For this approach Louis coined the term "numerical method" and his views are even today of great concern. Being aware of the problems arising in applying his numerical method, Louis stated: "The only reproach which can be made to the Numerical Method is that it offers real difficulties in its execution. For, on the one hand, it neither can, nor ought to be applied to any other than exact observations and these are not common and on the other hand, this method requires much more labour and time than the most distinguished members of our profession can dedicate to it. But what signifies this reproach, except that the research of truth requires much labour and is beset with difficulties."

These words are of great help in allowing us to understand why medical science had to wait until the late 1940s for clinical trials to be included among its tools of current use. It was in fact in 1946 when the Medical Research Council published the results of the trial carried out to evaluate the effect of streptomycin in tuberculosis of the lung. Thanks to the pioneering work of Bradford Hill, this trial incorporated the principles of randomization and replication which were being applied by researchers in agriculture, biology and industry through the work of Fisher, whose book *Design of Experiments* had been published some ten years earlier.

Randomization is the only tool that researchers have in order to ensure that groups of subjects being assigned different treatments differ only by chance in their characteristics and prognosis.

The base of the pyramid of table 2.1 is made up of meta-analysis (overview) of clinical trials, that is the process of combining results of different trials replicated with similar, though not identical, protocols, by means of statistical methods. This subject will be discussed in detail in Chapter 11.

Among the many books dealing with fundamentals of clinical trials available today we recall, without pretending to be exhaustive, those of Schwartz, Flamant and Lellouch (1970), Pocock (1983), Friedman, Furberg and DeMets (1985) and Meinert (1986). The reader is referred to these textbooks for the theoretical background and the practical issues in the design and conduct of clinical trals and in the analysis of their results. The present chapter has humbler aims, namely to introduce some general principles and to discuss some problems that, in spite of the considerable improvements in the use of statistical techniques for clinical trials in recent years, remain a common cause of difficulty in the design of clinical trials and in the interpretation of their results.

2.2 SELECTION OF PATIENTS AND TRIAL DESIGN

Suppose that the trial aims at answering a purely biological question, for example to assess if a drug is effective in reducing the solid mass of, say, advanced breast cancer; the study population would ideally be selected to be homogeneous so as to minimize the between-patients variability and to provide a sensitive and quick outcome. The *explanatory* approach (Schwartz *et al.*, 1970) one adopts in this trial aims at clarifying the state of knowledge concerning the *efficacy* of the drug tested without immediately extending the conclusions to any target population. However, a number of trials carried out nowadays aim at evaluating major therapies (such as beta-blockers in myocardial infarction) or treatment policies (such as demolitive or conservative surgery followed by adjuvant chemotherapy or hormonotherapy in breast cancer) and are frequently organized at a national or international level. In planning these trials it is wise to adopt the *pragmatic* approach by which "... the appropriate type of patient is determined in general terms by the population to which the results of the trial will be applied" (Schwartz *et al.*, 1970). Accordingly the basic scheme of the simplest clinical trial, completely randomized for the comparison of two treatments, is as shown in figure 2.1. The new therapy, say B, is usually compared with the best standard treatment (A) for that particular disease; in the example of figure 1.3 the standard surgical treatment was mastectomy. However, sometimes it is desirable to assess whether the new treatment has any effect at all. In this situation, in order to avoid any possibility that the patient's knowledge of his treatment may bias the comparison, the reference group A receives a *placebo* or dummy treatment which is like the treatment to be tested except for the absence of the active principle (blind study). The well-known

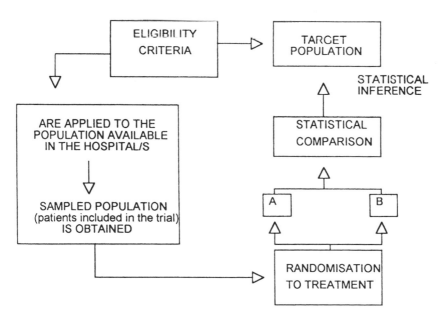

Figure 2.1 Basic scheme of a completely randomized controlled clinical trial (pragmatic approach).

"placebo effect" is due to the fact that, because of suggestion, the patients assuming the placebo often respond better than those receiving no treatment at all. Furthermore,in order to avoid a possible bias due to expectation of the outcome, even the observer may be "blinded" as regards the treatment assigned (double-blind study).

The eligibility criteria enable the investigator to restrict accrual to patients whose diagnosis has been properly made, who do not show contra-indications for any of the treatments being administered, and who have a sensible expectation of continued follow-up for the end-points under study. However, it is sensible not to adopt too restrictive criteria in order to reach conclusions which may have an important impact on medical practice. An in-depth discussion on the implications of wide or narrow eligibility criteria for selecting patients for clinical trials is given by Yusuf *et al.* (1990).

The pragmatic approach is the natural background of the large, simple clinical trials advocated by Richard Peto. These have been carried out in the last 15 years in cancer and cardiovascular diseases where earlier trials had excluded large advantages for new treatments and where small differences in outcome are still clinically and socially relevant. As an example let us consider the GISSI-2 trial mentioned in the previous chapter; the authors inform us that about one-third of the patients with acute myocardial infarction (AMI) admitted to 233 among the 250 Italian coronary care units were eligible for the trial. Since at the end of the trial 8.8% hospital mortality was observed in the study population as a whole, the proven beneficial impact of the acute treatments for AMI can be predicted in populations whose AMI incidence rates are known.

In multicentre trials the difficulty of convincing participants to randomize patients to more than two or three treatments, and the sound policy of keeping the protocol as simple as possible in order to stand a good chance of completing the study, has led the investigators quite often to adopt simple two-armed randomized designs. Sometimes a minimum amount of extra effort might allow the investigator to resort to the most powerful designs such as, for example, a 2×2 factorial design. Only factorial designs, in fact, enable the interaction between two or more treatments to be evaluated beside the main effects of each of them. A 2×2 factorial design was adopted in the GISSI-2 trial: half of all patients were allocated randomly to receive streptokinase (SK) or altepase (t-PA) (thrombolytic agents) and starting 12 hours after beginning SK or t-PA infusion they were randomized to receive or not to receive heparin. After observing that the interaction between thrombolytic agents and heparin was statistically not significant, the authors were able to compare the benefits and risks of the two thrombolytic agents, and to evaluate the effect of heparin. These two latter conclusions are as reliable as those which would have been obtained from trials of equal size carried out separately for each clinical question. In nontechnical papers, Byar and Piantadosi (1985) and Byar (1990) introduced methodological issues for applying factorial designs in clinical research, whilst Stampfer *et al.*

(1985) thoroughly discussed pros and cons of the use of a 2×2 factorial design in a large preventive clinical trial.

2.3 RANDOMIZATION

Randomization is the knot of the scheme in figure 2.1; as pointed out in section 2.1 it aims at ensuring that at the outset the two groups A and B differ only by chance. The basic randomization procedure is referred to as *simple* or *unrestricted* randomization. This is equivalent to assigning treatment A or B to any patient by tossing a fair coin, with equal probability ($p = \frac{1}{2}$) of showing either side. Several clinical investigators feel uncomfortable using randomization and the main arguments against its use seem to be as follows:

(1) Clinical judgement can secure a better matching of patient baseline features than if the two groups are obtained by random allocation. However, as a matter of fact randomization does not necessarily produce a perfect balance of the known characteristics in the groups. It is merely the unique way to preventing bias in the composition of the trial groups and of properly founding the assumption of a random-error term underlying quantitative statistical measures such as tests of significance and confidence intervals (see Appendix A2.1). Anyway, the probability of covariates imbalance decreases as the sample size increases and is usually of little influence.

(2) Ethical considerations do not allow a physician to choose a patient's treatment by the toss of a coin. But it has been said that "Frequently we have no acceptable evidence that a particular established treatment does benefit patients and, whether we like it or not, we are then experimenting upon them. As F.H.K. Green has pointed out, 'where the value of a treatment, new or old, is doubtful, there may be a higher moral obligation to test it critically than to continue to prescribe it year-in-year-out with the support merely of custom or wishful thinking.' It is certainly not always recognized that it may be unethical to introduce into general use a drug that has been poorly or inadequately tested. The ethical problem is, indeed, not solely one of human experimentation; it can also be one of *refraining from human experimentation*. Further, in withholding a new treatment from some patients it is well to remember that all the risks do not lie on one side of the balance. What is new is certainly not always the best and, as the history of antibiotics and hormones and of many other modern drugs has shown, it is by no means always devoid of serious danger to the patient. It may, therefore, be far more ethical to use a new treatment under careful and designed observation, in comparison with patients not so treated, than to use it

widely and indiscriminately before its dangers as well as its merits have been determined" (Bradford Hill, 1977).

A common feature is shared by nearly all papers reporting the results of clinical trials: the first tables and/or figures present the baseline characteristics of the case-series in terms of averages and standard deviations for continuous variables (age, laboratory findings, weight) or of absolute or relative frequencies for categoric variables (sex, stages of the disease, number of metastatic lymph nodes). Furthermore the similarity of the treatment groups with respect to each variable is tested by means of the t test, distribution free tests or chi-square test. The statistically not significant results usually obtained lead the authors to conclude that the groups are comparable and that the baseline variables do not need to be considered further. The inappropriateness of this approach was enlightened by Altman (1985) who stated: "Statistical significance is immaterial when considering whether any imbalance between groups might have affected the results. It is wrong to infer from the lack of statistical significance that the variable in question did not affect the outcome of the trial. Such an interpretation is unwarranted and potentially misleading". In our experience the alternative approach Altman suggested is extremely helpful: "In most trials the groups will be very similar. When there are one or more variables with known or suspected prognostic importance which are not very closely balanced between the treatment groups, there are two possible approaches. The prognostic importance of the variable(s) in question can be investigated by studying the relation with the outcome variable, preferably within treatment groups. Unfortunately, here too the use of significance tests may be unhelpful. It is the strength of the association rather than the significance level (which also depends upon sample size) which is of importance. The alternative approach is to compare the results from analyses of the trial both with and without allowance being made for the variable(s) in question. If the two results are essentially the same, this indicates that the simple comparison of treatment groups is reasonable. If they differ, then the imbalance was important and adjustment beneficial."

In pragmatic trials the sampled, and coherently, the target populations of figure 2.1 are usually heterogeneous with regard to known prognostic factors. In order to ensure reasonable balance between treatment groups with respect to these variables, stratified randomization could be adopted. On the basis of one or more prognostic factors, relatively homogeneous strata are identified and patient assignment to treatment groups is then accomplished within each stratum. It is widely accepted that stratifying for more than two or three factors can be detrimental to a trial and that, anyway, stratification complicates the management of a trial and thus improves the chances for something to go wrong. On the other hand in large studies simple randomization will, on average, provide allocation of patients to the different treatment groups which are well balanced with respect to the joint distribution of the prognostic variables. Furthermore

Peto *et al.* (1976) argued that if during the analysis the patients are subdivided into a few strata defined retrospectively from the factors which really affect prognosis, "... there is hardly ever need for stratification at entry in large trials". On average, the loss of information in applying retrospective stratification instead of pre-stratification is rather minor (Mantel, 1984). However in small trials (less than 200 patients overall) some limited stratification may be desirable to prevent the occasional badly skewed design, and in multicentre trials, regardless of size, it is sensible to stratify by centre, thus allowing for a major source of heterogeneity.

A common concern is that unrestricted randomization, with or without stratification, may by chance produce a treatment imbalance, thus affecting the power of statistical tests. This event is prevented by balancing the group sizes in successive blocks of patients by means of the random permuted blocks design simply called blocked randomization. For example, in the case of two treatments A and B arranged in blocks of four, the first three blocks could be ABAB, BBAA and ABBA, whilst if the block size is six, the first three blocks could be ABABBA, BAABAB and BAABBA. The point is that the block lengths should be neither too long nor too short. If the block length is long compared with the sample size, one takes the risk of not completing the block, thus missing the objective; this may happen in a multicentre trial where randomization is stratified by centre and the block is longer than the minimum sample size a given centre can accrue. On the other hand a too short block length has the drawback that the clinician may be able to predict the assignment for patients allocated at or towards the end of each block, nullifying the ignorance requirement. A sensible compromise consists in using blocks of variable length unknown to the clinician, that is, in alternating blocks of length ranging, for instance, from four to eight.

Blocked randomization is not only useful in balancing the totals of patients assigned to different treatments. It also prevents undesirable treatment imbalances at different times of the recruitment period so that, if the baseline characteristics of patients change during the enrolment time, the chance of imbalance in baseline characteristics is also reduced.

In staggered-entry trials, other types of restricted randomization procedures have been introduced to reduce, at any stage of the trial, the rsik of imbalance in the number of patients assigned to different treatments. Efron (1971) introduced the biased-coin design which was subsequently modified by Wei (1978). The idea behind this is briefly outlined here. Suppose that in a trial some strata are defined on the basis of relevant prognostic characteristics and/or by centre. Furthermore, as commonly happens, patients' entry is serially scattered over time and the total number of patients who end up in each stratum cannot be known until recruitment is completed. For each new patient to be randomized in the corresponding stratum a biased coin is tossed if, at that point of recruitment, more patients have been assigned one treatment than another. The biased coin will have a probability higher than $\frac{1}{2}$ to show the side corresponding to the treatment under-represented. If treatments are balanced in these strata when the new patient has to be randomized, a fair coin is used.

For an exhaustive discussion of procedures and properties of randomization the reader is referred to the series of papers published in 1988 in *Controlled Clinical Trials* (pp. 287–382).

2.4 STATISTICAL INFERENCE

In medical statistics courses one learns that we collect representative samples of subjects, make measurements or observations on them, and then, by the statistical analysis of data, draw inferences about the populations from which the samples were selected. On the contrary a clinical trial is carried out in one or more hospitals by accruing all or a subset of the patients admitted during a given time period. Therefore one may wonder whether in the absence of a formal process of sampling from predefined populations, statistical inference is still valid. Following Armitage (1971) one can be reassured "... by arguing that the observations are subject to random, unsystematic variation which makes them appear very much like observations on random variables. The population formed by the whole distribution is not a real, well-defined entity, but it may be helpful to think of it as a hypothetical population which would be generated if an indefinitely large number of observations, showing the same sort of random variation as those at our disposal, could be made. This concept seems satisfactory when the observations vary in a patternless way. We are putting forward 'model', or conceptual framework, for the random variation, and propose to make whatever statements we can about the relevant features of this model, just as we wish to make statements about the features of a population in a strict sampling situation."

In the clinical trial context we can think of statistical inference as a two-step procedure. Because of random allocation of treatments the difference between the two group means behaves like the difference between two means of random samples drawn from the same population if treatments are alike, or from different populations if treatments have different effects. The sampling theory of differences is thus directly relevant and enables us to do the first step by a formal significance test, namely to assess whether treatments A and B differ in the N patients accrued in the study.

The second step implies generalizing the observed results to the invoked population from which those N patients are thought to be drawn, i.e. the target population. This is a process of abstraction from space- and time-specific empirical measurement of a treatment effect in a trial to the assessment of a treatment effect in an abstract disease. This scientific generalization is not based upon statistical formalism but is done only resorting to the *a priori* defined eligibility criteria and to a detailed description of the patient baseline characteristics which are judged relevant to the generalization itself.

2.5   RANDOMIZATION AND "INTENTION TO TREAT" PRINCIPLE

In order to face problems arising in the analysis of the results of clinical trials, it is helpful to expand the central part of figure 2.1 as shown in figure 2.2 in which patients treated with A (or B) are divided into two subsets according to whether or not they completed the assigned treatment.

The two approaches routinely adopted to analyse data collected in a clinical trial are defined by the objectives the trial pursues. The first one, proper to the explanatory context (Schwartz *et al.*, 1970), focuses on "clinical efficacy", that is, it aims at establishing the extent to which a treatment does more good than harm to patients who take it as counselled. Efficacy analysis compares findings in group (1) with those in group (3). In relatively small trials carried out in a laboratory-like setting, one can avoid patients falling in groups (2) and (4) by making every effort: "... to prevent noncompliers from entering the trial at all. This can be accomplished by asking for volunteers, then putting these volunteers through a 'faintness of heart' period in which they are asked to comply with a regimen at least as demanding as the one that will be used in the trial. Only those patients who meet rigorous standards of compliance in the pretrial period should then be permitted to enter the study" (Haynes and Dantes, 1987).

The second approach, proper to the pragmatic context, focuses on evaluating the effectiveness of treatment, i.e. the extent to which a treatment: "... when deployed in the field, does what it is intended to do for a defined population" (Last, 1988). In this context the size of groups (2) and (4) is usually not negligible and hence the above-mentioned comparison may be biased by differences in patients excluded from one treatment group as opposed to the other. In fact noncompliance, though unpredictably, tends to select patients for outcomes, mainly adverse, independent of any treatment effect. The alternative approach consists in carrying out the analysis according to the "intention-

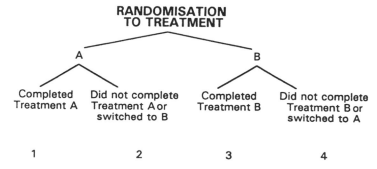

Figure 2.2   Expansion of the central part of figure 2.1 taking into account completness of treatment.

to-treat" principle and assessing the effect of each treatment whether or not the patient received the assigned treatment; thus findings in groups $(1) + (2)$ are compared with those in groups $(3) + (4)$. This approach ensures that selection bias is avoided because of the patient's data being analyzed as patients were randomized.

In our experience, however, analyses based upon the intention-to-treat principle are felt by some physicians as offending their intuition and sometimes they insist for comparisons suitable for evaluating "clinical efficacy". The results of the early trials of coronary artery bypass surgery (European Coronary Surgery Study Group, 1979) enable us to make clear the risk of drawing fallacious conclusions by following this approach. In this trial 768 men were accrued; 373 were assigned to medical and 395 to surgical treatment. One "surgical" patient was lost to follow-up. Among the patients allotted to surgical treatment, 26 did not undergo surgery, and among those allotted to medical treatment 50 patients were operated on. Two-year mortality from randomization was the end-point. Findings are summarized in the upper part of table 2.3. Altogether there were 50 deaths, 29 in the medical treatment and 21 in the surgical. The results of the analyses performed according to the two approaches discussed above are reported in the lower part of the same table.

Standard chi-square tests for 2×2 contingency tables are computed and the two analyses give conflicting results. Coherently with the intention-to-treat analysis the proper conclusion of the authors was that at two years there was no

Table 2.3 Two-year mortality in the coronary artery bypass surgery trial published by the European Coronary Surgery Study Group (1979) (upper part) and results of the analysis carried out according to the intention-to-treat principle and in terms of "clinical efficacy".

	Randomized to			
	Medical treatment		Surgery	
	Received medicine (1)	Switched to surgery (2)	Received surgery (3)	Switched to medicine (4)
Died	27	2	15	6
Survival at 2 years	296	48	354	20
Subtotal	323	50	369	26
Total	373		395	

Analysis by	Medical treatment	Surgery	x^2	p
Intention-to-treat	29/373(7.8%)	21/395(5.3%)	1.9	0.17
"Clinical efficacy"	27/323(8.4%)	15/369(4.1%)	5.6	0.018

significant difference in mortality between the two groups. Note that the mortality rate for deviants from the surgical protocol (six of 26 patients) was higher than for deviants from the medical group (two of 50). These findings cannot be taken as proving treatment effectiveness but are more likely to depend on the reasons which led to switching.

Let us consider a further example widely quoted in medical literature; it concerns the relationship of adherence to treatment and response in the long-term treatment of coronary heart disease (Coronary Drug Project Research Group, 1980). A cohort of 3892 men was randomized to receive either clofibrate (1103) or placebo (2789). The five-year mortality rates were respectively 20.0% and 20.9% ($z = -0.60$, $P = 0.55$, where z is the standardized Gaussian deviate for testing difference in mortality). For each patient the authors were able to estimate the cumulative adherence as the ratio of the number of capsules actually taken to the number that should have been taken according to protocol during the first five years of follow-up or until death; subjects were classified accordingly as either poor compliers ($< 80\%$) or good compliers ($\geqslant 80\%$). The results, for 3760 subjects, are reported in table 2.4.

Considering the clofibrate group alone, one can observe that five-year mortality among good compliers is substantially lower than among poor compliers. From the result of the analysis testing the difference between those two groups ($z = -3.60$, $P = 0.0003$), one could argue about the beneficial effect of clofibrate. However, overlapping findings arise even from the placebo group in which no pharmacological effect can be expected. Therefore one must conjecture that the observed differences are attributable to different characteristics of patients classified in the two subsets with good or poor compliance.

The lesson is that; "... irrespective of whether the object is efficacy or effectiveness, once in the study, all participants should be accounted for in the final analysis in the study group of which they were originally assigned, whether or not they have complied with prescribed treatments. In the efficacy study this permits valid comparison between the treatment and control groups whereas in the effectiveness study this is to maintain both validity and generalizability to the

Table 2.4 Five-year mortality in patients given clofibrate or placebo, according to compliance level (from Coronary Drug Project Research Group (1980), reproduced by permission of *New England Journal of Medicine*, **303**, 1038–41).

	Treatment group			
	Clofibrate		Placebo	
Compliance	No. of patients	% mortality $\pm$SE	No. of patients	% mortality $\pm$SE
$< 80\%$	357	24.6 ± 2.3	882	28.2 ± 0.8
$\geqslant 80\%$	708	15.0 ± 1.3	1813	15.1 ± 1.5
Total	1065		2695	

population from which the participants were selected. Only when these main, all patient, analyses have been done can the investigators afford to indulge in subgroup analyses – compliant patients only, for example – for whatever value they may have" (Haynes and Dantes, 1987).

2.6   STATISTICAL SIGNIFICANCE AND CLINICAL RELEVANCE

In the late 1980s several medical statisticians raised criticisms on the wide use of statistical tests as a unique inferential tool in medical literature and stressed the importance of "estimating with confidence"; even a book dealing with this topic was published (1989). Nevertheless, overemphasis on significance testing appears to persist today in medical literature.

In reporting the results of significance tests, particularly in tables referring to multiple regression analysis, clinicians tend to interpret statistical significance as synonymous with clinical relevance. As a matter of fact p values are only a guideline to the plausibility of the null hypothesis and should not be considered as proofs of treatment effects. Large p values suggest that data are plausibly consistent with the null hypothesis, whilst small p values lead the investigator to doubt the null hypothesis. Thus, to allow the reader to reach his own conclusions, the exact p values to a sensible number of significant digits together with the value of the test statistic should be given. Dichotomization into "significant" or "non-significant" with reference to the cutoff point, $p = 0.05$, is not acceptable.

The main issue in reporting trial results is the need to present findings directly on the scale of original measurement together with information on the inherent uncertainty attributable to sampling errors alone. In this way the reader is enabled to critically consider the *clinical relevance* of the results. This is the rationale for computing confidence intervals of any measure of effect of practical concern in clinical trials. On the one hand, a confidence interval conveys the same information as a p values, as far as statistical significance is concerned. A $(1 - \alpha) \times 100$ confidence interval which does not include the value specified by the null hypothesis is equivalent to a two-tailed test rejecting the null hypothesis with a significance level equal to α. Symmetrically, when the confidence interval includes the value specified by the null hypothesis the result is equivalent to a two-tailed test non-rejecting the null hypothesis. On the other hand a confidence interval is more informative than the p value in that its exact location and width suggest a good deal about the direction in which the truth lies and the adequacy of the sample size to pin it down.

Consider the clofibrate trial: the estimate of the difference in five-year mortality is $20.9\% - 20.0\% = 0.9\%$. Since the standard error of the difference of the two percentages is about 1.5%, by resorting to the Gaussian distribution, the approximate 95% confidence interval is $(-2.0\%, 3.8\%)$. This implies that the true difference could lie anywhere between these limits and that the data are likely to be consistent with a difference of zero. Furthermore, with such a narrow interval around zero, one may justifiably assume a conclusive negative finding with regard to clofibrate effect.

In commenting on the results of the two analyses reported in the lower part of table 2.3 we pointed out that the efficacy analysis led to the rejection of the null hypothesis while the intention to treat analysis did not. The 95% confidence interval of the difference of the two treatments computed for the efficacy analysis is found to be 0.66% to 7.93%. Since the lower limit is extremely near to zero, the clinician who is not willing to accept the conclusion of the intention to treat analysis could also be convinced that, even if statistically significant, the result for efficacy remains of questionable clinical relevance and thus further evidence is needed to assess the effectiveness of the surgical treatment.

2.7 SUBSET ANALYSIS

In previous sections it has been stressed that clinical trials aim at providing overall treatment comparisons. On the other hand the complexity of a chronic disease yields a setting in which patients with the same illness form groups which are remarkably heterogeneous in terms of baseline characteristics and prognosis. Therefore analysing patients' subsets is considered by several clinicians a natural part of the process enabling them to improve therapeutical knowledge by means of clinical trials. Unfortunately, improper approaches to statistical analysis and naive interpretations of the results are causes of confusion in therapeutical literature.

An approach frequently used to investigate how treatment effects vary among subsets consists in performing separate hypothesis tests for treatment differences within each subset ("subset-watching"). This approach should be avoided since it may lead to misleading results as will be clear from the following considerations. Suppose that we are dealing with a trial aiming at evaluating the effect of a new treatment for stage I melanoma of the skin. Three factors are known to influence prognosis: sex (m, f), minimum tumor thickness (three categories, say: 0–1.5 mm, 1.51–4.00 and >4.00 mm) and ulceration (present or absent). A total of 12 subsets of patients are defined and, consequently, 12 statistical tests will be computed to compare the new treatment to the standard one in all subsets using the significance level α (say $\alpha = 0.05$). The probability that, among s independent comparisons, at least one will be significant by chance alone is $1 - (1 - \alpha)^s$. Thus, as in our example, $1 - (1 - 0.05)^{12} \simeq 0.46$, we can state that even if the treatments are equivalent for all 12 subsets, there is a chance of some 46% that at least one test will be "statistically significant". The result given by the above formula must be considered an upper bound when the comparisons are not independent; nonetheless with many subsets this probability tends to one. The multiple comparison problem is made increasingly severe by the lack of power of the statistical tests carried out within subsets. In fact statistical significance is a function of the observed difference as well as the sample size, but this latter is computed to test the overall difference of the two treatment effects for the entire trial and not to detect subset effects.

In the previous example we considered three factors, the prognostic role of which is widely accepted in literature; the risk of wrong conclusions is increased even more if a subset analysis has been suggested by repeated delving into the data ("fishing expeditions", "data dredging").

The proper alternative consists in considering the treatment by subset interaction, i.e. how the difference between the effects of the two treatments varies among subsets of patients. In this case the multiple comparison problem is controlled by requiring that the overall interaction test be statistically significant. This test simultaneously assesses all the treatment subset interactions of potential concern to the clinician. However, the power of the interaction test will be low unless the trial has been planned to be sufficiently large for reliable subset analysis.

The difference between what we called "subset watching" procedure and tests for interaction is made clear by the examples given by Byar (1985) and reported in table 2.5. Two subsets of patients are considered here and the test statistic concerns the difference (Y) between the averages of the outcomes for the two treatments; the global test for the entire trial is significant at two standard errors $(2S)$ with a p value of 0.045. Now the data are supposed to be divided in equally sized subsets giving statistics Y_1 and Y_2 with standard errors $S\sqrt{2}$ for each

Table 2.5 Examples of apparent treatment–subset interaction; let $Y = \bar{X}_A - \bar{X}_B$ (from Byar, 1985).

	Statistic	S.E. of Y	p value (2 tail)
Unconvincing			
Overall test	$Y = 2S$	S	0.045
Subsets	$Y_1 = 3S$	$S\sqrt{2}$	0.034
	$Y_2 = 1S$	$S\sqrt{2}$	0.480
Interaction	$Y_1 - Y_2 = 2S$	$2S$	0.317
More convincing			
Overall test	$Y = 2S$	S	0.045
Subsets	$Y_1 = 4S$	$S\sqrt{2}$	0.005
	$Y_2 = 0$	$S\sqrt{2}$	1.000
Interaction	$Y_1 - Y_2 = 4S$	$2S$	0.045

subset. Consider the upper part of the table: examination of the results of the two tests computed within subsets might lead the reader to assess that the effects of the two treatments differ in the first subset, but not in the second one. On the contrary the interaction test is not statistically significant and one can conclude that findings in the two subsets as different as those observed would occur about one-third of the times by random variation alone ($p = 0.317$). In the lower part of the table, the difference between the two subsets is so marked that the interaction term is significant at $p = 0.045$. This result enables us to conclude that a difference between the treatment effects emerges in the first subset while little or no difference emerges in the second subset.

If we were dealing with the data of a trial, at this point we might argue about the biological plausibility of the above result in order to justify the conclusions and the subsequent therapeutic actions. However, testing for treatment–subset interaction deserves further comment because misinterpretation of the implications of testing such interaction is widespread in randomized clinical trials. Peto (1982) has distinguished *quantitative* and *qualitative* interactions. A quantitative interaction involves variation in the magnitude, but not in the direction, of the treatment differences among subsets, as shown in the upper graph of figure 2.3. The covariate reported on the horizontal axis is assumed to be dichotomous, with modalities: absent ($-$) and present ($+$) so that two subsets only are considered here. On the contrary a qualitative interaction implies a change in the direction of the treatment difference in different subsets, as shown in the lower graph of figure 2.3.

Note that interaction can be a matter of scale of measurement upon which the statistical model is based; monotone scale transformations may eliminate quantitative interactions. Furthermore, if the subsets are defined by important prognostic factors, differences between the treatment effects are expected to vary from one subset to another, enabling the investigator to anticipate quantitative interaction. It is therefore sensible to focus attention on qualitative interaction as only this is clinically important.

An interesting example of significant interaction of both types is given by Micciolo *et al.* (1985); by means of Cox's (1972) model they investigated the efficacy of chemotherapy (CMF) in breast cancer. Considering the event-free interval as response, they found a quantitative CMF × menopause interaction, whilst, considering the survival time as response, the same interaction appeared to be of qualitative type, suggesting that CMF benefited patients in pre-menopause but not those in menopause. The therapeutical message a clinician can draw from this result is in agreement with the conclusions later reached by the NIH consensus development conference on adjuvant chemotherapy for breast cancer, reported in the *Journal of the American Medical Association*, volume 254, pages 3461–3483.

Because significant results in all standard statistical tests for interaction do not give evidence on the presence of a qualitative interaction, some new approaches have been published. Shuster and van Eys (1983) present

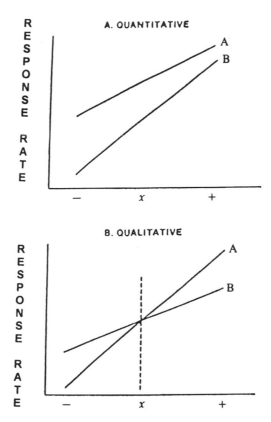

Figure 2.3 Two types of treatment-covariate interactions (see text for explanation).

a technique to divide a case series into three subsets: one region of superiority for each of the two treatments and one region of uncertainty. This method appears to be particularly useful in the case of a single continuous prognostic factor. Azzalini and Cox (1984) devise a test of qualitative interaction in a two-way analysis of variance and give approximate distributional results assuming that there are no treatment effects. Gail and Simon (1985) suggest a likelihood ratio procedure suitable for assessing the statistical significance of qualitative treatment × prognostic factors interactions with specification in advance of the set of disjoint subgroups of patients. Owing to its theoretical properties, this test compares favourably with its competitors; in addition it is easy to compute on all the response variables commonly used in clinical trials, and can be applied to published data. This test will be introduced in section 4.5.3.

"The obstacles to making reliable inferences about subset effects are thus considerable. Nevertheless, it is still appropriate to address subset questions provided we have scientific integrity and statistical sophistication to interpret and report the results sensibly, often admitting ambiguity and uncertainty" (Simon, 1988).

2.8* INTERIM ANALYSES IN CLINICAL TRIALS

2.8.1 Internal and external monitoring

It is well known that phase III randomized clinical trials for treatment of chronic diseases usually last for a considerable time, sometimes several years. Their duration may be partitioned into a recruitment stage, in which different treatments are administered, and a stage of continued observation after accrual has stopped. Because in the accrual period early information is already available, it is common practice to inspect the data gradually accumulating, for the following reasons:

(1) to control patient recruitment;
(2) to check the adherence of all investigators to the trial protocol with regard to the admission and exclusion criteria, the therapeutic regimen, and the discontinuation of treatments, etc. The early detection of possible deviations will enable the investigators to promptly modify their·behaviour and so prevent the study from losing its quality;
(3) to preserve the investigators' original interest and to satisfy their curiosity. In fact it is essential to keep them informed on how the study is progressing;
(4) to enable biostatisticians to check the quality of documentation and to evaluate the organization of data transfer in order to avoid undue delay. As discussed by Chlebowski *et al.* (1981), delayed reporting of data in key cases may stop accrual of patients.

Clearly these issues concern only the management aspects of the trial and can usually be handled with commonsense and effective administration; according to Armitage (1991) these are goals of the "internal" monitoring of a clinical trial which is usually performed by the Executive Committee on the basis of the reports prepared by the data coordinating centre.

On the other hand, in carrying out clinical trials with a long follow-up, it is considered unethical to make any inference about the treatment differences on mortality and other major response variables until the information on all patients has been gathered. In other words it appears unacceptable to resort to the "fixed sample size" approach, and interim analyses are usually accomplished at predetermined intervals on accumulating data in order to avoid patients being exposed to unsafe, inferior or ineffective treatments. These are goals of the "external" monitoring for which a group of experts, independent of the clinical investigators and sponsors of the trial, is usually responsible (data monitoring committee, DMC).

The interim analyses rely upon the group sequential approach (Pocock, 1977) which offers some of the theoretical advantages of the sequential techniques whose use in clinical trials is upheld by Armitage (1975) and Whitehead (1983). The strategy consists of analysing the accumulated data relative to primary and secondary end-points at periodic intervals (say every 6 or 12 months) during the course of the trial, and if a "substantial" treatment difference emerges from any of these analyses (hereafter called "looks"), prompt action can then be taken. This could mean stopping the accrual of patients prior to its scheduled end, if the above result is detected during the recruitment stage of the trial, or perhaps communicating results to other workers as soon as possible to prevent any use of an apparently inferior treatment.

2.8.2   Repeated Significance Testing and early rejection of H_0

From a methodological viewpoint the "group sequential" approach is a partially sequential procedure constructed on the basis of repeatedly performing fixed sample analysis in an organized way. To gain a deeper insight into this technique it is helpful to illustrate its main statistical aspects; this requires defining a formal stopping rule in terms of first kind error risk α and test power $(1 - \beta)$ and considering its relationship with the number of looks and the total (maximum) size of the trial.

Consider a homogeneous sample of patients sequentially entering into a clinical trial aiming at comparing the effects of the two treatments A and B. By a randomized permuted block design each consecutive set of $2n$ patients is assigned to the treatments so that n of them are allocated to A and B respectively. Assume that the response is a Gaussian random variable with a known variance σ^2 and unknown averages μ_A and μ_B and that $\bar{x}_{Aj}$ and $\bar{x}_{Bj}$ are corresponding observed mean responses in the jth set of n subjects. Then the

Table 2.6   Overall probability (%) of achieving results with a given nominal significance (%) after L repeated tests when there is no difference in the effects of two treatments.[a]

Nominal significance level	Number of repeated significance tests (L)								
	1	2	3	4	5	10	25	50	200
1	1	1.8	2.4	2.9	3.3	4.7	7.0	8.8	12.6
5	5	8.3	10.7	12.6	14.2	19.3	26.6	32.0	42.4
10	10	16.0	20.2	23.4	26.0	34.2	44.9	52.4	65.2

[a]Data are expressed as percentages. From McPherson (1974), reproduced by permission of *New England Journal of Medicine*. **290**, 501–2.

Table 2.7   Nominal significance level (α') and corresponding standardized normal deviate $z(\alpha, L)$ required for repeated two-tailed significance testing.[a]

L	$\alpha = 0.05$		$\alpha = 0.01$	
	α'	$z(\alpha, L)$	α'	$z(\alpha, L)$
1	0.0500	1.960	0.0100	2.580
2	0.0294	2.178	0.0056	2.772
3	0.0221	2.289	0.0041	2.873
4	0.0182	2.361	0.0033	2.939
5	0.0158	2.413	0.0028	2.986
6	0.0142	2.453	0.0025	3.023
7	0.0130	2.485	0.0023	3.053
8	0.0120	2.512	0.0021	3.078
9	0.0112	2.535	0.0019	3.099
10	0.0106	2.555	0.0018	3.117
11	0.0101	2.572	0.0017	3.133
12	0.0097	2.585	0.0016	3.147
15	0.0086	2.626	0.0015	3.182
20	0.0075	2.672	0.0013	3.224

[a]From Pocock (1977) by permission of the Biometrika Trustees.

empirical difference computed at the ith look is represented by

$$d(i) = \sum_{j=1}^{i} (\bar{x}_{Aj} - \bar{x}_{Bj}) \tag{2.1}$$

(where $i = 1, 2, 3, \ldots, L$; L = number of anticipated looks for a given trial), and is Gaussian with expectation $i\delta = i(\mu_A - \mu_B)$ and variance $2i\sigma^2/n$.

Therefore a two-tailed significance test of the null hypothesis $H_0 : \delta = 0$ computed at the ith look has probability $P_i = 2[1 - \Phi\{\sqrt{n}d(i)/\sigma\sqrt{2i}\}]$, where Φ is the standardized Gaussian distribution and can easily be read from the pertinent table. Considerations discussed in the previous section on the comparison among subsets are pertinent here; repeated testing at each look increases the probability of finding a statistically significant difference purely by chance (Armitage *et al.*, 1969). Table 2.6 illustrates the phenomenon of the inflation of the probability of first kind error for repeated two-tailed tests at 0.01, 0.05, and 0.10 levels of significance with equally sized samples of subjects receiving treatments A and B, respectively. It is assumed that, after achieving a given nominal level, the investigator stops the study. Suppose that in a given trial four looks have been anticipated; a significance level value of 5% under the usual conditions when a "once-and-for-all" test is computed at the end of the

Table 2.8 Different strategies aiming at making the final α'_L close to α by applying Bonferroni's inequality.[a]

α'_5	$z_{\alpha'_5}$	$\alpha'_i = (\alpha - \alpha'_5)/4$ $(i = 1, .., 4)$	z_{α_i}
0.045	2.005	0.001 25	3.227
0.040	2.005	0.002 50	3.024
0.035	2.108	0.003 75	2.900
0.030	2.170	0.005 00	2.807
0.025	2.242	0.006 25	2.375

[a]Two-tailed tests with $\alpha = 0.05$ and $L = 5$ looks.

Table 2.9 Nominal levels (α'_i) and their pertinent z values of three different stopping rules.[a]

Look number	Pocock		O'Brien and Fleming		Koepcke	
(i)	α'_i	z	α'_i	z	α'_i	z
1	0.015 81	2.413	0.000 01	4.417	0.000 11	3.867
2	0.015 81	2.413	0.001 26	3.225	0.006 14	2.740
3	0.015 81	2.413	0.008 45	2.634	0.025 26	2.238
4	0.015 81	2.413	0.022 56	2.281	0.025 26	2.238
5	0.015 81	2.413	0.041 34	2.040	0.025 26	2.238

[a]Two-tailed tests, $\alpha = 0.05$, five looks.

trial on the responses elicited in all patients would be increased to 12.6% (McPherson, 1974).

One may control the probability α of rejecting the above-mentioned null hypothesis on the whole set of tests when in fact it is true, by adopting more stringent nominal significant levels α' for the tests computed at each look. Table 2.7 (Pocock 1977) gives α' values and their corresponding Gaussian deviates $z(\alpha, L)$ for a number of looks from 1 to $L = 20$; they have been determined so that the overall risk of first kind error is $\alpha = 0.05$ or $\alpha = 0.01$. It is worth noting that given α and L, α' values do not depend on the number n of subjects whose responses are observed at each look becasue the differences $(\bar{x}_{Aj} - \bar{x}_{Bj})$ are random variables identically and normally distributed with variance $2\sigma^2/n$. At a glance at table 2.7, it appears that the testing procedure becomes more and more conservative as the number of looks increases.

In general, the two-tailed group sequential procedure is:

(a) reject H_0 and stop the trial at the ith look if

$$|d(i)/\sigma| > z(\alpha, L)\sqrt{2i/n} \qquad (2.2)$$

otherwise continue the trial, and

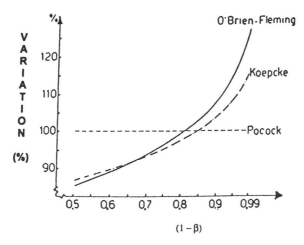

Figure 2.4 Comparison of three group sequential designs in terms of average sample number (ASN). Pocock's design is assumed as reference design. (Two-tailed tests with $\alpha = 0.05$ and five looks.) Ordinate: percentage variations of ASN; abscissa: power of the design.

(b) "accept" H_0 if at the last planned look

$$|d(L)/\sigma| \leqslant z(\alpha, L)\sqrt{2L/n} \qquad (2.3)$$

In interpreting the results of a clinical trial in which repeated significance testing at a constant nominal level α' has been adopted, a dilemma may arise when none of the consecutive tests reaches α' but the final p value is less than the overall significance level α. Some different suggestions aiming at making the final α'_L close to α could be given. For instance, after choosing a pertinent value of α'_L close to α, one could determine the remaining values of α' by resorting to Bonferroni's inequlity, that is, $\alpha'_i = (\alpha - \alpha'_L)/(L-1)$ for each $i \neq L$. For two-tailed tests with $\alpha = 0.05$, $L = 5$ and different values of significance level of the final analysis, table 2.8 reports the corresponding standardized Gaussian deviates to be adopted in the four *ad interim* analyses. The first of the five alternative designs is similar to that suggested by Haybittle (1971).

O'Brien and Fleming (1979) developed a strategy with α'_i increasing from one look to the next, with corresponding standardized Gaussian deviates proportional to $\sqrt{L/i}$. Their results were presented in terms of χ^2 for the comparison of two proportions but can be easily carried over the Gaussian responses. A procedure that takes advantage of both Pocock's (1977) and O'Brien and Fleming's (1979) strategies has been discussed by Koepcke *et al.* (1982). He adopts an increasing nominal significance level in the first looks and a constant level in the last ones. An example concerning five looks is given in table 2.9, which also reports the nominal values of Pocock's design and of O'Brien and

Fleming's design. Koepcke's procedure is close to the one reported in the last row of table 2.8.

The most relevant feature of a sequential design is known to be the extent to which it allows the trial to stop early when the alternative hypothesis is true. From this viewpoint, the most informative statistical quantity is the average sample number (ASN), i.e. the mean number of observations required to reach a decision. The performance of the three strategies in terms of ASN is easily seen from figure 2.4 in which the ASN of Pocock's design is taken as the reference value. For low power designs $[0.50 \leqslant (1 - \beta) \leqslant 0.70]$ O'Brien and Fleming's (1979) strategy appears to be the best, while for high power designs $(1 - \beta \geqslant 0.85)$ Pocock's strategy appears to be preferable. In the middle, that is for designs with power from 0.7 to 0.85, Koepcke's strategy compares favourably with the other two.

2.8.3 Asymmetric group sequential boundaries and early "acceptance" of H_0

Whichever of the previous strategies one adopts, the procedure specified by (2.2) and (2.3) enables a trial to be stopped early, rejecting the null hypothesis. However, clinical trials are often planned to test whether a new treatment is *better* than the standard treatment; investigating whether the new treatment is worse than the reference one may be utterly inappropriate. DeMets and Ware (1980) have shown that in one-sided settings one can take advantage of stopping rules which enable early termination and "acceptance" of H_0 at each look. In this case we consider $H_0 : \delta \leqslant 0$ and $H_a : \delta > 0$, and it is sensible to assume that "acceptance" of H_0 may be suggested before the statistic reaches a value as extreme as the one needed to reject H_0. Hence asymmetric group sequential

Table 2.10 Upper boundary parameter, z_U, for several values of the lower boundary parameter, z_l, with $\alpha = 0.05$ and 0.01 for various maximum numbers of tests L (from De Mets and Ware 1980, reproduced by permission of the Biometrika Trustees)

	$\alpha = 0.05$				$\alpha = 0.01$			
L	$z_l = 0.5$	$z_l = 1.0$	$z_l = 1.5$	$z_l = 2.0$	$z_l = 0.5$	$z_l = 1.0$	$z_l = 1.5$	$z_l = 2.0$
2	1.875	1.876	1.876	1.876	2.532	2.532	2.532	2.532
3	1.990	1.992	1.993	1.993	2.636	2.637	2.637	2.637
4	2.062	2.067	2.068	2.068	2.703	2.705	2.705	2.705
5	2.113	2.120	2.122	2.122	2.751	2.754	2.754	2.754
6	2.152	2.161	2.163	2.164	2.789	2.792	2.792	2.793
7	2.182	2.194	2.197	2.197	2.817	2.822	2.823	2.823
8	2.207	2.221	2.225	2.225	2.841	2.846	2.848	2.848
9	2.229	2.244	2.248	2.249	2.861	2.868	2.870	2.870
10	2.247	2.263	2.269	2.270	2.879	2.886	2.888	2.889

boundaries are obtained and at each look:
reject H_0 and stop the trial if:

$$d(i)/\sigma > z_u(\alpha, L)\sqrt{2i/n}$$

or accept H_0 and stop the trial if:

$$d(i)/\sigma < -z_1\sqrt{2i/n}$$

otherwise continue the trial.

The two boundaries (z_u and z_1) are given in table 2.10. The value of z_1 is selected by the investigator to reflect the strength of the negative evidence he requires before "accepting" H_0 and stopping the trial. Coherently with Pocock's (1977) approach, the value of $z_u(\alpha, L)$ is such that the overall significance level after L tests is α. In a further paper, DeMets and Ware (1982) suggested a modification of the procedure which combines the advantage of utilizing the O'Brien–Fleming upper boundary with a preselected lower boundary.

2.8.4 Lan–DeMets procedure

From previous subsections the two requirements of the group sequential analysis emerge: the number of interim analyses must be specified in advance and they must be performed at equal increments of information. The procedure developed by Lan and DeMets (1983) generalizes the group sequential approach, allowing the investigator to escape the two above-mentioned restrictions and to resort to properly built boundaries; this procedure relies upon the "α spending function". In essence, let us assume that the trial be completed by time T, scaled arbitrarily such that $T = 1$. The $\alpha(t)$ function allocates the amount of type I error that the investigator can "spend" at each analysis carried out at time $t (0 \leqslant t \leqslant 1)$ and is such that $\alpha(0) = 0$ and $\alpha(1) = \alpha$. Note that the parameter t defines the position of the interim analyses during the trial, and if n_i patients have been accrued before the ith analysis, $t_i = n_i/N$ (where N is the target sample size); thus the final analysis corresponds to $T = 1$. Furthermore if, by construction, the increment $\alpha(t_i) - \alpha(t_{i-1})$ is the amount of the significance level the investigator can utilize at time t_i, for the whole trial the risk of type I error is α regardless of the data or the number of interim analyses. Specifically:

$$\alpha(t) = P_0\{\tau \leqslant t\} \qquad 0 \leqslant t \leqslant 1$$

where τ is the first boundary crossing of a Brownian motion process. At time points $0 < t_1 < t_2 < \ldots < t_i \leqslant 1$ the boundary values b_i are chosen such that under H_0 and given $b_1, b_2, \ldots, b_{i-1}$:

$$P_0\{|W_1| < b_1, |W_2| < b_2, \ldots, |W_{i-1}| < b_{i-1}, |W_i| > b_i\} = \alpha(t_i) - \alpha(t_{i-1})$$

where W_i represents the Brownian process, $W(t)$, at $t = t_i$. The evaluation of the boundary values requires numerical integration, and a Fortran program available from the authors gives the value $b_i = z_i \sqrt{t_i}$ so that the user can easily make the translation to the standardized statistic for $z_i = z_u(t_i) = -z_l(t_i)$. Furthermore it is worth noting that Pocock or O'Brien–Fleming procedures can be reformulated in terms of "spending function" so that the nominal levels of the first kind error probability are approximated by the following two analytical functions:

$$\alpha_P = \alpha \ln\{1 + (e - 1)t\}$$

$$a_{O'B-F}(t) = 2 - 2\phi(1.96/\sqrt{t})$$

where ϕ denotes the standardized Gaussian cumulative function.

2.8.5   Stochastic curtailed sampling

The procedures considered so far base the recommendation to stop the trial on the current evidence about the size of the relevant parameter, in relation to properly defined critical values. An alternative approach, broadly known as *stochastic curtailment*, has been suggested by Lan *et al.* (1982). By taking into account information collected at a given interim analysis, it attempts to predict the final results that would be obtained if the trial were allowed to complete its planned course until *T*. Clearly the future data are unknown and the final results are subject to random uncertainty. However, if at the *i*th look it becomes known that, with a high probability, the future data could not change the conclusion reached, it is sensible to stop the trial at this point.

Formally, let *R* denote the rejection region, $\bar{R}$ denote the acceptance region and $S(t)$ the test statistic computed at time *t*. Lan *et al.* (1982) consider curtailing in favour of the alternative hypothesis $H_a: \delta \neq 0$ in terms of the conditional probability:

$$P\{S(T) \in R | S(t), H_0\} = \gamma_0$$

where $S(T)$ is the value of the test statistic $S(.)$ predicted at the scheduled end of the trial, given $S(t)$ and assuming that the null hypothesis H_0 is true; of course, the investigator must utilize a sufficiently large γ_0 in order to give the ultimate decision some degree of certainty. Similarly, for sufficently large γ_1, the procedure considers curtailing in favour of $H_0: \delta = 0$ in terms of

$$P\{S(T) \in \bar{R} | S(t), H_a\} = \gamma_1 \tag{2.4}$$

where $S(T)$ is the value of the statistic predicted at T, given $S(t)$ and assuming that H_a is true.

By this procedure both the type I and type II error probabilities of the final analysis are inflated at most by α/γ_0 and β/γ_1 respectively. However, for γ_0 and γ_1 sufficiently large the inflation is negligible; for example, for $\gamma_0 = 0.95$ and $\alpha = 0.05$, $\alpha/\gamma_0 = 0.052\,63$. But the choice of γ_0 and γ_1 near 1 makes the procedure conservative and early stopping unlikely. In order to overcome this drawback, Lan *et al.* (1984) suggested evaluating the conditional probabilities (2.4) for a series of alternative hypotheses δ that may be of concern for the clinician. Rather naturally, this suggests resorting to a Bayesian approach whereby prediction is based on the posterior probability for δ (Spiegelhalter and Freedman, 1988; Spiegelhalter *et al.*, 1986; Freedman and Spiegelhalter, 1992).

2.8.6 Confidence intervals

In section 2.6 the importance of "estimating with confidence" was stressed in order to aid the physician to interpret the final results of a clinical trial in terms of clinical relevance. The same need emerges in reporting results after a trial was stopped on the basis of conclusions reached in an interim analysis. One must at once realize that when a stopping boundary is crossed, say the upper one, the observed difference and the confidence interval calculated in the usual way (for instance, $\bar{d} \pm z_{1-\alpha/2} \text{SE}(\bar{d})$) (naive confidence interval) tend to be biased upward. The justification for this is that crossing a stopping boundary at, say, the ith look and not earlier, is ignored in the previous formula. Tsiatis *et al.* (1984) provided a numerical method for computing exact confidence limits after the trial has been stopped according to the Pocock or O'Brien–Fleming plan. Their intervals are pulled towards zero when compared with the naive ones; this is particularly evident if the treatment difference is highly significant. Kim and DeMets (1985) developed a method for estimation and confidence interval computation for the Lan–DeMets (1983) flexible boundaries. As an alternative to group sequential testing, Jennison and Turnbull (1989) discussed a technique by which confidence intervals can be obtained at all looks of a trial with a given probabilistic assurance that *all* the intervals contain the true value. Not surprisingly these latter are unusually wide, and if the investigator's primary concern is in obtaining the confidence interval after stopping at the ith look, they are unduly conservative.

2.8.7 Survivorship data

So far our considerations of group sequential methods have focused on comparison of treatment means as indicated by (2.1). However, the application of group sequential methods can be extended to comparing proportions and survival curves (Pocock, 1977), in the latter case assuming that the response time is distributed according to an exponential function (see section 5.5). Then one can resort to the stasistic

$$y(k) = \sqrt{\frac{k}{2}} \log \frac{\hat{\lambda}_A(k)}{\hat{\lambda}_B(k)}$$

where $\hat{\lambda}_A(k)$ and $\hat{\lambda}_B(k)$ are the observed death rates (per person year of exposure) in treatment groups A and B, respectively, computed immediately after the kth death (counting both treatments together). The statistic $y(k)$ behaves approximately like the cumulative sum of k independent, normally distributed variables with expectation $\delta = \sqrt{k/2} \log(\lambda_A/\lambda_B)$ and $\sigma^2 = 1.0$, under the null hypothesis $\lambda_A = \lambda_B$ and $\delta = 0$. By Monte Carlo techniques Breslow and Haug (1972) showed that the Gaussian approximation works extremely well also for small k; therefore the group sequential boundaries at *equally spaced number of deaths* appear to be applicable. In this latter case, $k = in$ enables us to resort to (2.1), since now n is the prespecified *number of deaths* between two looks. Canner (1977), who also assumed exponentiality, presented and evaluated a statistical procedure to analyse accumulating data *at fixed time intervals* rather than at the occurrence time of events. By means of computer simulation studies, which provide for different mortality rates and different accrual patterns, the author showed that group sequential boundaries can be used in survival analysis for test statistics that are nearly normal, under a variety of conditions. These include cases where the assignment of patients to treatments can differ by a ratio of as much as $1:3$ during the course of the trial.

In clinical trials it is common practice to analyse survival data by methods which make no distributional assumption on failure time and the appropriate test statistics will be commented upon in Chapter 4. Let us consider the Mantel–Haenzel test (Mantel and Haenzel, 1959) which appears to be the most widely used. It suffices to say here that the computation of this test statistic and its comparison with the selected boundary at each increment of $2n$ death has been shown to be justified both asymptotically (Tsiatis, 1981a) and for smaller sample sizes (Gail, 1982). These comments can be extended to the other tests mentioned in section 4.6 except to the Gehan test (Slud and Wei, 1982).

At the beginning of this section it was assessed that in several clinical trials interim analyses are accomplished periodically (say each year), not at equal increments of information. DeMets and Gail (1985) showed that the operating characteristics (i.e. probabilities of type I and II error) of the Mantel–Haenzel test are quite robust for unequal increments of information even though the boundaries are computed under the assumption of equal increments.

In concluding this subsection, one must remember that there will be a special need for caution in applying any procedure of *ad interim* analysis to survival data. In fact, if the assumption of proportional hazard does not hold and the two survival curves cross, the result of some *ad interim* analyses could enable the investigator to conclude prematurely in favour of a given treatment, while with further follow-up no clear treatment difference would become apparent.

2.8.8 Some caveats

It is timely to state that although the previously considered rules are tools suitable for informing the investigators whether an early termination of the trial is possible, it would be inappropriate to reach such a decision mechanically on the basis of a predetermined boundary applied to one particular end-point of the trial. In fact, the problem of decision making in clinical trials is extremely complex and requires the organizers to consider simultaneously several issues and to weight them properly in a subjective decision process. The issues to be considered are as follows:

(1) results of the analyses performed on outcome measures other than those used as the end-point of the monitoring procedure (these include measures of incidence and severity of side-effects and of quality of life during the treatment and follow-up periods);

(2) relationships between the effect of treatments and differential baseline distributions of prognostic factors in the two groups A and B to sort out possible biases;

(3) investigating whether early termination is attributable to some particular subgroups or to the whole sample;

(4) considering, if applicable, the evidence from similar trials being carried out.

APPENDIX A2.1   RANDOMIZATION AND TREATMENT COMPARISON

This appendix shows that randomization is the only tool suitable for avoiding confusion in comparing two or more treatments. Zelen's (1985) approach will be followed. We define the following variables:

t = survival time or any other outcome;

z = treatment (A or B);

$\mathbf{x}$ = prognostic factor(s) affecting outcome;

and the following conditional probability distributions:

$p(t|z, \mathbf{x})$ = probability distribution of outcome conditional on assigned treatment and prognostic factors;

$p(t|z)$ = probability distribution of outcome conditional on treatment;

$p(\mathbf{x}|z)$ = probability distribution of prognostic factors conditional on treatment.

The effect of the new treatment is assessed by comparing $p(t|B)$ with $p(t|A)$; under the null hypothesis of equal effectiveness, $p(t|B) = p(t|A) = p(t)$. Probability theory enables us to write

$$p(t|z) = \sum_{\mathbf{x}} p(\mathbf{x}|z)p(t|z, \mathbf{x})$$

where the summation is over the whole set of prognostic factors. Assuming that the outcome does not depend on treatment, but on prognostic factors only, that is $p(t|z, \mathbf{x}) = p(t|\mathbf{x})$, the previous expression may be rewritten

$$p(t|z) = \sum_{\mathbf{x}} p(\mathbf{x}|z)p(t|\mathbf{x})$$

It is worth noting that the right-hand side of the equation is still dependent on treatment even if we assumed the outcome to be independent of treatment. This result applies to all studies which do not randomize treatments and proves why they are biased even if the investigator sums over all the known prognostic factors. The randomized studies enjoy the important property that the prognostic factors admit the same distribution regardless of treatment, i.e. $p(\mathbf{x}|z) = p(\mathbf{x})$. As a consequence

$$p(t|z) = \sum_{\mathbf{x}} p(\mathbf{x})p(t|\mathbf{x}) = p(t)$$

which shows that, as expected under the null hypothesis, the distribution of outcome does not depend on treatment. Under the alternative hypothesis of different effect between the two treatments one has

$$p(t|z) = \sum_{\mathbf{x}} p(\mathbf{x})p(t|z, \mathbf{x})$$

3

Estimation of Survival Probabilities

3.1 INTRODUCTION

In 1963 Freireich *et al.* published the results of a multicentre controlled clinical trial on remission maintenance in children with acute lymphoblastic leukemia. The study was designed to test whether patients who had achieved complete remission, after some induction treatment, could benefit from further treatment. Patients in complete remission were randomized to receive maintenance therapy with 6-mercaptopurine or placebo. The study was stopped according to a sequential-type rule after 42 patients were randomized. In the placebo group all patients had a relapse, while in the 6-MP group 12 patients were still in remission when the study was stopped and analysed. The observed remission or censored times, recorded in weeks, and rearranged in ascending order, were, by treatment group:

Placebo (21 patients): 1, 1, 2, 2, 3, 4, 4, 5, 5, 8, 8, 8, 8, 11, 11, 12, 12, 15, 17, 22, 23

6-MP (21 patients): 6, 6, 6, 6*, 7, 9*, 10, 10*, 11*, 13, 16, 17*, 19*, 20*, 22, 23, 25*, 32*, 32*, 34*, 35*

where * indicates censored times.

An analysis with classical methods would consider the estimation and comparison of the mean duration of remission in the two groups. In the placebo group, the calculation of the arithmetic mean or, alternatively, of the harmonic mean, which has the advantage of being less influenced by longer durations, is straightforward. On the contrary, in the 6-MP group, censored data create a major problem. The mean and standard error cannot be computed unless patients still in remission at the end of follow-up are assigned hypothetical times of relapse (greater than the observed censored time). This adjustment of censoring introduces an element of arbitrariness.

The same disturbing feature may arise when another location measure, the median, is adopted. While the median can be straightforwardly calculated as 8 weeks in the placebo group, it cannot be calculated in the 6-MP group unless we assign arbitrary relapse times at least to the smaller censored times. It is worth noting that if observed censored times would have fallen after the 11th ranked duration only, the median could have been estimated without arbitrary adjustments even in the 6-MP group. However, we do not wish to rely on statistics where applicability depends on the pattern of censoring.

An alternative approach to the analysis of these data could possibly look at counts of failures. This implies the calculation and comparison of the proportion of relapses in the two groups, namely 1 in the placebo group and $9/21 = 0.43$ in the 6-MP group. Each proportion estimates the probability of relapse in the corresponding group. Nonetheless, we would like to account for the time at which relapses occur, and not only for the binary information "relapsed or not relapsed". For example, we could reasonably wish to investigate if treatment delays relapses, besides preventing some. Furthermore, it could be argued that if subjects whose observation is censored at 6, 9 or 10 weeks were followed up for an additional period, some more relapses would possibly be observed. In fact, it is likely that more failures would be seen as the time of observation grows longer.

Clearly, the description and comparison of numbers of failures needs to be done by referring to some unit period of observation. To achieve this, we use the simple idea of subdividing the observation time into a sequence of time intervals, that may become as short as the smallest time unit (one week, in this example, or more frequently one day). Then, the proportion of failures at each time unit will be used to describe the data in terms of "survival curve" (sections 3.2 to 3.5).

An alternative way of accounting for the observation time when counting the number of failures occurring in a group of subjects is to calculate failure rates. These have been commonly applied by epidemiologists to study mortality in a population and are easily extended to study a failure process in a clinical setting (section 3.6).

To conclude, methods for survival analysis which are presented in this chapter are basic methods which (i) accommodate censoring; (2) account for different periods of observation on each experimental unit; and (3) account for the time at which events occur.

3.2 THE PRODUCT LIMIT ESTIMATE

3.2.1 Basic concepts

Suppose we wish to estimate the probability of surviving one year from diagnosis for subjects who were diagnosed a given disease. The product limit method is based on the simple consideration that in order to survive one year from the beginning of the observation (i.e. diagnosis), the subject has to survive every day from the first to the 365th. It is natural to estimate the probability of surviving a given day by calculating the proportion of patients, among those alive just before that day, who survive until the next day. All patients entered in the study contribute to the estimation of the probability of surviving the first day:

$$\hat{p}(1) = \frac{\text{number of patients entered} - \text{number of failures on the first day}}{\text{number of patients entered}}$$

However, in any following day, the number of alive patients at the denominator will be possibly less than the initial number. For example, at day 180 from the start of the observation, none of the patients who had a failure or were censored on one of the previous days convey any information on the probability of surviving day 180. Only subjects who are alive and under observation just before day 180, including those whose failure time is 180, are useful for estimating $p(180)$; these are called patients "at risk" and the probability $p(180)$ is estimated by

$$\hat{p}(180) = \frac{\text{number of patients at risk on day 180} - \text{number of failures on day 180}}{\text{number of patients at risk on day 180}}$$

How should we combine these probabilities, called conditional probabilities, to obtain the probability of surviving one full year? We accumulate step by step the probabilities of surviving each day by multiplying them, that is the probability $P(2)$ of surviving two days is $P(2) = p(1) \times p(2)$. By multiplying $P(2) \times p(3)$ we will obtain the probability of surviving three days and so on up to $P(365)$.

Actually, it is clear from the calculation of the probability of surviving each day that we do not need to calculate $\hat{p}(180)$ if on day 180 no events were observed. Since all patients at risk on day 180 survived until the next day, it is simply $\hat{p}(180) = 1$. So, the cumulative probability $\hat{P}(t)$ only changes at those time points where some failures occur.

The simple method we have outlined may be usefully applied to the data given in the previous section to calculate the probability of being in remission at different time points for subjects in the two treatment groups. In this example, the event of interest is relapse instead of death and time is reported in weeks instead of days, but the same arguments apply. To ease computations, all times, censored and non-censored, are put in ascending order and the corresponding numbers of censored observations, failures, and patients at risk are counted as in table 3.1, columns (1) to (4). The probability of being in remission is calculated at each distinct failure time only, as shown in columns (5) and (6) of the same table.

Consider the 6-MP group where the first three failures are reported to occur at six weeks from remission over 21 accrued patients.

Table 3.1 Computation procedure for the remission probability in a controlled clinical trial on leukemia in childhood. Data from Freireich et al. (1963).

Treatment	Time to relapse (week) $t_{(j)}$ (1)	No. of censored data (2)	No. of failures d_j (3)	No. of patients at risk n_j (4)	Conditional probability of remission $\hat{p}_j = (n_j - d_j)/n_j$ (5)	Cumulative probability of remission $\hat{P}(t)$ (6)
Placebo	1		2	21	$19/21 = 0.90476$	0.9048
	2		2	19	$17/19 = 0.89474$	$0.9048 \times 0.8947 = 0.8095$
	3		1	17	$16/17 = 0.94118$	$0.8095 \times 0.9412 = 0.7619$
	4		2	16	$14/16 = 0.87500$	$0.7619 \times 0.8750 = 0.6667$
	5		2	14	$12/14 = 0.85714$	$0.6667 \times 0.8571 = 0.5714$
	8		4	12	$8/12 = 0.66667$	$0.5714 \times 0.6667 = 0.3810$
	11		2	8	$6/8 = 0.75000$	$0.3810 \times 0.7500 = 0.2857$
	12		2	6	$4/6 = 0.66667$	$0.2857 \times 0.6667 = 0.1905$
	15		1	4	$3/4 = 0.75000$	$0.1905 \times 0.7500 = 0.1429$
	17		1	3	$2/3 = 0.66667$	$0.1429 \times 0.6667 = 0.0952$
	22		1	2	$1/2 = 0.50000$	$0.0952 \times 0.5000 = 0.0476$
	23		1	1	$0/1 = 0.00000$	$0.0476 \times 0.0000 = 0.0000$
6-MP	6	1	3	21	$18/21 = 0.85714$	0.8571
	7		1	17	$16/17 = 0.94118$	$0.8571 \times 0.9412 = 0.8067$
	9	1				
	10	1	1	15	$14/15 = 0.93333$	$0.8067 \times 0.9333 = 0.7529$
	11	1				
	13		1	12	$11/12 = 0.91667$	$0.7529 \times 0.9167 = 0.6902$
	16		1	11	$10/11 = 0.90909$	$0.6902 \times 0.9091 = 0.6275$
	17	1				
	19	1				
	20	1				
	22		1	7	$6/7 = 0.85714$	$0.6275 \times 0.8571 = 0.5378$
	23		1	6	$5/6 = 0.83333$	$0.5378 \times 0.8333 = 0.4482$

The conditional probability of being in remission beyond the sixth week is $18/21 = 0.8571$. Since this is the first observed failure time, the estimated value corresponds to the cumulative probability of remission $\hat{P}(6)$. Note that, by convention, the patient whose time is censored at six weeks is still considered at risk at that time but is dropped from the risk set immediately after. Therefore, the set at risk at week 7 is $21 - (3 + 1) = 17$ and the corresponding conditional probability of remission is $16/17 = 0.9412$. Consequently, $\hat{P}(7) = (18/21) \times (16/17) = 0.8067$. The value of $\hat{P}(t)$ is successively "updated" in an analogous way at each distinct relapse time.

In the placebo group, no censoring is present, since all 21 subjects have a relapse during the observation period. Therefore, the probability that remission duration is 23 weeks or more is estimated to be zero. The corresponding percent cumulative probability in the 6-MP group is some 45%, and no other event occurs after week 23, up to week 35, with five patients still in complete remission.

The estimated remission probability $\hat{P}(t)$ may be graphed against time t. The resulting curve is a step function that has a new step, i.e. changes value, at each distinct failure time. The survival curve starts from value one, at time zero, since all patients are in remission at their entry time. The curve does not change at censored times, but censored observations influence the height of the steps by "eroding" the set of patients at risk. The graphical representations of the cumulative probability of remission estimated in the two treatment groups are represented in figure 3.1.

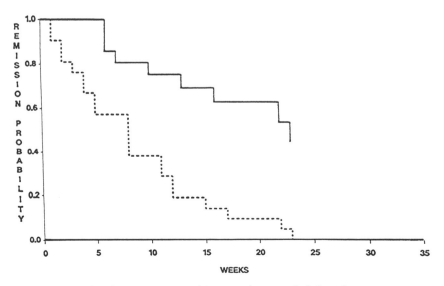

Figure 3.1 Graphical representation of the cumulative probability of remission estimated in table 3.1 for patients treated with 6-MP (——) and placebo (----).

These results suggest a superiority of treatment over placebo that can be properly tested and estimated (see Chapter 4), taking into account all the time points, and not just the last point estimate. The fact that, for the 6-MP group, the value $\hat{P}(t) = 0.4482$ remains unchanged from week 23 up to 35 must be considered with great caution. The "tail" of a survival curve, when only a few patients are at risk, is very unstable (see section 3.4); the graph of the survival curve is displayed here up to the last event only.

3.2.2   Formal definition of the product limit estimate

Suppose failure and censored times $\{t_1, t_2, \ldots, t_j, \ldots, t_N\}$ are known on N subjects in a random sample.

Let $t_{(1)} < t_{(2)} < \cdots < t_{(j)} \cdots < t_{(J)}$, $J \leqslant N$, be the distinct ordered failure times observed among N patients, where now (j) indicates the jth ordered time and not the jth subject. Let d_j $(j = 1, \ldots, J)$ be the corresponding numbers of failures with $d_j \geqslant 1$. Multiple failures occurring at the same time $t_{(j)}$ are called tied observations. The number n_j of patients "at risk" is the number of subjects actually observed when failures occur, that is the number of subjects who have either failure or censored times greater than or equal to $t_{(j)}$. The reasonable convention is adopted that censored times, due to either loss or withdrawal, follow failures in case the recorded times coincide. The sample conditional probability of surviving beyond time $t_{(j)}$, $p(t_{(j)})$, is more simply indicated by p_j, according to the notation in table 3.1. The estimate of p_j is

$$\hat{p}_j = \frac{n_j - d_j}{n_j} = 1 - \hat{q}_j \tag{3.1}$$

where $\hat{q}_j = d_j / n_j$ is the estimated conditional probability of failure at $t_{(j)}$.

The proportion $P(t)$ in the population surviving at t is estimated by the product

$$\hat{P}(t) = \prod_{j|t_{(j)} \leqslant t} \hat{p}_j \tag{3.2}$$

The product in (3.2) runs over the distinct failure times only, so that the value of $\hat{P}(t)$ only changes at time points $t_{(j)}$, $j = 1, \ldots, j$, when at least one new failure occurs. By definition of conditional probability, time units at which no failures occur contribute with unit factors at the estimation of $P(t)$ and can therefore be neglected in (3.2). Consequently, the value of the estimate of $P(t)$ at $t_{(j-1)}$, $\hat{P}(t_{(j-1)})$, will vary at $t_{(j)}$ by a factor $\hat{p}_j$ as follows: $\hat{P}(t_{(j)}) = \hat{P}(t_{(j-)}) \times \hat{p}_j$. This last is the recursive formula adopted in table 3.1 (column (6)) for the calculation of $\hat{P}(t)$.

The probability of surviving beyond the start of observation is defined to be $\hat{P}(0) = 1$. If the greatest observed lifetime t' corresponds to a failure time, then $\hat{P}(t') = 0$ and $\hat{P}(t) = 0$ for $t > t'$. If t' corresponds to a censored time, $\hat{P}(t)$ with $t > t'$ can be regarded as lying between 0 and $\hat{P}(t')$, but is not more closely defined.

The corresponding graph of $\hat{P}(t)$ is a right continuous step function: the value of $\hat{P}(t)$ at each failure time $t_{(j)}$ is the one on the right of $t_{(j)}$.

The estimator (3.2) of $P(t)$ for censored data is called the product limit estimator. Since it was derived by Kaplan and Meier (1958), is it often called the Kaplan–Meier (K–M) estimator. Despite censored data, the K–M method allows the user to estimate the proportion $P(t)$ of subjects in the population whose survival time would exceed t (in the absence of such censoring), provided that survival is independent of the censoring process. This important assumption was discussed in section 1.3. The product limit method provides a minimum variance unbiased estimator of $P(t)$ since it is the distribution, unrestricted as to form, which maximizes the likelihood of the observation (Appendix A3.1).

It is worth noting that the K–M estimator reduces to the usual binomial estimate of the probability of surviving when censoring is absent. Consider the group treated with placebo in table 3.1, where every subject relapses. According to (3.2), $\hat{P}(t)$ is a product of subsequent proportions $\hat{p}_j$ (column (5), table 3.1) in which all integers, except the first denominator N and the last numerator considered, cancel out. In other words, $\hat{P}(t)$ reduces to the observed proportion of survivors beyond t over all entered patients. For example, since 14 patients out of 21 are still in remission beyond the fourth week, the cumulative probability $P(4)$ is estimated by the proportion 14/21. The use of the product of conditional probabilities to estimate the cumulative probability is therefore superfluous when no censoring is present. It is instead the only way to account for censoring when estimating survival curves.

An alternative formulation of (3.2) often given in the case of continuous time, i.e. in absence of ties, is based on ranks. In a sample of N individuals, relabel the failures and censored times in order of increasing magnitude $t_{(1)} < t_{(2)} < \cdots < t_{(j)} \cdots < t_{(N)}$. The index j, $j = 1, \ldots, N$, is the rank corresponding to the subject whose observation time is the jth one in the sequence. If $t_{(j)}$ represents a failure time for a particular value of j, the corresponding conditional probability of surviving $t_{(j)}$, given alive just before $t_{(j)}$, is estimated by $\hat{p}_j = (N - j)/(N - j + 1)$.

The estimate of $P(t)$ is then

$$\hat{P}(t) = \prod_{j | t_{(j)} \leq t} \frac{N - j}{N - j + 1} \tag{3.3}$$

provided that the product only runs through those times $t_{(j)} \leq t$ which are failure times and that $\hat{P}(0) = 1$.

Again, if no censoring is present, and therefore ranked times are all failures, the product in (3.3) simplifies to a single proportion and the complete formulation of $\hat{P}(t)$ is as follows:

$$\hat{P}(t) = \begin{array}{ll} 1 & \text{for } t < t_{(1)} \\ \dfrac{N-j}{N} & \text{for } t_{(j)} \leqslant t < t_{(j+1)} \\ 0 & \text{for } t > t_{(N)} \end{array} \tag{3.4}$$

Beside describing survival in the sample, $\hat{P}(t)$ may also be regarded as an empirical estimate of the population distribution. The term "empirical" is used here as synonymous of non-parametric: it indicates that the estimate is obtained without any recourse to assumptions of functional form for the distribution.

The product limit estimator can be easily modified to account for left truncation (Cnaan and Ryan, 1989).

3.3 LIFE TABLES: FUNDAMENTALS AND CONSTRUCTION

The life table method was originally developed by demographers and actuaries to describe the lifetime of a population. A population life table depicts the length of life of a hypothetical cohort, followed from birth to death, that is assumed to experience the same mortality as that estimated from the observed population. The estimated age and sex-specific death rates are applied to the hypothetical cohort, usually of 100 000 or 1 000 000 individuals. For each age class, the number of events and the survivors entering the following age class are calculated on the basis of the corresponding age-specific death rate estimated in the population. From life table data it is possible to calculate the expected age of death of an individual of a given age, the probability of surviving from one age to another and other related quantities (Elandt-Johnson and Johnson, 1980, pp. 83–127).

Since the 1950s the method has been found to be easily adaptable to deal with grouped mortality data with censored observations. Especially since Cutler and Ederer's paper in 1958, the life table method has been used to estimate the probability of survival from a sample. Cutler and Ederer gave life table estimates of the yearly survival probabilities of Connecticut residents with localized kidney cancer diagnosed in the period 1946–1951. The sample may be thought to represent a specific cohort. This could be defined as a group of individuals who experienced the same initial event (i.e. diagnosis of kidney cancer) and were characterized by some common features (i.e. residence in Connecticut).

Historically, the product limit method was derived from the life table method. The product limit method needs the time at which each subject experiences a failure or is censored. On the contrary, the life table method applies to grouped data. In this case, observation time is subdivided into intervals of fixed length; the number of subjects alive at the beginning of each interval and the number of failures and censored data (due to loss or withdrawal) occurring during each interval are the only "ingredients" needed to estimate survival.

However, the basic concepts underlying the life table estimate of the survival curve are the same as those discussed in section 3.2.1 for the product limit: the probability of surviving a long period of time is computed by multiplying the conditional probabilities of surviving each of the intervals constituting it.

3.3.1 The life table estimate

Given a sample of N individuals, let the time axis be partitioned into j intervals $I_j = [t_j, t_{j+1})$, $j = 0, 1, \ldots, J-1$. (The notation [) is equivalent to the notation $t_j \leqslant t < t_{j+1}$.) The length of the jth interval is $h_j = t_{j+1} - t_j$. The times t_0 and t_J are defined to be respectively zero (the origin of the time axis) and an upper limit of the observation times. For the jth interval I_j, let

n'_j = number of survivors at the beginning of the jth interval;
c_j = number of censored observations in the jth interval;
d_j = number of failures which occurred in the jth interval.

It follows that $n'_0 = N$ and $n'_{j+1} = n'_j - c_j - d_j$ for $j = 0, \ldots, J-1$. The number c_j of censored observations is the sum of the number lost to follow-up and the number withdrawn alive during the jth interval.

In the absence of censored data ($c_j = 0$, $j = 0, \ldots, J-1$), an obvious estimate of q_j, the conditional probability of death in the interval $I_j = [t_j, t_{j+1})$, given alive at t_j, would be d_j/n'_j. However, when censoring is present, this would generally underestimate q_j, since subjects who withdraw alive or are lost to follow-up in I_j might have failed during the interval, had they not been censored.

To define the number of patients at risk during the interval I_j we assume that, on average, censoring occurs at the mid-point of the interval I_j. This means that each individual with censored time corresponds to half an individual alive at the beginning of the interval and exposed to risk for the full interval. Therefore the "set at risk", defined as n_j = "effective number of initial exposed to risk", is obtained by subtracting from the number n'_j of subjects entering the interval half of those with censored observations. The actuarial estimator of q_j is

$$\hat{q}_i = \frac{d_j}{n'_j - \frac{1}{2}c_j} = \frac{d_j}{n_j} \tag{3.5}$$

This estimator was introduced without using probability theory and provides a rather crude adjustment for the calculation of the set at risk n_j. Nonetheless, it is sensible in most situations, especially when censoring is limited and occurs evenly within the intervals. In a probabilistic framework, this adjustment derives from taking the mean number of censored subjects as being at risk after assuming a uniform distribution of withdrawal and loss during the interval. If it were known that all censored data occurred either at the beginning of the interval or right at the end of it, the appropriate estimator of q_j would be respectively $\hat{q}_j = d_j/[n'_j - (w_j + l_j)]$ or $\hat{q}_j = d_j/n'_j$. No qualitative difference is assumed to exist between censoring due to withdrawal or loss, as discussed in section 1.3. The actuarial estimator of q_j considers that the subjects who failed in the interval I_j would have been exposed to risk for the entire interval, had they not failed.

Once q_j is estimated by (3.5), the conditional probability of surviving the interval I_j is $\hat{p}_j = 1 - \hat{q}_j$.

Let us denote by P_j the cumulative probability of surviving all intervals preceding I_j, that is, the probability of surviving up to time t_j. This quantity will be estimated by the product of the conditional probabilities of surviving each interval before I_j:

$$\hat{P}_j = \hat{p}_0 \times \hat{p}_1 \times \hat{p}_1 \times \cdots \times \hat{p}_{j-1} = \hat{P}_{j-1} \times \hat{p}_{j-1}, \quad j = 1, \ldots, J \qquad (3.6)$$

The last product in (3.6) shows the recursive formula for the estimation of P_j: surviving to the start of the interval I_j is equivalent to surviving to the start of the interval I_{j-1} and then surviving the entire I_{j-1}. It is defined that $\hat{P}_0 = \hat{P}(0) = 1$. Note that no estimate is possible beyond time t_J, although $\hat{P}(t)$ with $t > t_J$ can be thought of as lying between $\hat{P}_J$ and 0.

The structure of the life table is described in table 3.2.

The life table method may be applied to survival data such as those presented in table 3.3 (columns (1)–(5)). A retrospective study conducted at the Mayo Clinic reported the survival experience of 2418 males with angina pectoris diagnosed from 1/1/1927 to 31/12/1936 (Parker *et al.*, 1946). For each yearly interval, the collected data were the number of patients alive and under observation at the beginning of the interval, the number of deaths and the number of censored times observed during the interval. Since this was a retrospective study, censored data arose because it was not always possible to update follow-up. The number of patients at risk (column (6)) is calculated by subtracting half of the censored patients from those entering the interval.

The values of $\hat{q}_j$ and $\hat{P}_j$ are given in columns (7) and (9). For illustration, suppose we wish to estimate the 10-year survival. We see that survival at nine years is $\hat{P}_9 = 0.3342$ and we calculate that the conditional probability of surviving the tenth year $(I_9 = [9, 10))$ is $\hat{p}_9 = 1 - \hat{q}_9 = 1 - 42/[427 - \frac{1}{2}(64)] = 353/395 = 0.8937$. Consequently, the probability of surviving 10 years is about 30%, given that $\hat{P}_{10} = 0.3342 \times 0.8937 = 0.2987$.

Note that the probability of surviving 15 years is $\hat{P}_{15} = 0.1429$. Thirty subjects are still alive at 15 years, but no follow-up is known on them. As time increases (to infinity) we would observe 30 deaths more. The survival curve will drop to zero but we have no data to estimate how this will occur in time.

Apart from the adjustments we need to introduce because of censoring, the life table is essentially equivalent to a nonparametric method for estimating

Table 3.2 Structure of the life table

j	$I_j = [t_j, t_{j+1})$	h_j	n'_j	c_j	d_j	n_j	$\hat{q}_j$	$\hat{p}_j$	$\hat{P}_j = \hat{P}(t_j)$
0	$[t_0, t_1)$	$t_1 - t_0$	$n'_0 = N$	c_0	d_0	n_0	$\hat{q}_0$	$\hat{p}_0$	$\hat{P}_0 = 1$
1	$[t_1, t_2)$	$t_2 - t_1$	n'_1	c_1	d_1	n_1	$\hat{q}_1$	$\hat{p}_1$	$\hat{P}_1 = \hat{p}_0$
$\cdots$	$\cdots$	$\cdots$	$\cdots$	$\cdots$	$\cdots$	$\cdots$	$\cdots$	$\cdots$	$\cdots$
j	$[t_j, t_{j+1})$	$t_{j+1} - t_j$	n'_j	c_j	d_j	n_j	$\hat{q}_j$	$\hat{p}_j$	$\hat{P}_j = \hat{p}_0 \times \hat{p}_1 \times \cdots \times \hat{p}_{j-1}$
$\cdots$	$\cdots$	$\cdots$	$\cdots$	$\cdots$	$\cdots$	$\cdots$	$\cdots$	$\cdots$	$\cdots$
$J-1$	$[t_{J-1}, t_J)$	$t_J - t_{J-1}$	n'_{J-1}	c_{J-1}	d_{J-1}	n_{J-1}	$\hat{q}_{J-1}$	$\hat{p}_{J-1}$	$\hat{P}_{J-1} = \hat{p}_0 \times p_1 \times \cdots \times \hat{p}_{J-2}$
J	$t_J +$	—	n'_j	c_j	0	—	—	—	$\hat{P}_J = \hat{p}_0 \times p_1 \times \cdots \times \hat{p}_{J-1}$

Table 3.3 Life table estimation of $P(t)$ on data from Parker *et al.* (1946) by yearly intervals.

Years after diagnosis j (1)	Interval $[t_j, t_{j+1})$ I_j (2)	No. entering interval n'_j (3)	No. censored (Lost to follow-up) c_j (4)	Deaths d_j (5)	Set at risk n_j (6)	Conditional probability of death $\hat{q}_j$ (7)	Conditional probability of survival $\hat{p}_j$ (8)	Cumulative probability of survival $\hat{P}_j$ (9)
0	[0, 1)	2418	0	456	2418.0	0.1886	0.8114	1.0000
1	[1, 2)	1962	39	226	1942.5	0.1163	0.8837	0.8114
2	[2, 3)	1697	22	152	1686.0	0.0902	0.9098	0.7170
3	[3, 4)	1523	23	171	1511.5	0.1131	0.8869	0.6524
4	[4, 5)	1329	24	135	1317.0	0.1025	0.8975	0.5786
5	[5, 6)	1170	107	125	1116.5	0.1120	0.8880	0.5193
6	[6, 7)	938	133	83	871.5	0.0952	0.9048.	0.4611
7	[7, 8)	722	102	74	671.0	0.1103	0.8897	0.4172
8	[8, 9)	546	68	51	512.0	0.0996	0.9004	0.3712
9	[9, 10)	427	64	42	395.0	0.1063	0..8937	0.3342
10	[10, 11)	321	45	43	298.5	0.1441	0.8559	0.2987
11	[11, 12)	233	53	34	206.5	0.1646	0.8354	0.2557
12	[12, 13)	146	33	18	129.5	0.1390	0.8610	0.2136
13	[13, 14)	95	27	9	81.5	0.1104	0.8896	0.1839
14	[14, 15)	59	23	6	47.5	0.1263	0.8737	0.1636
15	15+	30	0	0				0.1429

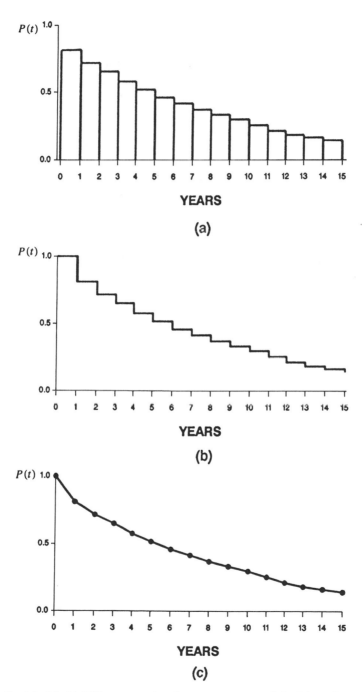

Figure 3.2   (a), (b), (c) Different graphical representations of the life table estimate of survival of subjects in table 3.3

a cumulative survival distribution. The corresponding histogram consists of a series of vertical bars, one for each interval, whose height represents the relative frequency of observations equal to or greater than the upper limit of the interval, i.e. the frequency of subjects surviving the interval. Figure 3.2(a) shows the histogram corresponding to data in table 3.3. However, a graphical representation in the form of a survival curve is preferred since it is visually more appealing and clearly asserts that the probability of surviving has value one at the start of observation. A curve based on life table estimates may be represented in two different ways, neither of which is as immediate as the graphical representation of the K–M estimate.

Figure 3.2(b) shows the survival curve corresponding to table 3.3 as a step function. What is represented on the y-axis is, in fact, the cumulative proportion surviving at the beginning of the interval: it takes value $\hat{P}_0 = 1$ at time t_0, and changes to value $\hat{P}_1$ at time t_1, and to value $\hat{P}_j$ at time t_j. The first step at t_1 represents the decrease in survival observed during the first interval, the second step at t_2 that observed in the second interval, and so on.

The same data are represented in figure 3.2(c) by plotting the cumulative probability of surviving against the starting points of time intervals and by connecting them with straight lines. This makes the curve easy to look at. However, it must be kept in mind that the straight lines do not represent survival at intermediate points in the intervals, which is not estimated in life tables.

3.3.2   Comparison of the life table and product limit estimates

More and more often in clinical studies individual data are available both on time to failure and on time to last seen alive. These data are in what is called "continuous" form, rather than grouped. This usually leads to the use of the K–M estimator for the survival curve. However, in large data sets, with many failures, the life table method may be preferred if a simpler and more concise representation of cumulative survival is wanted. The life table may be constructed provided that the continuous data are grouped by time intervals, the length of which, usually the same in grouped data sets, may nonetheless be varied. When the individual censoring times are known, an estimator of the effective number of initial subjects exposed to risk n_j which generalizes the denominator of (3.5) may be adopted. Indicating by θ_{ij} the portion of time an individual i, lost or withdrawn, has been at risk in the interval I_j, the number n'_j of subjects entering interval $I_j = [t_j, t_{j+1})$ may be adjusted as follows to obtain the set at risk n_j:

$$n_j = n'_j - \sum_{i=1}^{c_j} (1 - \theta_{ij})$$

This expression reduces to the n_j given in (3.5) if θ_{ij} is assumed to be $\frac{1}{2}$ for all individuals with censored data. However, the simple actuarial estimator $\hat{q}_j$ given in (3.5) is a good and robust estimator of q_j. For a review of different estimators of q_j proposed in the literature and their statistical properties, see Elandt-Johnson and Johnson (1980, pp. 162–172).

As an illustration, the K–M estimator (3.2) and the actuarial estimator (3.6) of $P(t)$ are applied to the same data set. This includes continuous data on survival of 349 women with early breast cancer, who underwent radical surgery (Halsted mastectomy) according to the protocol outlined in figure 1.3. Ten individual records randomly selected from the entire data set are shown in table 3.4. The entry into the study corresponds to the date of surgery (column (2)). The study of long-term survival was conducted by fixing the cut-off date of 31/12/91, more than 10 years after the end of the recruitment period. Follow-up was updated during the first semester of 1992 by consulting records from the Registry Office for those subjects who were last seen before the cut-off date. The resulting information is shown in columns (3) and (4).

The random variables (T, δ), the observation time and the indicator of failure, respectively, are represented in column (5). When $\delta = 1$, T is the difference, in years, between dates of death and surgery; when $\delta = 0$, T is the difference between the cut-off date and surgery, unless the last update is before 31/12/91. In such cases (observations 3 and 4) time is computed to the last follow-up date. Among censored data, subjects 3 and 4 and subjects 2, 7, 8 and 10 are lost to follow-up and withdrawn alive, respectively.

Figure 3.3 shows the survival curves of women who underwent radical surgery obtained with the two approaches. Clearly, the more irregular picture is that representing the K–M estimate, with a step at each of the 103 deaths observed in the sample. The life table estimate is shown by the step function with yearly intervals. It is correct to "read" this curve at the start of each interval,

Table 3.4 Raw data on survival of 10 patients from a trial on treatment of early breast cancer (figure 1.3).

Observation no. (1)	Date of surgery (2)	Date of last contact (3)	Date of death (4)	(T, δ) (years) (5)
1	27/06/73	.	14/04/84	(10.81, 1)
2	17/07/73	27/01/92	.	(18.49, 0)
3	28/01/74	18/11/91	.	(17.82, 0)
4	29/04/76	08/01/91	.	(14.70, 0)
5	17/01/77	.	28/09/85	(8.70, 1)
6	19/04/78	.	15/12/88	(10.67, 1)
7	24/07/78	12/05/92	.	(13.45, 0)
8	26/07/78	08/06/92	.	(13.43, 0)
9	23/04/80	.	08/03/85	(4.88, 1)
10	30/04/80	11/03/92	.	(11.68, 0)

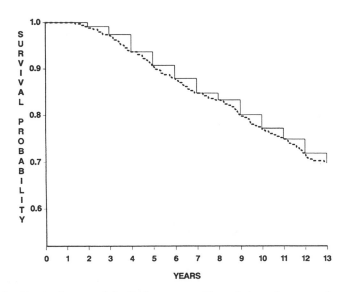

Figure 3.3   Survival curve of the 349 women with early breast cancer who underwent radical surgery according to the protocol outlined in figure 1.3: K–M estimate (----) and life table estimate (——).

while the horizontal lines through the interval are just a graphical device. If we connect the estimates at the starting points of the intervals with straight lines as in figure 3.2(c), and read the values corresponding to time points inside the intervals, we assume a linear decrease which may be unrealistic.

If the survival estimate is needed in more detail, even adopting the life table method, then it suffices to define shorter time intervals, for example monthly intervals. Actually, thinking of time intervals that become so small that each includes only one death leads to the concept introduced by Kaplan and Meier. Long-term survival following radical surgery in early breast cancer would probably be presented by giving survival estimates and standard errors at few time points, say at 5, 10 and 15 years. Yearly survival estimates obtained with the two approaches are practically superimposed as is clear also from the graphs in figure 3.3. For example, the estimated 10-years percent survival is 77.22 and 77.23 respectively with the life table and the K–M estimator.

3.4   MEASURES OF PRECISION OF THE SURVIVORSHIP PROBABILITY ESTIMATES

Both the K–M and life table estimates of $P(t)$ discussed in the previous sections are subject to random variation like any statistic; thus it is desirable that they are reported together with information on their inherent imprecision worded in terms of standard error or of $(1 - \alpha)$ 100% confidence interval. In reports dealing with results of clinical trials, however, confidence intervals of $P(t)$ must be regarded with a great deal of caution for the considerations already stated in section 2.4. In a clinical trial, in fact, the random allocation of treatments to patients justifies the process of statistical inference (test of significance and confidence limits) on the difference of treatment effects, and only an informal procedure allows the researcher to generalize the conclusions to the advocated population. On the other hand, for any useful inferences to be made on $P(t)$, the assumption of random sampling is needed, but the trial is performed on a set of patients who can rarely be regarded as a random sample of any well-defined population. The clinical trial, particularly if carried out in one or only a few institutions and small sized, is a poor tool for describing populations suffering from a disease. In the examples which follow, attention will be focused on computation procedures only, disregarding the above-mentioned problems.

3.4.1 Standard errors and confidence intervals

In the absence of censoring, the estimate of $P(t)$ given by (3.2) reduces to the usual binomial estimate, the standard error of which is estimated to be (see Kish, 1965, p. 46):

$$SE[\hat{P}(t)] = \sqrt{\frac{\hat{P}(t)[1 - \hat{P}(t)]}{N - 1}}$$

The $(1 - \alpha)$100% confidence interval may be obtained by the following Gaussian approximation to the binomial distribution:

$$\hat{P}(t) \pm z_{1-\alpha/2}SE[\hat{P}(t)] \tag{3.7}$$

where $z_{1-\alpha/2}$ is the standardized Gaussian deviate.

This interval is symmetric about the point estimate. The sampling distribution, however, is asymmetric except for $\hat{P}(t)$ around 0.5 and the symmetrical interval may be inaccurate. It is not unusual, for instance, to obtain a small but unstable estimate for $P(t)$ with an associated interval estimate yielding a lower bound below 0.0, outside the admissible range. Theoretically, in order to reflect the asymmetry of the sampling distribution, the limits should be asymmetric around the estimate of $P(t)$ and within the admissible range. This aim can be pursued by resorting to the statistic

$$z = \frac{\hat{P}(t) - P(t)}{\sqrt{\frac{P(t)[1 - P(t)]}{N}}}$$

from which a quadratic equation in $P(t)$ is easily derived. Therefore the $(1 - \alpha)$ 100% confidence limits of $P(t)$ are obtained as solutions of the following equation (see Armitage and Berry, 1987, pp. 117–118):

$$[P_L(t), P_U(t)] = \frac{(2N\hat{P}(t) + z_{1-\alpha/2}^2) \pm z_{1-\alpha/2}\sqrt{4N\hat{P}(t)[1 - \hat{P}(t)] + z_{1-\alpha/2}^2}}{2(N + z_{1-\alpha/2}^2)} \quad (3.8)$$

These limits have been shown to be accurate even for small values of $P(t)$.

With reference to the Placebo group of table 3.1, the estimate of the standard error of $\hat{P}(15)$ is

$$SE[\hat{P}(15)] = \sqrt{\frac{0.142\,86 \times 0.857\,14}{20}} = 0.078\,25$$

and the 90% confidence interval of $P(15)$ by means of (3.8) is

$$[P_L(15), P_U(15)] = \frac{42 \times 0.142\,86 + 2.706\,02 \pm 1.645\sqrt{4 \times 21 \times 0.142\,86 \times 0.857\,14 + 1.645^2}}{2(21 + 1.645^2)}$$

$$= (0.058\,57, 0.308\,69)$$

In the presence of censoring, the asymptotic estimate of the variance of the life table estimator of survival at t_j, P_j, was first proposed by Greenwood (1926):

$$var_G(\hat{P}_j) = \hat{P}_j^2 \sum_{l=0}^{j-1} \frac{d_l}{(n_l - d_l)} \quad (3.9)$$

and is derived in Appendix A3.2. This formula has the advantage that it can be calculated interactively; in other words, the calculation of $var_G(\hat{P}_j)$ requires adding one more term to the sum needed for $var_G(\hat{P}_{j-1})$.

From table 3.3, the probability of surviving at two years from diagnosis is estimated to be $\hat{P}_2 = 0.7170$ and by (3.9) the estimate of its standard error is

$$SE_G(\hat{P}_2) = 0.7170\sqrt{\frac{456}{2418 \times (2418 - 456)} + \frac{226}{1942.5 \times 1716.5}} = 0.009\,18$$

The estimate $SE_G[\hat{P}_j]$, when inserted in (3.7), gives confidence limits suffering the same drawbacks we previously commented upon.

Rothman (1978) suggested resorting also in this case to the quadratic equation (3.8) after replacing N with the "effective sample size" N_j^*. This latter is obtained by equating the Greenwood formula to the variance of a proportion and is (Cutler and Ederer, 1958):

$$N_j^* = \frac{(1 - \hat{P}_j)}{\hat{P}_j \sum_{l=0}^{j-1} \frac{d_l}{n_l(n_l - d_l)}} \qquad (3.10)$$

Note that N_j^* does not correspond to the set at risk n_j but is an indicator of the information available in the intervals preceding t_j in terms of sample size under observation. In the previous example:

$$N_2^* = \frac{1 - 0.7170}{0.7170\left(\dfrac{456}{2418 \times 1962} + \dfrac{226}{1942.5 \times 1716.5}\right)} = 2408.19$$

This result enables us to assess that the estimate $\hat{P}_2 = 0.7071$ is as reliable as one based on a case series of 2408 patient starting the trial with *all of them* followed until death or survival for two years.

Note that the Greenwood formula is accurate only asymptotically, for increasingly large values of n_j. It underestimates the true variance of $\hat{P}_j$ in the life table tail which is often the region of greatest medical concern. Consequently the effective sample size computed by (3.10) is overestimated and, coherently, the coverage of the confidence interval is less than the expected $(1 - \alpha)100\%$. To compensate for this, Peto *et al.* (1977) suggested defining the effective sample size as the number of patients at risk at time t_j divided by the survival probability at t_j:

$$N_j^{**} = \frac{n_j'}{\hat{P}_j} \qquad (3.11)$$

This effective sample size, when used in the usual formula for the binomial variance of a proportion, gives

$$var_P[\hat{P}_j] = \frac{\hat{P}_j^2(1 - \hat{P}_j)}{n_j'} \qquad (3.12)$$

The pertinent confidence interval may be computed by inserting N_j^{**} instead of N in (3.8) (Simon and Lee, 1982).

The formula (3.11) has been justified by arguing that, in order to observe n_j' patients at t_j where the survival probability is $\hat{P}_j$, there must be initially at least $n_j'/\hat{P}_j$ patients. If more patients were accrued, the life table estimate of P_j would be somewhat more reliable than this. Peto *et al.* (1977) state: "This estimate is therefore usually conservative, although the actual degree of conservatism is often surprisingly slight, and is counterbalanced by the fact that the formula

Table 3.5 Effective sample size according to Cutler and Ederer (1958) (N_j^*) and Peto et al. (1977) (N_j^{**}) and 90% confidence limits computed according to Rothman (1978) and Simon and Lee (1982). Data from table 3.3.

Years after diagnosis j (1)	No. entering interval n_j' (2)	Set at risk n_j (3)	Cumulative probability of surviving P_j (4)	Effective sample size N_j^* (5)	N_j^{**} (6)	90% confidence intervals according to Rothman L (7)	U (8)	90% confidence intervals according to Simon and Lee L (9)	U (10)
0	2418	2418.0	1.0000	2418.00	2418.00				
1	1962	1942.5	0.8114	2408.19	2366.81	0.797 97	0.824 13	0.797 97	0.824 13
2	1697	1686.0	0.7170	2392.79	2334.46	0.701 66	0.731 85	0.701 53	0.731 98
3	1523	1511.5	0.6524	2371.84	2296.92	0.636 22	0.668 23	0.636 02	0.668 43
4	1329	1317.0	0.5786	2350.68	2253.03	0.561 84	0.595 18	0.561 57	0.595 45
5	1170	1116.5	0.5193	2306.52	2034.27	0.502 34	0.536 22	0.501 97	0.536 58
6	938	871.5	0.4611	2226.23	1730.58	0.444 08	0.478 21	0.442 98	0.479 32
7	722	671.0	0.4172	2085.61	1470.91	0.400 12	0.434 48	0.397 85	0.436 81
8	546	512.0	0.3712	1937.41	1277.68	0.353 98	0.388 76	0.350 73	0.392 14
9	427	395.0	0.3342	1765.95	1074.66	0.316 81	0.352 05	0.312 86	0.356 24
10	321	298.5	0.2987	1537.43	911.22	0.281 10	0.316 91	0.276 26	0.322 15
11	233	206.5	0.2557	1292.80	683.52	0.237 84	0.274 42	0.232 68	0.280 17
12	146	129.5	0.2136	1083.85	516.59	0.195 46	0.232 94	0.188 97	0.240 49
13	95	81.5	0.1839	910.09	360.64	0.165 34	0.204 04	0.157 53	0.213 56
14	59	47.5	0.1636	692.49	209.94	0.144 43	0.184 76	0.134 09	0.198 12
15	30		0.1429			0.122 41	0.166 17	0.107 71	0.187 18

deals appropriately with the increasing uncertainty that should properly be expected as one goes along the long flat region with which many life tables finish."

Table 3.5 reports N_j^*, N_j^{**} and 90% confidence intervals computed according to Rothman (1978) and Simon and Lee (1982) on data from table 3.3. Columns (5) and (6) show that N_j^* and N_j^{**} start to diverge markedly after six years from diagonosis because of the evident loss to follow-up from the fifth year from diagnosis onwards. The Rothman and Simon–Lee confidence intervals are practically equal up to 10 years from diagnosis, after which the Rothman intervals tend to be narrower, though only slightly, than the Simon and Lee ones.

With continuous failure time, as is assumed in computing the K–M esttimator, a proper derivation of the Greenwood variance (3.9) requires that expression (3.2) be thought of as a stochastic process in t. Under mild conditions on censoring it may be shown that (3.2) is asymptotically a Gaussian process; this allows us to compute approximate $(1 - \alpha)100\%$ confidence intervals for the survivorship function $P(t)$ by means of (3.8). Coherently with (3.2) the Greenwood formula is now

$$var_G[\hat{P}(t)] = \hat{P}^2(t) \sum_{j|t_{(j)} \leqslant t} \frac{d_j}{n_j(n_j - d_j)} \tag{3.9a}$$

and the effective sample size is now

$$N^*(t) = \frac{1 - \hat{P}(t)}{\hat{P}(t) \sum_{j|t_{(j)} \leqslant t} \frac{d_j}{n_j(n_j - d_j)}} \tag{3.10a}$$

In order to compute $N^{**}(t)$ and $var_P[\hat{P}(t)]$, $(n_j - d_j)$ must be substituted for n_j' in (3.11) and (3.12) respectively.

Table 3.6 is the counterpart of table 3.5 for data from the 6-MP group in table 3.1. The width of the confidence intervals shows how the uncertainty increases as one moves to the final part of the curve. This example includes a limited set of patients and, unlike the previous example where only slight differences were seen between the Rothman and Simon–Lee approaches, confidence intervals are now quite different in width.

Owing to the underestimation of the variance of $\hat{P}(t)$ due to the Greenwood formula, the Rothman procedure is expected to give intervals with confidence lower than 90% whilst the contrary is expected by adopting the Simon–Lee procedure. Dorey and Korn (1987) showed that a satisfactory compromise may be reached by using the Cutler–Ederer effective sample size, $N^*(t)$, for obtaining the upper confidence limit and the Peto *et al.* effective sample size, $N^{**}(t)$, for the lower confidence limit. This means defining confidence intervals whose lower

Table 3.6 Effective sample size according to Cutler and Ederer (1958) (N_j^*) and Peto *et al.* (1977) (N_j^{**}) and 90% confidence limits computed according to Rothman (1978) and Simon and Lee (1982). Data from table 3.1.

| Time | Events | Set at risk | Cumulative probability of surviving $\hat{P}(t)$ | Effective sample size | | 90% confidence intervals according to | | | |
| | | | | | | Rothman | | Simon and Lee | |
$t_{(j)}$ (1)	d_j (2)	n_j (3)	(4)	N_j^* (5)	N_j^{**} (6)	L (7)	U (8)	L (9)	U (10)
6	3	21	0.857 14	21.00	21.00	0.691 33	0.941 43	0.691 33	0.941 43
7	1	17	0.806 72	20.63	19.83	0.632 08	0.910 24	0.628 22	0.911 58
10	1	15	0.752 94	20.04	18.59	0.571 08	0.874 62	0.563 76	0.877 85
13	1	12	0.690 20	18.74	15.94	0.500 22	0.832 19	0.484 29	0.840 91
16	1	11	0.627 45	17.97	15.94	0.435 08	0.786 47	0.423 84	0.794 07
22	1	7	0.537 82	15.12	11.16	0.337 73	0.726 43	0.310 05	0.750 83
23	1	6	0.448 18	13.65	11.16	0.254 32	0.659 19	0.238 35	0.678 24

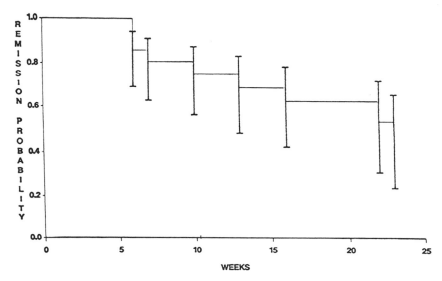

Figure 3.4 Remission probability curve for the 6-MP group of table 3.1, with Dorey–Korn 90% confidence intervals.

limits are the values reported in column (9) of tables 3.5 and 3.6 while the upper limits are reported in column (8). The curve of cumulative remission probability in the 6-MP group, together with Dorey–Korn confidence intervals, is drawn in figure 3.4.

3.4.2* Confidence bands

In several medical reports the upper (lower) confidence limits computed at each point estimate are connected to each other to draw the upper (lower) confidence limit for the entire survival curve. This procedure should be avoided because the $(1 - \alpha)100\%$ confidence refers to each of the J estimates of $P(t)$ but not to all of them simultaneously. Therefore a proper measure of reliability of the entire survival curve is a simultaneous confidence band covering the whole set of J survival probabilities.

In the absence of censoring, the desired band for the survival curve can be computed by means of standard distribution free methods suitable for constructing confidence bands for the population cumulative distribution (see Bradley, 1968). The sampling distribution of the maximum observed absolute deviation,

$$_N D = \max |\hat{P}(t) - P(t)|$$

used in the Kolmogorov–Smirnov test is the core of the procedure. Assume that

$_ND_\alpha$ is the critical value of $_ND$ corresponding exactly to a significance level α. Then $\Pr(\max|\hat{P}(t) - P(t)| < _ND_\alpha) = 1 - \alpha$; this means that there is an *a priori* $(1 - \alpha)$ probability that the sample step function $\hat{P}(t)$ will always be between the two curves whose ordinates are the bracketed values in the following inequality: $[P(t) - _ND_\alpha] < \hat{P}(t) < [P(t) + _ND_\alpha]$. But, since $\hat{P}(t)$ lies within $\pm _ND_\alpha$ of $P(t)$, it turns out that $P(t)$ is included within

$$\hat{P}(t) \pm _ND_\alpha \tag{3.13}$$

Therefore we can attach a confidence level of $(1 - \alpha)$ to the statement that $P(t)$ stays in the interval (3.13) for all $t \leqslant t_j$. For $\alpha = 0.20, 0.10, 0.05, 0.01$ and different values of N, the values of $_ND_\alpha$ are given in Appendix A3.3. This table is an abridged version of table XIII by Bradley (1968).

Let us now consider the patients accrued in the placebo group of the Freireich *et al.* (1963) trial. From Appendix A3.3, for $N = 21$ and $\alpha = 0.10$, $_{21}D_\alpha = 0.258\,58$; therefore, by using $\hat{P}(t)$ given in the last column of table 3.1 it is easy to compute the 90% confidence band: $\hat{P}(t) \pm 0.258\,58$. In the absence of censoring, the 90% confidence interval for $P(t)$ at each time point in which a relapse was recorded was computed by means of (3.8) with $N = 21$. The survival curve estimated for this example, with the 90% confidence band together with the 90% confidence limits of $P(t)$ for each time point in which a relapse occurred, are shown in figure 3.5. It appears evident that connecting

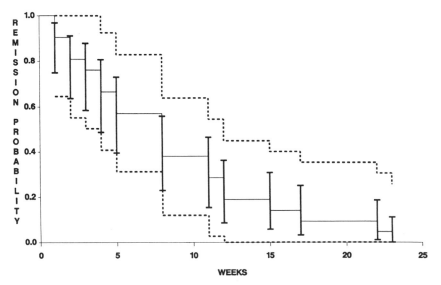

Figure 3.5 Remission probability curve for the Placebo group of table 3.1, with 90% confidence band.

the upper (lower) limits of $\hat{P}(t)$ leads to a dramatic underestimation of the width of the confidence region of the whole survival curve.

Hall and Wellner (1980) extended the above approach to make allowance for censoring. After defining:

$$C_N(t) = N \frac{var_G[\hat{P}(t)]}{\hat{P}^2(t)} = N \sum_{j|t_{(j)} \leqslant t} \frac{d_j}{n_j(n_j - d_j)}$$

$$\bar{K}_N(t) = [1 + C_N(t)]^{-1}$$

they proved that analogously with (3.13) the intervals

$$\hat{P}(t) \pm {}_ND_\alpha \left[\frac{\hat{P}(t)}{\bar{K}_N(t)} \right] \quad \text{for all } t \leqslant t_j \tag{3.14}$$

form an asymptotic $(1 - \alpha)100\%$ confidence band for the survival curve, based on the K–M estimator. Nair (1984) used a Monte Carlo simulation to investigate the achieved confidence levels of the bands under different sizes of the study $(N = 25, 50, 100)$ and with several extents of censoring. He was able to show that the coverage of the confidence band given by (3.14) is satisfactory even with a sample size as low as $N = 25$ and up to 50% censoring.

The computation procedure of the Hall–Wellner 90% confidence band for the survival curve of the 6-MP group of table 3.1 is shown in table 3.7. Despite the sample size ($N = 21$ is slightly less than the smallest sample size tested by Nair (1984)), the 6-MP group data have been utilized here only because this now familiar example is easy to process with a hand calculator.

From Appendix A3.3, ${}_{21}D_{0.10} = 0.258\ 58$; overall the censoring is some 30%. The factor $[\hat{P}(t)/\hat{K}_N(t)]$ making allowance for censoring increases as one moves towards the tail of the curve. Let us consider $t_{(j)} = 23$ weeks; in the absence of censoring, the band would be $0.448\ 18 \pm 0.258\ 58 = (0.189\ 60, 0.706\ 76)$. Owing to the observed censoring process, the critical value is increased, say, 1.30 times and the confidence band is widened to $0.448\ 18 \pm 0.258\ 58(1.297\ 00) = (0.112\ 80, 0.783\ 56)$. As a consequence the difference between the width of the Hall–Wellner band and the width of the band one obtains by the improper procedure of connecting the confidence limits computed at each $t_{(j)}$ is even larger than that shown in figure 3.5.

Note that Hall and Wellner assess that the correction factor inserted in (3.14) tends to make the bands somewhat conservative when the term $[1 - \bar{K}_N(t_j)]$ drops below 0.75. To compensate for this they provide a table of values which, divided by $\sqrt{N}$, must be substituted for ${}_ND_\alpha$ in (3.14). An abridged version of their table 1 is given in Appendix A3.4.

In the previous example $[1 - \bar{K}_N(t_j)] = 0.654\ 45$; by linear interpolation in the first row of table in Appendix A3.4 we obtain the critical value = 1.194. Since

Table 3.7 Computation procedure of Hall–Wellner 90% confidence band for the survival curve of the 6-MP group. Data from table 3.1.

Remission time (weeks) $t_{(j)}$	No. of events d_j	Set at risk n_j	$\sum_{t_{(j)} \le t} \dfrac{d_j}{n_j(n_j - d_j)}$	$C_N(t)$	$\hat{P}(t)$	$\bar{K}_N(t) = [1 + C_N(t)]^{-1}$	$\dfrac{\hat{P}(t)}{\bar{K}_N(t)}$	90% confidence band	
(1)	(2)	(3)	(4)	(5) = 21 × (4)	(6)	(7) = [1 + (5)]$^{-1}$	(8)	L (9)	U (10)
6	3	21	7.93651×10^{-3}	0.16667	0.85714	0.85714	1.00000	0.59856	1.00000
7	1	17	11.61298×10^{-3}	0.24387	0.80672	0.80394	1.00346	0.54725	1.00000
10	1	15	16.37488×10^{-3}	0.34387	0.75294	0.74412	1.01185	0.49130	1.00000
13	1	12	23.95064×10^{-3}	0.50296	0.69020	0.66535	1.03735	0.42196	0.95844
16	1	11	33.04155×10^{-3}	0.69387	0.62745	0.59036	1.06283	0.35262	0.90228
22	1	7	56.85107×10^{-3}	1.19387	0.53782	0.45582	1.17990	0.23272	0.84292
23	1	6	90.18440×10^{-3}	1.89387	0.44818	0.34555	1.29700	0.11280	0.78356

$1.194/\sqrt{21} = 0.2606$ is almost equal to the Kolmogorov–Smirnov critical value of $0.258\,58$, the use of this latter is justified.

3.5   CONSTRUCTION OF A SURVIVAL CURVE: SOME COMMENTS

Methods for calculating survival curves are easily applied, often by means of routine computer programs. However, seriously biased estimates may be obtained if calculations are done without carefully considering the definition of (1) time of entry into the study (entry point); (2) events of interest (end-point); and (3) modalities of follow-up

For the reader's comprehension and critical reading of study results, it is essential that survival plots are accompanied by notes on how the three points above have been handled. For the same reason, it is necessary, when reporting a survival curve, to describe some features of the analysed data set, such as number of patients included and excluded, reasons for exclusion, amount and cause of censoring.

3.5.1   Definition of the entry point

The definition of the "event" which gives rise to the observation influences the count of deaths and of patients at risk. In a non-randomized study this "event" must be clearly defined, depending on the study aims. For example, in describing the results achieved with a specific therapeutic protocol in the care of childhood leukemia, the entry may be chosen to be the date of either diagnosis or beginning of treatment. Coherently with this choice, data to be analysed will include or exclude patients who die very early, before treatment has even been started. The choice of the entry point should be the date of diagnosis if the study aims at describing the overall impact of a treatment on patients' survival. This, in fact, also depends on treatment being suitable for delivery as soon as possible after diagnosis.

In any randomized study, a patient enters the trial at the moment he is randomly assigned one of the treatments of choice. Therefore, date of randomization is the starting point in the calculation of survival or observation time, and not date of disease onset, of diagnosis or of beginning of treatment. For example, if dates of randomization and of start of treatment do not coincide, deaths between these two dates must be counted as events in the assigned treatment arm. Because of this, it is recommended, when planning a trial, that the randomization be scheduled as near as possible to the start of treatments to be compared. This will allow a straightforward evaluation and interpretation of trial results.

The definition of the entry point may be easily mistaken when survival is calculated on patients who are reclassified on the basis of information collected after treatment has started. The easiest situation to deal with is that of a study which, by design, fixes a precise time point when the information of interest is collected on all survivors. For example, in modern treatment of leukemia, response to the so-called "induction treatment" is evaluated at a fixed time point from diagnosis, say, six weeks. The correct date of entry for calculating survivorship of respondent patients is the date of response evaluation, and not that of diagnosis or beginning of treatment. If these last were chosen, the curves would show a spurious plateau at 100% in the first six weeks, due to the fact that only survivors at that time could possibly be classified for response.

However, in many situations, a time point for reclassification cannot be fixed *a priori*, either because it does not make sense to do so or because it is difficult to apply in practice. In these situations the problem of defining the entry point cannot be solved straightforwardly as before. Since, most commonly, patients are reclassified with the aim of comparing survival in the resulting groups, the correct approach to the analysis will be discussed in section 4.7.

3.5.2   Definition of the end-point

The definition of the end-point depends on the objective of the study. Death from any cause or from a specific cause is the end-point for the analysis of mortality in clinical and epidemiological studies. In clinical studies, researchers are also interested in evaluating the ability of treatment to prevent or delay recurrent disease as well as death. Consequently, the end-point is recurrence or death, whichever occurs first. Each patient will have a failure time or, if none of the events occurred, a time at which he is alive and event free. The K–M or the life table estimator applied to these data give the so-called *event free survival* curve. This is also termed *disease free survival* curve when, at the entry point, patients may be considered "free of disease". This is the case for event free survival calculated from date of surgery in women with breast cancer. However, it is improper to take this term literally, since it is reasonable to think that patients whose cancer recurs were never, in fact, disease free. The more general term "event free survival" must necessarily be used in other situations. For example, patients with leukemia do not receive surgery and certainly have a disease present at the start of treatment (usually chemotherapy). In such studies it is usual to not only include recurrence or death as an end-point, but also to define any patient who has not achieved remission by the end of the initial treatment (induction therapy) as having failed at that time. Note that the failure time for individuals who do not achieve remission is somewhat arbitrary, and the "event free" survival curve will have a jump at the end of the induction period. The term "disease free" survival is often reserved for the subset of patients who achieved remission by the end of the induction period. If this is done, it is important also to report the proportion of these patients.

Whereas the date of death is usually known exactly, the recorded date of recurrence is often dependent on follow-up schedules. In practice it will reflect the date on which recurrence is detected, which will be sometime after the actual date of recurrence. It is important therefore that all patients are consistently examined at fixed times after entry into the study.

A more intensive investigation of the clinical course could detect recurrence earlier. This should be borne in mind when planning follow-up schedules in different groups of patients to be compared in terms of event free or disease free survival. These analyses are often useful but, particularly in comparative studies of different treatments, overall survival should always be plotted and compared to ensure an honest evaluation of treatment effect. Since this usually requires a longer follow-up, it is often neglected in favour of earlier results on event free survival. However, on many occasions, for example in cancer patients, higher response rates or lower recurrence rates do not necessarily translate into a gain for survival.

For overall survival it is intended here that the end-point of the analysis is death from any cause. In studying cancer patient survival, an issue which is often debated is whether cancer deaths only should be considered. What is not satisfactory about this analysis is that it may be biased by errors in the classification of causes of death. Furthermore, many intensive chemotherapic regimens in cancer patients reduce the so-called "cancer death" rates, but are not related to an improvement in the overall survival. Whether this is due to the "cost" of treatment in terms of toxic deaths or to the above-mentioned classificatory problem, overall survival alone indicates in practice if treated patients have a better prognosis. As with all general rules, that of considering overall survival does not mean that analysis by specific causes of death are not proper in some situations. For example, one of the objectives of the GISSI-2 (figure 1.1) trial was to compare the effect on early mortality (in-hospital and six-months mortality) of different treatments of evolving myocardial infarction. In this case, the analysis of mortality due to myocardial infarction can be considered reliable since, in such a short term, patients were routinely examined by the same team of investigators.

Sometimes, more complicated end-points are defined in order to reflect the natural history of the disease under study. Consider, for example, the paper by Fisher *et al.* (1989); they report the results of a trial aimed at comparing three treatments: total mastectomy, lumpectomy with irradiation and lumpectomy without irradiation in breast cancer patients. In defining their end-point the authors assess: "The occurrence of tumor in the same breast after lumpectomy was not designated as an end-point event in determining disease-free survival, since patients who initially underwent total mastectomy were not a risk for an ipsilateral breast tumor. All patients assigned to the three groups were followed with respect to disease-free survival, distant-disease-free survival and overall survival. Recurrences of tumor in the chest wall and the operative scar, but not in the ipsilateral breast, were classified as local treatment failures. Tumors in the internal mammary, supraclavicular or ipsilateral axillary nodes were classified as regional treatments failures. Tumors in all other locations were considered distant treatment failures. The events considered in our analysis of *disease-free survival* were first recurrences of disease, second cancers, and deaths without recurrence of cancer. The patients classified as having any distant disease were those in whom a distant metastasis was the first recurrence, those with a distant metastasis after a local or regional recurrence, and those with a second cancer (including a tumor in the other breast). 'Overall survival' refers to survival with or without recurrence of disease."

For tracing the disease-free survival curve the authors properly consider, beside all types of recurrence, any second cancer and deaths without recurrence. This is done to make allowance for tumors which are possibly linked to the primary breast tumor and for possible mistakes in diagnosing the cause of death. However, the end-point used to study distant-disease survival deserves some comments: deaths without disease recurrence are treated as censored observations. Since we are here in the presence of "competing risks", estimating time to distant failure by the product limit or the life table methods leads to a hypothetical construction that does not have a useful interpretation in terms of the probability of occurrence of events.

In concluding, whatever end-point one adopts, the decision as to which patients meet the criteria defining the end-point must be made explicit in the study protocol and applied without knowledge of the treatment group to which the patients were assigned.

Statistical problems arising in the simultaneous analysis of different types of events, such as deaths from different causes or relapses in different sites, will be discussed in Chapter 9.

3.5.3   Modalities of follow-up

This section discusses errors that can be easily introduced in any analyses of failure times if quality and modalities of follow-up are not properly considered.

In data collection, the attempt to closely monitor patient status often causes "bad news to come first", which creates a source of potential bias. As an example, we use data from a multicentre study on acute leukemia which enrolled, between 1 August 1989 and 31 July 1990, 25 newly diagnosed patients. They had to be routinely seen at fixed intervals while on treatment. In addition, relapses and deaths had to be notified to the data centre as soon as they were known to the clinicians. Data on survival available for these patients one year later, on 31 July 1991, are shown in table 3.8, column (3). Suppose these

Table 3.8 Survival data on 25 patients enrolled between 1 August 1989 and 31 July 1990. Raw follow-up data available and updated at 31 July 1991 (columns (3) and (4)) and calculated survival time in days (column (5)), with status indicator ($\delta = 0$: alive; $\delta = 1$: dead).

| Patient identification number (1) | Date of diagnosis (entry) (2) | Follow-up data | | Time from diagnosis and status at cut-off date 31/07/91 (T, δ) (5) |
		Available on 31/07/91 (3)	Updated around 31/07/91 (4)	
1	04/08/89	18/12/90 alive	02/08/91 alive	726.0
2	05/08/89	15/04/91 dead	15/04/91 dead	618.1
3	11/08/89	03/01/91 alive	02/08/91 alive	719.0
4	25/08/89	19/09/89 dead	19/09/89 dead	25.1
5	08/09/89	25/01/91 alive	31/07/91 alive	691.0
6	15/09/89	05/02/91 dead	05/02/91 dead	508.1
7	29/09/89	18/04/91 alive	08/08/91 alive	670.0
8	12/10/89	25/03/91 alive	10/08/91 alive	657.0
9	30/10/89	15/04/91 alive	31/07/91 alive	639.0
10	05/11/89	10/05/91 alive	08/08/91 alive	633.0
11	11/12/89	01/04/91 dead	01/04/91 dead	476.1
12	27/12/89	31/01/91 alive	31/01/91 lost	400.0
13	30/12/89	18/07/91 alive	18/07/91 alive	565.0
14	21/01/90	18/03/91 alive	31/07/91 alive	556.0
15	30/01/90	18/07/90 alive	20/07/91 alive	536.0
16	22/02/90	03/01/91 alive	12/08/91 alive	524.0
17	07/03/90	29/12/90 alive	09/08/91 alive	511.0
18	07/04/90	16/07/91 dead	16/07/91 dead	465.1
19	11/05/90	21/03/91 alive	20/08/91 alive	446.0
20	26/05/90	15/05/91 dead	15/05/91 dead	354.1
21	03/06/90	29/10/90 dead	29/10/90 dead	148.1
22	29/06/90	17/07/91 alive	17/07/91 alive	383.0
23	06/07/90	29/03/91 alive	12/08/91 alive	390.0
24	13/07/90	11/07/91 dead	11/07/91 dead	363.1
25	24/07/90	10/03/91 alive	31/07/91 alive	372.0

data are fed into a computer and analysed as they are, without seeing that dates on which patients were last seen alive and disease-free are on average less promptly updated than dates of death. The resulting K–M survival curve (continuous line), shown in figure 3.6, will necessarily overestimate the probability of death. In order to avoid bias in estimation of survival, these data need to be uniformly updated at a given date. A request to the clinical centres for follow-up on 31 July 1991 gives data in column (4) of table 3.8.

The correct analysis is that done at the cut-off date of 31 July 1991. The observation time T is calculated in column (5), where the value of δ indicates whether a patient is alive ($\delta = 0$) or dead ($\delta = 1$). The resulting survival curve is shown in figure 3.6 (broken line).

The same problem may arise with a type of follow-up scheme called anniversary follow-up, which is being commonly used. Suppose that data on patients status are collected and reported at fixed intervals from diagnosis, for example at months 3, 6, 12, 18, 24 and annually thereafter. However, patients with clinical problems may be seen more often by clinicians. In practice their follow-up will tend to be selectively updated, while, as in the previous example, less updated information would be available on patients without overt symptoms. In order to

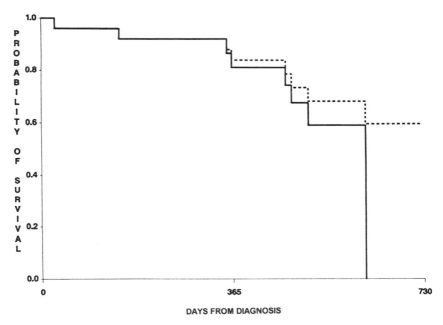

Figure 3.6 Survival curves for 25 patients in table 3.8 calculated from data available at 31 July 1991 (——) and at the cut-off date of 31 July 1991 (-----) after correct updating at follow-up.

avoid bias, information from the last routine follow-up only should be taken into account in the analysis, while subsequent events should be ignored.

Another peculiar source of bias is the way recurrences are reported in the anniversary follow-up scheme. Peto (1984) shows a worked example of this. Suppose, for simplicity, that 100 patients have to be seen for recurrence at annual visits and that recurrence is never diagnosed between these visits. The centre schedules 25 visits each day for four days at about one year from entry or previous follow-up. If recurrent disease is detected, it is reported at the date of the visit. If these dates were fed into a computer, a standard program for K–M estimate of survival would give the results shown in table 3.9. However, since four recurrences were detected overall, the correct estimate of disease free survival at one year would be 96%. These data should be analysed after recording all visit dates to exactly 365 days from entry so that a standard application of the K–M method would give the correct estimate. In fact recurrence dates are not known and these data only mean that four recurrences occurred on 100 patients at risk for the entire year. The life table method could be successfully applied to these grouped data, leading to the correct estimate of the probability of recurrence at one year. This problem does not arise when mortality only is estimated, since exact dates will be recorded for deaths occurring between follow-up visits.

An alternative to the anniversary follow-up scheme is that of updating follow-up at a certain calendar date for every live patient. Most commonly, especially in cooperative studies, a cut-off date for the analysis is decided in advance, for example at about one year from the last patient entry, usually at the end of one month, as in table 3.8. Collaborating centres are requested to update follow-up of all patients at that date. If they are notified of the cut-off data a few months in advance, it is likely that they will be able to arrange the examination of live patients within a few days around that date and to retrace the exact date of death for subjects who are found to be no longer alive. The entire data set would be available within a relatively short period following the cut-off date. The analysis could then be carried out on survival times calculated

Table 3.9   Recurrent disease detected in 100 patients at four visits of annual follow-up and resulting incorrect K–M estimate of disease free survival; the correct estimate is 96% disease free at one year (from Peto, 1984, *Cancer Chemical Trials*, reproduced by permission of Oxford University Press).

Days from entry to annual visit	No. of patients at risk	No. of patients examined	No. of recurrences detected	Incorrect K–M curve
363	100	25	1	$1.000 \times 99/100 = 0.9900$
364	75	25	1	$0.9900 \times 74/75 = 0.9768$
365	50	25	1	$0.9768 \times 49/50 = 0.9573$
366	25	25	1	$0.9573 \times 24/25 = 0.9190$

as in table 3.8, column (5). It is worth noting that deaths occurring after the cut-off date should not be included as events, since this would lead to slightly overestimated mortality.

3.5.4 Reporting and interpreting a survival curve

A list is given here of some suggestions for reporting and interpreting a survival curve which should prevent misleading or wrong conclusions to be drawn from published results:

(1) The graph of a survival curve should be properly plotted, according to whether it was obtained with the K–M or the life table method, as shown in sections 3.2 and 3.3. In some contexts it may be more effective to plot the cumulative incidence of failure instead of the survival curve. The estimated cumulative incidence is $1 - \hat{P}(t)$ or $1 - \hat{P}_j$ where $\hat{P}(t)$ and $\hat{P}_j$ are the K–M and the life table estimates of the cumulative probability of survival given in (3.2) and (3.6) respectively. For example, the life experience of the 349 women whose survival curve was represented in figure 3.3 may be described in terms of cumulative death incidence as in figure 3.7.

(2) A statement clarifying which patients have been analysed and which

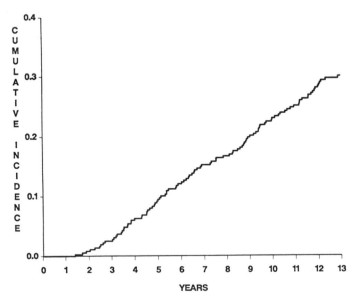

Figure 3.7 Cumulative death incidence estimated according to K–M for the 349 women with early breast cancer who underwent radical surgery (see protocol in figure 1.3).

have been possibly excluded or "withdrawn" is needed. Patients may be excluded only when they are found to be ineligible by reviewing the clinical information. It may happen, for example, that the formulated diagnosis is contradicted when the complete set of clinical tests done at presentation becomes available. However, exclusions justified by lack of eligibility criteria should be of concern if the review of clinical records is not being done blindly and homogeneously on all patients. Exclusions based on events subsequent to entry are to be avoided because they never satisfy the above requests. Patients are "withdrawn" from the analysis when they are included in the sample but are subsequently dropped at the occurrence of some event. This happens, for example, when the observation time is censored at the occurrence of a major deviation from treatment, thus excluding all subsequent follow-up information. What is often unclear is that this type of approach usually produces biased results. For example, if treatment is suspended for patients in very bad clinical conditions, the estimation of treatment effect would be biased since deaths occurring after the deviation will not be accounted for in the analysis. It is therefore recommended that, in any case, the primary analysis should be done not allowing for exclusions and "withdrawns". In particular, in randomized trials, exclusions and "withdrawns" following randomization are not acceptable, as discussed in section 2.5.

(3) The entry point and the event (or events) which compose the end-point should be clearly stated for any plotted survival, event free or disease free survival curve.

(4) The date to which analysis is referred or, equivalently, follow-up is updated, should be clearly stated, together with the dates of start and end of recruitment. The difference between dates when last enrolment occurred and follow-up was updated define the minimum observation time potentially guaranteed to any patient. For example, if patients with breast cancer are first examined after three months from surgery, the analysis should be done allowing for a minimum follow-up period of three months, at least. This aspect is often neglected in interim analyses performed during the enrolment period, where patients who entered the study at a date very close to that of the analysis are included. Often medical publications report, as an additional indicator of completeness of follow-up, the median follow-up time. This should not be confused with the median survival time estimated on a survival curve (see point 7 below). The median follow-up time should be calculated on the distribution of potential observation times elapsed from date of entry to that of last update, as explained in detail by Korn (1986). When the recruitment pattern needs to be described, graphical devices, for example histograms of annual recruitment or *ad hoc* charts, are by far the most effective.

(5) The percentage of observations in the data set which are censored because they come from patients who were lost to follow-up should be reported. As

discussed in section 1.2.2, censoring due to loss may occur for reasons which are related to the patient's condition or to treatment effect. Thus, this type of censoring does not satisfy the assumption of "non-informativeness" which is required for estimating survival. The presence of a relatively high percentage of patients lost to follow-up may therefore lead to biased estimates. For this reason, every effort should be made to keep the percentage of lost to follow-up very low, under 5%, as a rule of thumb, and preferably much less in studies where a low incidence of failures is expected. A high percentage of loss in a prospective study justifies some doubts on the quality of its conduction and results, and even more so if censoring due to loss is not balanced among the treatment groups to be compared.

(6) Reading a survival curve is very easy and immediate. Nonetheless, it should be borne in mind that the survival curve shows the pattern of mortality in time, not the details. This means that any conclusion based on the fine detail of a curve is likely to be wrong. In particular, when the tail of a survival curve presents a flat region, this does not necessarily mean that the risk of death is negligible for patients who are still alive. In fact, only if there is a large number of patients at risk along the flat region, may this provide a reasonable estimate of the proportion of patients who have been cured. This type of finding should always be evaluated in the light of previous knowledge of the natural history of the disease under study. A final flat region, for example, is expected in childhood leukemia, and in some other types of cancer, where long term survivors are likely to be cured. On the other hand, with a few patients at risk, a survival curve may drastically drop at the occurrence of one event, as in figure 3.6, for the continuous line curve. These drops do not convey any sensible information on survival, as is to be expected since very few patients are followed up at that point. To avoid misinterpretation of the tail of the curve, a simple device is to write below it the number of subjects at risk at regular intervals. An alternative is not to plot the curve beyond the point where there are only a few patients at risk, say 10 as a rule of thumb.

(7) The median survival time may be roughly read on a survival curve as the time point which corresponds to the value $\hat{P}(t) = 0.5$. It estimates the time period beyond which 50% of the patients are expected to survive in the population under study. In diseases with favourable prognosis the survival probability may not reach the value 0.5 across the study period, preventing the estimate of the median time. When possible, the estimated median survival time is often reported. However, it should be regarded with great caution, since it can be very inaccurate, unless the survival curve drops steadily and rapidly in the region around the 50% value of survival, say between 30% and 70%. For the calculation of the median confidence limits see Simon and Lee (1982) and Slud *et al.* (1984).

3.6 EVENT RATES

3.6.1 Basic concepts

The use of rates is motivated by the need to account for the fact that the number of deaths observed in a set of subjects strictly depends on how long these subjects have been observed for. This also motivates the construction of survival curves. However, the concepts of rate and conditional probability are related, but different. The rate of death in a sample expresses the total number of observed deaths related to the total time all subjects have been observed for. The longer we observe a subject, the more we are likely to see a failure and the more we want him to "weight" in the evaluation.

The concept of rate might be best understood, on an intuitive basis, by analysing the standard rate estimate adopted by epidemiologists.

Suppose we wish to estimate the mortality rate λ experienced by the ten women with early breast cancer whose data are in table 3.4. Each subject contributes to the rate estimate $\hat{\lambda}$ with the amount of time, in some chosen unit, she has been under observation. The number of events is related to this total observation time, and not just to the number of subjects observed, as follows:

$$\hat{\lambda} = \frac{\text{total number of events (deaths)}}{\text{total observation time (person–time units)}}$$

From figure 3.8, representing the survival experience of each woman in table 3.4, it is clear that:

total number of deaths = 4
total observation time in person–years
$= 10.81 + 18.49 + 17.82 + 14.70 + 8.70 + 10.67 + 13.45 + 13.43$
$+ 4.88 + 11.68 = 124.63.$

Consequently the estimated mortality rate is:

$$\hat{\lambda} = \frac{4}{124.63} = 0.032$$

This means that, for every 1000 women at risk for an entire year, 32 deaths are expected to occur.

In estimating the mortality rate of women whose life experience is represented in figure 3.8, we implicitly assume that the rate is constant during 18 years of observation. The single value $\hat{\lambda}$ refers to all time points during that long interval. This is quite unsatisfactory, since it is unlikely that this constancy reflects the real mortality process. We could instead subdivide the follow-up period into intervals and assume that the rate is constant during each of them but not

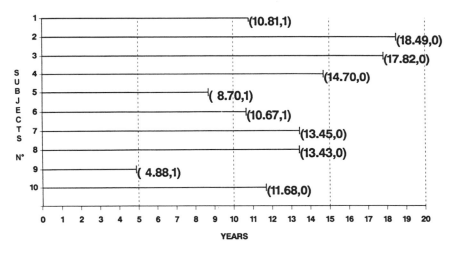

Figure 3.8 Graph of the follow-up experience of the 10 subjects in table 3.4. For each subject, in parentheses, the observation time (in years) and the indicator of failure (1 = dead; 0 = alive) are reported.

necessarily across them. Then estimation of interval specific rates proceeds as before, provided that failures and total observation times are partitioned among the time intervals. As an example, in figure 3.8 consider the time axis subdivided into five-years intervals. During the first five years of follow-up, one death occurs at 4.88 years and the remaining nine subjects survive through the entire interval. If we indicate with λ_1 the rate in this first interval, $\hat{\lambda}_1 = 1/$ $(4.88 + 9 \times 5) = 1/49.88 = 0.020$. In the following interval the mortality rate, estimated on nine subjects at risk, is $\hat{\lambda}_2 = 1/(3.70 + 8 \times 5) = 0.023$. In the third interval, eight subjects contribute to the estimation, and their observation time is written as the denominator of $\hat{\lambda}_3$ which becomes:

$$\hat{\lambda}_3 = \frac{2}{0.81 + 5 + 5 + 4.70 + 0.67 + 3.45 + 3.43 + 1.68} = 0.081$$

Since no deaths occur in the last interval, $\hat{\lambda}_4 = 0$. The four figures obtained for $\hat{\lambda}$ suggest that the risk of death is not constant over 18 years. Either when the number of subjects at risk is small or few events occur, $\hat{\lambda}$ may vary a lot for little change in the data (i.e. a few more events) or with the choice of the length of the intervals.

3.6.2 Estimate of rates and their standard errors in life tables

It comes natural to think of extending the use of rates for the analysis of data in a life table. Such data are grouped by time intervals and a model allowing for the rates to change from interval to interval, while still remaining constant within intervals, could be usefully adopted.

When data are grouped, like those in table 3.3, and exact times of death, loss and withdrawal are not given, some kind of approximation is needed to evaluate the number of person–years at risk in each interval. The standard approach is to assume that deaths and censored observations, on average, contribute for half the interval to the total observation time in the interval where deaths and censoring take place.

According to the notation adopted in the life table (see table 3.2),

$$\hat{\lambda}_j = \frac{d_j}{[n_j' - \frac{1}{2}(c_j + d_j)] \times h_j} = \frac{d_j}{(n_j - \frac{1}{2}d_j)h_j} \tag{3.15}$$

The denominator in (3.15) may also be seen as the product of the length of the interval by the average number of survivors at the mid-point of the interval (assuming a uniform distribution of deaths and censored data).

Yearly rates estimated for data in table 3.3 are shown in table 3.10.

To investigate the difference between rate and probability, consider the expression (3.15) as compared to the formula $\hat{q}_j = d_j/n_j$ for estimating the conditional probability of death in the interval I_j: the two look very similar, especially when the length of the interval is $h_j = 1$ (year), for every j, as in this example. However, probability should not be confused with rate. The ratio d_j/n_j estimates the proportion of deaths which occur in a fixed period of time (here one year), d_j being a subset of n_j. The rate, instead, gives the "intensity" of death per year, since the denominator represents person–years of observation during the interval and not just persons. Consider the second interval in table 3.10. In estimating q_j we consider that, among the 39 subjects lost to follow-up during the interval, 19.5 subjects only are initially exposed to risk of dying in the second year. Coherently, we adjust the number of subjects entering the interval by subtracting 39/2 from 1962 (which gives 1942.5 as in column (6) in table 3.3). Differently, in estimating λ, we approximately calculate the observation time for censored subjects as well as for those who die, and obtain 1829.5 person–years. While the probability is a pure number, the rate is dimensioned as the inverse of time.

The estimator (3.15) is often called the actuarial estimator and may equivalently be given as

$$\hat{\lambda}_j = \frac{2\hat{q}_j}{h_j(1 + \hat{p}_j)}$$

by resorting to the life table estimators $\hat{q}_j$ and $\hat{p}_j$ introduced in section 3.3.1.

The actuarial estimator $\hat{\lambda}_j$ is the central death rate for interval I_j, and is plotted against time at the midpoint of I_j. For estimates in table 3.10, see figure 3.9. The

Table 3.10 Yearly rate estimates for $\hat{\lambda}_j$ on data from Parker *et al.* (1946) (see table 3.3).

Years after diagnosis j (1)	Interval $[t_j, t_{j+1})$ l_j (2)	No. entering interval n_j^* (3)	No. of censored observations c_j (4)	No. of deaths d_j (5)	Person–years (6)	$\hat{\lambda}_j$ (7)	SE$(\hat{\lambda}_j)$ (8)
0	[0, 1)	2418	0	456	2190.0	0.2082	0.0097
1	[1, 2)	1962	39	226	1829.5	0.1235	0.0082
2	[2, 3)	1697	22	152	1610.0	0.0944	0.0076
3	[3, 4)	1523	23	171	1426.0	0.1199	0.0092
4	[4, 5)	1329	24	135	1249.5	0.1080	0.0093
5	[5, 6)	1170	107	125	1054.0	0.1186	0.0106
6	[6, 7)	938	133	83	830.0	0.1000	0.0110
7	[7, 8)	722	102	74	634.0	0.1167	0.0135
8	[8, 9)	546	68	51	486.5	0.1048	0.0147
9	[9, 10)	427	64	42	374.0	0.1123	0.0173
10	[10, 11)	321	45	43	277.0	0.1552	0.0236
11	[11, 12)	233	53	34	189.5	0.1794	0.0306
12	[12, 13)	146	33	18	120.5	0.1494	0.0351
13	[13, 14)	95	27	9	77.0	0.1169	0.0389
14	[14, 15)	59	23	6	44.5	0.1348	0.0549
15	15 +	30	—	—	—	—	—

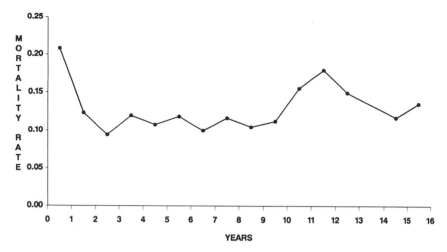

Figure 3.9 Estimated yearly mortality rate from data in table 3.10.

points representing $\hat{\lambda}$ in different intervals are connected with lines just to help interpretation. The mortality rate tends to remain constant for a relatively large interval between 1 and 10 years from diagnosis, while it is higher in the first year after diagnosis, and in later years, where estimates seem, however, quite unstable.

Gehan (1969) showed that the standard error of λ_j is estimated by

$$SE(\hat{\lambda}_j) = \frac{\hat{\lambda}_j}{\sqrt{n_j \hat{q}_j}} \sqrt{1 - \left(\frac{h_j \hat{\lambda}_j}{2}\right)^2} \qquad (3.16)$$

Like Greenwood's formula, (3.16) is a large-sample approximation and, assuming that the sampling distribution of $\hat{\lambda}_j$ is approximately Gaussian, enables us to obtain approximate confidence limits. Consider $\hat{\lambda}_2 = 0.0944$ in table 3.10; from columns (6) and (7) of table 3.3 we have: $n_2 = 1686$ and $\hat{q}_2 = 0.0902$ respectively. Since $h_j = 1$ we obtain

$$SE(\hat{\lambda}_2) = \frac{0.0944}{\sqrt{1686 \times 0.0902}} \sqrt{1 - \left(\frac{0.0944}{2}\right)^2} = 0.007\,65$$

Column (8) of table 3.10 reports the standard errors of the estimated $\hat{\lambda}_j$ given in column (7). The standard error of $\hat{\lambda}_j$ increases with the decrease of the number of patients exposed to risk, and at 12 years from diagnosis the precision of the estimated mortality rate is about a quarter of the corresponding estimate at one year. The approximate 95% confidence interval of λ_2 is $0.0944 \pm 1.96 \times 0.007\,65 = (0.0794, 0.1094)$.

3.6.3* The relationship between probability and rate

Let us now consider time intervals that tend to become so short that we can call them "instants". Then, in any of these intervals, the instantaneous rate may be seen as the ratio of the conditional probability of dying in that instant to the length of the instant itself. To exemplify, consider intervals of small but finite length, such as days which are very short compared to the length of a clinical study. It is unlikely that more than one individual will have failed or been censored on a given day t, so the number of person-days at risk will be well approximated by the number at risk at the beginning of day t. Consequently, the denominator in (3.15) would reduce to the number of subjects "entering" the day t multiplied by the length $h = 1$ of the day. Apart from the dimensionality (inverse of time), this ratio coincides with that suitable for estimating the conditional probability of failure on day t. This is the consequence of the formal definition of the instantaneous death rate $\lambda(t)$ as $\lambda(t) = q(t)/dt$, where $q(t)$ indicates the conditional failure probability in the interval from t to $t + dt$ of infinitesimal length $h = dt$.

In words, rates represent the intensity of failures, in terms of relative rapidity of occurrence of failures in time. The term relative means that the rapidity, at time t, is measured conditionally on the subjects at risk of presenting the failure at t. Conversely, it is $q(t) = \lambda(t) \times dt$.

Suppose we know a subject experiences a constant failure rate λ. We wish to calculate the cumulative probability $P(t)$ that he will survive a fixed period of time t (from the start of observation to day t), given that he will not be lost during that period. We begin by dividing the period t into many instants, each of length dt, so that the number of instants will be $N = t/dt$. The conditional probability of dying in each instant dt is λdt. The cumulative probability $P(t)$ of surviving beyond t is the product of the conditional probabilities of surviving each of the N instants in t, that is $P(t) = (1 - \lambda dt)^N$. The logarithm of the cumulative survival is

$$\log P(t) = N \log(1 - \lambda dt)$$

As dt tends to zero, $\log(1 - \lambda dt)$ tends to $-\lambda dt$; since $N = t/dt$, we obtain

$$\log P(t) = -N\lambda dt = -\lambda t \tag{3.17}$$

The quantity λt represents the sum of a constant instantaneous rate λ over the time period t and is called the cumulative failure rate. By exponentiating, we obtain:

$$P(t) = \exp(-\lambda t) \tag{3.18}$$

Suppose we describe the failure process over J years by means of yearly rates λ_j as in a life table setting. In each of the years, (3.18) holds and, since survival probabilities multiply, we obtain

Cumulative probability of surviving J years
$$= \exp[-(\lambda_0 h_0 + \lambda_1 h_1 + \cdots + \lambda_{J-1} h_{J-1})] \qquad (3.19)$$

where the sum in parenthesis is the cumulative failure rate over J years. This relationship suggests a method for estimating the survival probability which is alternative to that introduced in section 3.3. Table 3.11 shows the survival estimate P_j obtained by (3.19) for data from Parker *et al.* (1946); these results are very similar to those in table 3.3.

The relationship (3.19) holds (in the limit) even if one allows the rates to vary at each instant. In practice, instantaneous rates would be estimated as zero at every instant at which no failures occur and as $d_j/n_j dt$ at the instants $t_{(j)}$ at which d_j subjects fail, among n_j subjects at risk. The additive contribution of each instant to the cumulative failure rate is therefore either zero or $(d_j/n_j dt) \times dt = d_j/n_j$. Therefore the estimated cumulative failure rate is represented by an increasing step function with jumps at each distinct failure time:

$$\hat{\Lambda}(t) = \sum_{j|t_j \leqslant t} \hat{\lambda}(t_j) = \sum_{j|t_j \leqslant t} \frac{d_j}{n_j} \qquad (3.20)$$

Table 3.11 Estimate of cumulative survival probability based on mortality rate estimates for data from Parker *et al.* (1946). Yearly rates are from table 3.8.

Years after diagnosis j (1)	Yearly rates $\hat{\lambda}_j$ (2)	Estimated Cumulative rates (3)	Cumulative survival probability $\bar{P}_j$ (4)
0	0.208 22	0.2082	1.0000
1	0.123 53	0.3318	0.8120
2	0.094 41	0.4262	0.7176
3	0.119 92	0.5461	0.6530
4	0.108 04	0.6541	0.5792
5	0.118 60	0.7727	0.5199
6	0.100 00	0.8727	0.4618
7	0.116 72	0.9894	0.4178
8	0.104 83	1.0943	0.3718
9	0.112 30	1.2066	0.3348
10	0.155 23	1.3618	0.2992
11	0.179 42	1.5412	0.2562
12	0.149 38	1.6906	0.2141
13	0.116 89	1.8075	0.1844
14	0.134 82	1.9423	0.1641
15	0.000 00	0.0000	0.1434

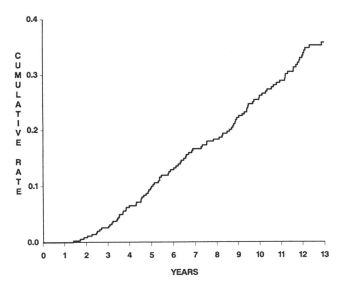

Figure 3.10 Graph of the cumulative mortality rate (Aalen estimator) for a set of 349 women with early breast cancer who underwent radical surgery according to the protocol in figure 1.3.

and this is known as the Aalen estimator of the cumulative failure rate (Aalen, 1976). This can only be calculated when individual data are known. It is the same context in which Kaplan–Meier survival curves arose. However, the two curves convey different information. The cumulative failure rate curve drawn for a relatively large number of subjects gives a sense of how the failure rates vary with time. As an example, figure 3.10 shows this curve for 349 women with early breast cancer who underwent radical surgery according to the protocol in figure 1.3.

Periods in which the instantaneous mortality rate keeps fairly constant correspond to linear increments of the cumulative hazard curve. Conversely a non-linear increment will be observed where the rate increases or decreases. Figure 3.10 shows that mortality sharply increases during the third year from diagnosis from a relatively low initial rate, as reflected by the increase in the gradient of the curve around $2\frac{1}{2}$ years. It then maintains a fairly constant rate in the following years. For example, we can roughly estimate the rate, assumed constant, in the period from three to six years, as this is the slope of the line approximating the curve in that period. The estimate is: $(0.13 - 0.03)/(6 - 3) = 0.033$. Looking at how mortality rates vary in time is a useful complement to the information given by the survival curve (figure 3.3) and helps in the description and understanding of the data.

As in the life table context, relationship (3.19) may provide an estimate of the cumulative probability of surviving based on the estimated cumulative death rate (3.20). Owing to the mathematical fact that, for large values of n_j, $\exp(-1/n_j)$ is approximately equal to $1 - 1/n_j$, the survival probability estimated by means of (3.19) will approximate the K–M estimate, at least in large samples, with virtually no tied observations.

APPENDIX A3.1   THE PRODUCT LIMIT METHOD: THEORY

The product limit is a non-parametric estimate of the cumulative survival probability $P(t)$ or equivalently of the survival function $S(t)$ as defined in Chapter 5. We can obtain this estimator on the basis of a likelihood approach without formulating a specific class of distributions for the random variable T.

Let $t_{(j)}$, with $j = 1, 2, \ldots, J$, be the ordered failure times, let c_j be the number of censored times in the interval $[t_{(j)}, t_{(j+1)})$ and let t_{ij}^* be one of the c_j times: $t_{ij}^* \in [t_{(j)}, t_{(j+1)})$, $i = 1, 2, \ldots, c_j$. In particular, if we define $t_{(0)} = 0$ and $t_{(J+1)} = \infty$, censored times lower than the first failure time $t_{(1)}$ are $t_{i,0}^* \in [t_{(0)}, t_{(1)})$ for $i = 1, 2, \ldots, c_0$ and censored times greater than t_J are $t_{i,J}^* \in [t_J, t_{(J+1)})$, $i = 1, 2, \ldots, c_J$. If the censoring process is assumed to be non-informative, for the censored individuals we know only that the survival time is greater than the censored time. Therefore the contribution to the likelihood for an individual with time t_{ij}^* is

$$\Pr\{T > t_{ij}^*\} = S(t_{ij}^*)$$

An individual with failure time $t_{(j)}$ will contribute to the likelihood with the quantity $S(t_{(j)} - 0) - S(t_{(j)})$ where $S(t_{(j)} - 0)$ indicates $\lim_{x \to 0^-} S(t_{(j)} + x) = \Pr\{T \geq t_{(j)}\}$ and $S(t_{(j)}) = \Pr\{T > t_{(j)}\}$. By (5.19) the contribution to the likelihood is the value of the density function at time $t_{(j)}$.

The likelihood function can therefore be computed as the product of the following contributions:

$$\prod_{i=1}^{c_0} S(t_{i,0}^*)$$

$$\times [S(t_{(1)} - 0) - S(t_{(1)})]^{d_1} \prod_{i=1}^{c_1} S(t_{i,1}^*)$$

$$\vdots$$

$$\times [S(t_{(J)} - 0) - S(t_{(J)})]^{d_J} \prod_{i=1}^{c_J} S(t_{i,J}^*)$$

and may be written as

$$L = \prod_{i=0}^{c_0} S(t_{i,0}^*) \times \prod_{j=1}^{J} \left\{ [S(t_{(j)}^{c_0} - 0) - S(t_{(j)})]^{d_j} \prod_{i=1}^{c_j} S(t_{i,j}^*) \right\}$$

The maximum likelihood estimator cannot be continuous from the left, otherwise all the contributions related to failure times and consequently L would be null.

Since $S(t)$ is a monotonic decreasing function, the greatest value that can be chosen for any censored time contribution $S(t_{(j)}^*)$ is the probability $S(t_{(j)})$ of surviving beyond the lower extreme of the interval $[t_{(j)}, t_{(j+1)})$ to which $t_{i,j}^*$ belongs. The c_0 censored times occurring before $t_{(1)}$ will each contribute a factor $S(t_{(0)}) = 1$.

Also, in order to maximize L, the value of S before $t_{(j)}$, $(S(t_{(j)} - 0))$, should be chosen to be as high as possible and that at $t_{(j)}$ and beyond $(S(t_{(j)}))$ as low as possible; the choice of these values must respect the fact that the function decreases at each new failure. So for $S(t_{(j)} - 0)$ we can go back to pick up the value of $S(t_{(j-1)})$ as the greatest possible one. According to the notation adopted in Chapter 3, L may be written as

$$L = \prod_{j=1}^{J} [P(t_{(j-1)}) - P(t_{(j)})]^{d_j} P(t_{(j)})^{c_j}$$

where the factor $P(t_{(0)})^{c_0}$ having a value of 1 has been omitted. In order to transform the difference contained in L into a product we note that

$$p_j = \frac{P(t_{(j)})}{P(t_{(j-1)})}$$

for $j = 1, 2, \ldots, J$, and that

$$q_j = 1 - p_j = \frac{P(t_{(j-1)}) - P(t_{(j)})}{P(t_{(j-1)})}$$

The likelihood can be written as a function of p_j and q_j as follows:

$$L = \prod_{j=1}^{J} \left\{ (p_1 \cdots p_{j-1} q_j)^{d_j} (p_1 \cdots p_j)^{c_j} \right\}$$

$$= \prod_{j=1}^{J} \left\{ (p_1 \cdots p_{j-1})^{d_j + c_j} p_j^{c_j} q_j^{d_j} \right\}$$

$$= \prod_{j=1}^{J} q_j^{d_j} (1 - q_j)^{n_j - d_j}$$

The likelihood achieves its maximum at $\hat{q}_j = d_j / n_j$ which is the product limit estimate of q_j and $\hat{P}(t) = \hat{S}(t) = \prod_{j | t_{(j)} \leq t} (1 - d_j / n_j)$. If the greatest observation time is a censored time $t^* > t_{(J)}$ then, by definition, $S(t^*) = S(t_{(J)})$ and, since the likelihood function does not depend on $S(t)$ for $t > t^*$, no estimation is possible beyond t^*.

A formula for the asymptotic variance of the product limit estimator may be obtained by regarding L as the usual likelihood function in the parameters $q_j (j = 1, \ldots, J)$. It turns out that the asymptotic variance obtained after calculating the second derivative of L in q_j is $var(\hat{q}_j) = d_j(n_j - d_j)n_j^{-3}$ and consequently, by applying the delta method (see Appendix A3.2),

$$var[\ln \hat{S}(t)] = \sum_{t_{(j)} \leqslant t} (1 - \hat{q}_j)^{-2}$$

$$var(1 - \hat{q}_j) = \sum_{t_{(j)} \leqslant t} \frac{d_j}{n_j(n_j - d_j)}$$

$$var[\hat{S}(t)] = [\hat{S}(t)]^2 \times \sum_{t_{(j)} \leqslant t} \frac{d_j}{n_j(n_j - d_j)}$$

This estimator is coherent with that given by Greenwood (1926) in (3.9) for the life table and reduces to the binomial estimator $\hat{S}(t)[1 - \hat{S}(t)]n^{-1}$ in the absence of censoring. Several studies have investigated properties of the product limit estimator, deriving it as the limiting case of the life table estimator when the number of time intervals tends to infinity and considering different hypotheses on the censoring process. Some of the references are Breslow and Crowley (1974), Johansen (1978) and Miller (1983).

APPENDIX A3.2   DELTA METHOD

Using the large sample approximation formula in Kendall and Stuart (1958, p. 231), the variance of a function of random variables $p_l (l = 0, 1, \ldots, j - 1)$ is

$$var[g(\hat{p}_0, \hat{p}_1, \ldots, \hat{p}_{j-1})] \simeq \sum_{l,k=0}^{j-1} \left\{ \frac{\partial g}{\partial \hat{p}_l} \frac{\partial g}{\partial \hat{p}_k} cov(\hat{p}_l, \hat{p}_k) \right\}$$

where, according to (3.6):

$$g(\hat{p}_0, \hat{p}_1, \ldots, \hat{p}_{j-1}) = \hat{P}_j = \hat{p}_0 \times \hat{p}_1 \times \cdots \times \hat{p}_{j-1}$$

It has been shown by Chiang (1960) that $cov(\hat{p}_l, \hat{p}_k) = 0$, $l \neq k$. Furthermore $\partial \hat{P}_j / \partial p_l = \hat{P}_j / \hat{p}_l$ and the conditional, on set $\{n_l\}$, variance of $\hat{p}_l$ is estimated as $var(\hat{p}_l) = \hat{p}_l(1 - \hat{p}_l)/n_l$. Thus

$$var(\hat{P}_j | n_0, n_1, \ldots, n_{j-1}) \simeq \hat{P}_j^2 \sum_{l=0}^{j-1} \hat{p}_l^{-2} \frac{\hat{p}_l(1 - \hat{p}_l)}{n_l}$$

$$= \hat{P}_j^2 \sum_{l=0}^{j-1} \frac{d_l}{n_l(n_l - d_l)}$$

since $1 - \hat{p}_l = d_l/n_l$.

APPENDIX A3.3   PERCENTILES OF THE KOLMOGOROV–SMIRNOV STATISTIC

Values of $_ND_\alpha$ suitable for computing $(1-\alpha)100\%$ confidence intervals of cumulative survival probability are given in the following table.

N	\(1-\alpha\) 0.80	0.90	0.95	0.99
1	0.900 00	0.950 00	0.975 00	0.995 00
2	0.683 77	0.776 39	0.841 89	0.929 29
3	0.564 81	0.636 04	0.707 60	0.829 00
4	0.492 65	0.565 22	0.623 94	0.734 24
5	0.446 98	0.509 45	0.563 28	0.668 53
6	0.410 37	0.467 99	0.519 26	0.616 61
7	0.381 48	0.436 07	0.483 42	0.575 81
8	0.358 31	0.409 62	0.454 27	0.541 79
9	0.339 10	0.387 46	0.430 01	0.513 32
10	0.322 60	0.368 66	0.409 25	0.488 93
11	0.308 29	0.352 42	0.391 22	0.467 70
12	0.295 77	0.338 15	0.375 43	0.449 05
13	0.284 70	0.325 49	0.361 43	0.432 47
14	0.274 81	0.314 17	0.348 90	0.417 62
15	0.265 88	0.303 97	0.337 60	0.404 20
16	0.257 78	0.294 72	0.327 33	0.392 01
17	0.250 39	0.286 27	0.317 96	0.380 86
18	0.243 60	0.278 51	0.309 36	0.370 62
19	0.237 35	0.271 36	0.301 43	0.361 17
20	0.231 56	0.264 73	0.294 08	0.352 41
21	0.226 17	0.258 58	0.287 24	0.344 27
22	0.221 15	0.252 83	0.280 87	0.336 66
23	0.216 45	0.247 46	0.274 90	0.329 54
24	0.212 05	0.242 42	0.269 31	0.322 86
25	0.207 90	0.237 68	0.264 04	0.316 57
30	0.190 32	0.217 56	0.241 70	0.289 87
35	0.176 59	0.201 85	0.224 25	0.268 97
40	0.165 47	0.189 13	0.210 12	0.252 05
45	0.156 23	0.178 56	0.198 37	0.237 98
50	0.148 40	0.169 59	0.188 41	0.226 04
55	0.141 64	0.161 86	0.179 81	0.215 74
60	0.135 73	0.155 11	0.172 31	0.206 73
65	0.130 52	0.149 13	0.165 67	0.198 77
70	0.125 86	0.143 81	0.159 75	0.191 67
75	0.121 67	0.139 01	0.154 42	0.185 28
80	0.117 87	0.134 67	0.149 60	0.179 49
85	0.114 42	0.130 72	0.145 20	0.174 21
90	0.111 25	0.127 09	0.141 17	0.169 38
95	0.108 33	0.123 75	0.137 46	0.164 93
100	0.105 63	0.120 67	0.134 03	0.160 81
>100	$1.07/\sqrt{N}$	$1.14/\sqrt{N}$	$1.22/\sqrt{N}$	$1.63/\sqrt{N}$

*From Bradley (1968)

APPENDIX A3.4 CRITICAL VALUES FOR THE HALL–WELLNER CONFIDENCE BANDS

Critical values to be used to compute the Hall–Wellner $(1 - \alpha)100\%$ confidence band when $[1 - \bar{K}_N(t_j)] < 0.75$. Each entry of the table must be divided by $\sqrt{N}$.

$1 - \alpha$	0.10	0.25	$1 - \bar{K}_N(t_j)$ 0.40	0.50	0.60	0.75
0.90	0.599	0.894	1.062	1.133	1.181	1.217
0.95	0.682	1.014	1.198	1.273	1.321	1.354
0.99	0.851	1.256	1.470	1.552	1.600	1.626

*From Hall and Wellner (1980). Reproduced by permission of the Biometrika Trustees.

4

Non-parametric Methods for the Comparison of Survival Curves

4.1 INTRODUCTION

In clinical research one is concerned not only with estimating the cumulative probability of surviving but, more often, with the comparison of the life experience of two or more groups of subjects differing for a given characteristic or randomly allocated to different treatments. Unlike reliability analysis in the technological field where the physical and/or chemical properties of the production process can suggest the theoretical failure function (exponential, Weibull, etc.) to fit the experimental data, in the biomedical field it is extremely difficult to have a priori knowledge to make reliable hypotheses on the underlying theoretical survival functions; thus the non-parametric approach is usually adopted also to compare survival curves. Among the various non-parametric tests one can find in the statistical literature, the Mantel–Haenzel (M–H) test (1959), currently called the "log-rank" test, will be described in the following section. Originally it was developed for comparing two sets of proportions over several confounding factors in case-control studies. In a later paper Mantel (1966) extended the applicability of the M–H test to the analysis of survival times.

Though random allocation of patients of either treatment aims at making the two groups initially equivalent in all respects relevant to the enquiry, it must be recognized that this mechanism works in a probability sense, that is in the long run (section 2.3). Therefore it seems sensible to make allowance for the known sources of variability by grouping patients into homogeneous categories or "strata" and comparing the effect of the two treatments within each stratum. The M–H test can be easily extended to analyse stratified data as shown in section 4.5.

Nowadays the K–M method for estimating survival curves and the M–H test for comparing two estimated survival curves are the most frequently used statistical tools in medical reports on survival data.

4.2   THE TWO SAMPLE MANTEL-HAENZEL (LOG-RANK) TEST

The rationale of the M–H test may be understood on a heuristic basis: this appears to be a feature extremely appealing for the clinical investigator.

Suppose that N_A patients have been allocated at random to treatment A and N_B patients to treatment B, $N = N_A + N_B$ being the total number of patients accrued in the study. In order to obtain a procedure providing a running comparison of the two failure processes as they develop in time, all failures recorded in the two groups are considered jointly. The distinct failure times in the two groups together are ranked in ascending order; let $t_{(1)} < t_{(2)} < \ldots$ $< t_{(j)} < \ldots t_{(J)}$ be the ordered times and d_j the number of failure times equal to $t_{(j)}$, i.e. multiplicity of $t_{(j)}$. Further let n_j be the set of patients exposed to the risk of failing just before $t_{(j)}$. The information on the two failure processes can be depicted by a series of 2×2 contingency tables, one for each of the J failure times. The contingency table corresponding to the time $t_{(j)}$ is given in table 4.1.

Of the whole set of patients at risk before $t_{(j)}$ suppose that three-fifths belong to group A and two-fifths to group B. Under the hypothesis that the two treatments be equally effective, three-fifths of the deaths observed at $t_{(j)}$ are expected to belong to group A and two-fifths to group B. The number of observed deaths in group A may be smaller or greater than the expected value and this holds for each of the J contingency tables. However, if the two treatments A and B are equivalent, at the last failure time, $t_{(J)}$ in a given treatment, say A, the total number of observed deaths will tend to be equal to the sum of the deaths expected, for the treatment, in all the J tables considered. The possible difference between these two totals suggests that treatment A is more or less effective than treatment B and the researcher can resort to a pertinent statistical test to evaluate such a difference.

More formally, let H_0 and H_a indicate the null and the alternative hypotheses respectively:

$$\left| \begin{array}{l} H_0 : \lambda_A(t) = \lambda_B(t) \\ H_a : \lambda_A(t) = \theta \lambda_B(t) \end{array} \right. \tag{4.1a}$$

where $\lambda(t)$ represents the failure rate at time t and θ is the unknown constant of

Table 4.1 2×2 contingency table at failure time $t_{(j)}$.

Treatment	Dead at time $t_{(j)}$	Alive at time $t_{(j)}$	Set at risk just before $t_{(j)}$
A	d_{Aj}	$n_{Aj} - d_{Aj}$	n_{Aj}
B	d_{Bj}	$n_{Bj} - d_{Bj}$	n_{Bj}
Totals	d_j	$n_j - d_j$	n_j

proportionality of the rates in two groups. If $\theta < 1$, treatment A is more effective than B, whilst the contrary is true for $\theta > 1$.

Alternatively, by considering the relationship between the hazard function and the survival function which will be discussed in section 5.3, the previous hypothesis may be written

$$\left|\begin{array}{l} H_0:P_A(t) = P_B(t) \\ H_z:P_A(t) = [P_B(t)]^\theta \end{array}\right. \tag{4.1b}$$

The marginals d_j, n_{Aj} and n_{Bj} of table 4.1 are random variables depending upon the entire history before $t_{(j)}$ in term of occurrence of failures and censored observations.

As in the Kaplan–Meier method, individuals in either group with survival times equal to or greater than $t_{(j)}$ but less than the next larger known failure time contribute to the risk set at $t_{(j)}$ but are omitted from the risk set at $t_{(t+1)}$. Mantel and Haenzel (1959) suggested considering the distribution of the observed cell frequencies conditional on the observed marginal totals under the null hypothesis of equally effective treatments. This implies, in turn, considering the distribution of the frequency of just one cell, say d_{Aj}, since the other frequencies are implicitly determined by the fixed marginals.

Under the null hypothesis, the distribution of d_{Aj} is shown to be hypergeometric (see Appendix A4.1) and the (conditional) expected value of d_{Aj} is

$$E(d_{Aj}) = n_{Aj} \times \frac{d_j}{n_j} \tag{4.2}$$

The (conditional) variance of d_{Aj}, and also of the difference, $d_{Aj} - E(d_{Aj})$, is

$$var(d_{Aj}) = \left[n_{Aj} \times \frac{d_j}{n_j}\left(1 - \frac{d_j}{n_j}\right) \right]\left(\frac{n_j - n_{Aj}}{n_j - 1}\right) = \frac{n_{Aj}n_{Bj}d_j(n_j - d_j)}{n_j^2(n_j - 1)} \tag{4.3}$$

This latter is computed by multiplying two terms: the first one, in square brackets, is the estimate of the variance of a binomial random variable and the second term, in parentheses, is the correction factor for sampling from a finite population of size n_j.

Statistical theory proves that, under the null hypothesis, the ratio

$$X_j^2 = \frac{[d_{Aj} - E(d_{Aj})]^2}{var(d_{Aj})} \tag{4.4}$$

is approximately distributed as chi-square with one degree of freedom, provided that neither n_{Aj} nor n_{Bj} be too small. The statistic taking into account the information addressed by all tables is

$$U = \sum_{j=1}^{J} [d_{Aj} - E(d_{Aj})] \tag{4.5}$$

which appears to be especially sensitive to the consistent difference between failure probabilities related to treatments A and B (see section 4.4). Conditionally on fixed d_j, n_{Aj} and n_{Bj}, the variance of U under the null hypothesis is obtained by adding the J statistics (4.3) computed at each of the ordered failure times. Thus, the test statistic as suggested by Mantel and Haenzel is

$$Q_{M-H} = \frac{\left\{ \sum\limits_{j=1}^{J} [d_{Aj} - E(d_{Aj})] \right\}^2}{\sum\limits_{j=1}^{J} var(d_{Aj})} = \frac{U^2}{var(U)} \tag{4.6}$$

which, like (4.4), is approximately distributed as chi-square with one degree of freedom.

The computation procedure of the M–H test will be shown by using the Freireich *et al.* (1963) data presented in table 3.3.

When the times of the two groups are ordered together, one obtains the sequence of failure times $t_{(j)}$: 1, 2, 3, 4, 5, 6, 7, 8, 10, 11, 12, 13, 15, 16, 17, 22, 23. For each of the 17 times $t_{(j)}$, the data are arranged in a 2×2 contingency table as shown in table 4.2; this is convenient for computing the ingredients of the M–H test: $E(d_{Aj})$ according to (4.2), and $var(d_{Aj})$, according to (4.3). More-over this table reports other quantities used to compute statistics which will be commented upon later on.

As far as the first contingency table is concerned, $E(d_{A1}) = 2 \times 21/42 = 1.00$; $var(d_{A1}) = 21 \times 21 \times 2 \times 40/(42^2 \times 41) = 0.4878$. These calculations are then repeated at each $t_{(j)}$.

Therefore from the totals of columns (2), (8) and (9) of table 4.2 it is straightforward to get

$$Q_{M-H} = \frac{(9.00 - 19.250\,50)^2}{6.256\,96} = 16.97$$

From the chi-square distribution with 1 d.f. the corresponding probability is $p < 0.0001$ and one can sensibly reject the null hypothesis of equivalence of the two treatments.

But how to move from this finding, important in terms of "statistical significance", to an interpretation of the trial results in terms of "clinical relevance"?

Since the hypotheses (4.1a) are given in terms of relative failure rate, an estimate of θ appears to be a plain measure of the new treatment effectiveness compared with that of the standard one or of placebo. This statistic summarizes, over the entire life experience, the failure rate in the new treatment compared to that in the reference treatment, making allowance for the follow-up lengths.

Table 4.2 Computation procedure of the M–H test, of the estimate of the proportionality constant θ and its confidence interval. Data from Freireich *et al.* (1963) (see text).

Time of failure $t_{(j)}$ (1)	Failures d_{Aj} (2)	d_{Bj} (3)	d_j (4)	Patients at risk n_{Aj} (5)	n_{Bj} (6)	n_j (7)	$E(d_{Aj})$ (8)	$var(d_{Aj})$ (9)	R_j (10)	S_j (11)	W_j (12)
1	0	2	2	21	21	42	1.00000	0.48780	0.	1.00	0.47619
2	0	2	2	21	19	40	1.05000	0.48596	0.	1.05000	0.47250
3	0	1	1	21	17	38	0.55263	0.24723	0.	0.55263	0.24723
4	0	2	2	21	16	37	1.13514	0.47723	0.	1.13514	0.46019
5	0	2	2	21	14	35	1.20000	0.46588	0.	1.20000	0.44571
6	3	0	3	21	12	33	1.90909	0.65083	1.09091	0.	0.62810
7	1	0	1	17	12	29	0.58621	0.24257	0.41379	0.	0.24257
8	0	4	4	16	12	28	2.28571	0.87075	0.	2.28571	0.73469
10	1	0	1	15	8	23	0.65217	0.22684	0.34783	0.	0.22684
11	0	2	2	13	8	21	1.23810	0.44807	0.	1.23810	0.41270
12	0	2	2	12	6	18	1.33333	0.41830	0.	1.33333	0.37037
13	1	0	1	12	4	16	0.75000	0.18750	0.25000	0.	0.18750
15	0	1	1	11	4	15	0.73333	0.19556	0.0	0.73333	0.19556
16	1	0	1	11	3	14	0.78571	0.16837	0.21429	0.0	0.16837
17	0	1	1	10	3	13	0.76923	0.17751	0.0	0.76923	0.17751
22	1	1	2	7	2	9	1.55556	0.30247	0.11111	0.66667	0.32099
23	1	1	2	6	1	7	1.71429	0.20408	0.0	0.71429	0.20408
Totals	9	21	30				19.25050	6.25696	2.42793	12.67840	5.97110

4.3 ESTIMATES OF THE RELATIVE HAZARD RATE θ

The quantity U as given by (4.5), divided by its variance, was suggested by Peto *et al.* (1976) as an approximate estimate of the natural logarithm of θ, namely

$$\hat{\xi} = \frac{U}{var(U)} \tag{4.7}$$

This anticipates a result deriving from the Cox model; $\hat{\xi}$ is the one-step approximation of the maximum likelihood estimator of $\log \theta$ obtained by applying the Newton–Raphson technique (section 6.4.1).

This statistic (4.7) is asymptotically normally distributed with variance equal to

$$var(\hat{\xi}) = [var(U)]^{-1} \tag{4.8}$$

Therefore approximate $100(1 - \alpha)\%$ confidence limits of $\log \theta$ are $\hat{\xi} \pm z_{1-\alpha/2}/\sqrt{var(U)}$ and by exponentiating these latter one obtains those of θ.

From totals of columns (2), (8) and (9) of table 4.2 the estimate of θ is

$$\exp\left(\frac{9 - 19.2505}{6.25696}\right) = 0.194$$

with 95% confidence limits

$$\exp(-1.638\,26 \pm 1.96/\sqrt{6.256\,96}) = (0.089, 0.425)$$

These results enable the investigator to conclude, with a 95% confidence of being right, that the risk of relapsing in children treated with 6-MP is about five times less than in placebo treated children, ranging from a minimum of about two times $(1/0.425)$ to a maximum of about eleven times $(1/0.089)$. These three statistics are extremely informative for the chinician trying to evaluate the benefit–risk ratio of the therapy under study and to predict results in future patients.

Alternatively, in accordance with Mantel and Haenzel (1959), θ may be estimated as follows:

$$\hat{\theta}_{M-H} = \frac{\sum\limits_{j=1}^{J} d_{Aj}(n_{Bj} - d_{Bj})/n_j}{\sum\limits_{j=1}^{J} d_{Bj}(n_{Aj} - d_{Aj})/n_j} = \frac{\sum\limits_{j=1}^{J} R_j}{\sum\limits_{j=1}^{J} S_j} = \frac{R_+}{S_+} \tag{4.9}$$

This estimator may be written as a weighted average of the odds ratios of failure (ω_j) estimated in each contingency table:

$$\hat{\theta}_{M-H} = \frac{\sum\limits_{j=1}^{J} w_j \hat{\omega}_j}{\sum\limits_{j=1}^{J} w_j}$$

where

$$\hat{\omega}_j = \frac{d_{Aj}(n_{Bj} - d_{Bj})}{d_{Bj}(n_{Aj} - d_{Aj})}$$

$$w_j = \left(\frac{1}{n_{Aj}} + \frac{1}{n_{Bj}}\right)^{-1} [(1 - p_{Aj})p_{Bj}]$$

p_{Aj} and p_{Bj} being estimates of conditional probability of failure at $t_{(j)}$ under treatments A and B respectively.

The first factor of w_j shows that the weight function assigns higher weight to tables with larger n_{Aj} and/or n_{Bj}; that is, this factor can be considered to weight approximately by the precision of the odds ratio in each table. Furthermore let us assume, as in the previous example, that B is the placebo or the standard treatment group; because of the multiplicative factor p_{Bj} in the weight function, the M–H procedure will attach more "importance" to estimates of the odds ratio from tables showing a greater failure frequency in the reference group.

An alternative estimator of θ (Anderson and Bernstein, 1985) has the form

$$\hat{\theta}^{*}_{M-H} = \frac{\sum\limits_{j=1}^{J} d_{Aj} n_{Bj}/n_j}{\sum\limits_{j=1}^{J} d_{Bj} n_{Aj}/n_j} \tag{4.9a}$$

This may be seen as the ratio of the weighted averages of the estimated failure rates in the two groups, with weights $w_j = n_{Aj} n_{Bj}/n_j$.

If there is a constant of proportionality common to all the $t_{(j)}$ times, the statistics (4.9) and (4.9a) estimate this constant (see conditions (1) and (2) of section 4.4). If there is no common proportionality constant, these statistics give a weighted average of the separate constants provided that condition (3) of section 4.5 is satisfied.

According to Sato (1990) approximate $100(1 - \alpha)\%$ confidence limits for θ can be obtained as the two solutions of the quadratic equation

$$(R_+ - \theta S_+)^2 / \theta W_+ = z^2_{(1 - \alpha/2)}$$

where

$$W_+ = \sum_{j=1}^{J} W_j = \sum_{j=1}^{J} [d_{Aj}(n_{Bj} - d_{Bj})(n_{Aj} - d_{Aj} + d_{Bj} + 1)$$
$$+ (n_{Aj} - d_{Aj})d_{Bj}(n_{Bj} - d_{Bj} + d_{Aj} + 1)]/n_j^2$$

By means of Monte Carlo experiments Sato showed that the coverage of this approximate confidence interval is remarkably close to the expected.

The explicit solution to the previous equation are

$$(\hat{\theta}_L, \hat{\theta}_U) = \frac{2R_+S_+ + z_{(1-\alpha/2)}^2 W_+ \pm \sqrt{(4R_+S_+ + z_{(1-\alpha/2)}^2 W_+)z_{(1-\alpha/2)}^2 W_+}}{2S_+^2} \quad (4.10)$$

With reference to the Freireich *et al.* (1963) trial the ingredients to estimate θ and to compute its confidence limits are given in the last three columns of table 4.2.

As far as the first contingency table is concerned, $d_{A1}(n_{B1} - d_{B1})/n_1 = 0.0$ and $d_{B1}(n_{A1} - d_{A1})/n_1 = 1.00$; moreover

$$\frac{d_{A1}(n_{B1} - d_{B1})[(n_{A1} - d_{A1} + d_{B1} + 1] + (n_{A1} - d_{A1})d_{B1}(n_{B1} - d_{B1} + d_{A1} + 1)}{n_1^2} = \frac{21 \times 2 \times (19 + 1)}{42^2} = 0.476\,19$$

From totals of columns (10), (11) and (12) of table 4.2 one obtains

$$\hat{\theta}_{M-H} = \frac{2.427\,93}{12.678\,40} = 0.192$$

and the 95% confidence limits of θ are

$$(\hat{\theta}_L, \hat{\theta}_U) = \frac{2 \times 2.427\,93 \times 12.6784 + 3.8416 \times 5.9711 \pm \sqrt{(4 \times 30.782\,27 + 3.8416 \times 5.9711) \times 22.938\,58}}{2 \times 12.6784^2} = (0.083, 0.443)$$

In this example the estimates of θ and of the confidence intervals computed with the two approaches lead to the same conclusions on treatment effect.

It is worth noting that if no ties occurs at $t_{(j)}$, $d_j = 1$ and $d_{Aj} - E(d_{Aj})$ is equal to $1 - n_{Aj}/n_j$ if failure is observed in group A or to $-n_{Aj}/n_j$ if failure is observed in group B, and they correspond to the contribution to R_+ and S_+ respectively. Moreover, since the variance (4.3) becomes $var(d_{Aj}) = n_{Aj} \times n_{Bj}/n_j^2$, it is easy to see that $var(d_{Aj})$ corresponds to W_j and calculation of the confidence interval of θ takes advantage of quantities already obtained for computing the Q_{M-H} statistic.

4.4   A FAMILY OF TWO SAMPLE RANK TESTS

Since the distribution function for the r.v. T is $1 - P(t)$, when there are no censored observations, the hypotheses (4.1b) can be tested by resorting to non-parametric rank tests (Lehmann, 1975), for example the Wilcoxon test or the Mann–Whitney U test. It may be shown that the M–H test belongs to a family of tests which is obtained as an extension to censored data of the standard non-parametric tests suitable for comparing two complete distributions. Other tests, of current use in survival analysis, such as the test suggested by Gehan (1965), by Tarone and Ware (1977) and by Prentice (1978), are members of this family.

Without making direct use of the theory of ranks (the interested reader is referred to Lawless (1982), pp. 412–430) the above-mentioned family of tests may be heuristically introduced as follows. Let us consider deaths $d_{A1}, d_{A2}, \ldots,$ $d_{Aj}, \ldots, d_{AJ}$ as Gaussian, independent, random variables with expectation and variance given by (4.2) and (4.3) respectively, and let $w_1, w_2, \ldots w_j, \ldots, w_J$ be known constants. The variable $\sum_{j=1}^{J} w_j[(d_{Aj} - E(d_{Aj})]$ is normally distributed and, thus, the statistic

$$Q = \frac{\left\{ \sum_{j=1}^{J} w_j[d_{Aj} - E(d_{Aj})] \right\}^2}{\sum_{j=1}^{J} w_j^2 \, var(d_{Aj})} \tag{4.11}$$

is distributed according to the chi-square density function with 1 d.f. under the null hypothesis that the survival times are equally distributed in the two treatment groups A and B. As a matter of fact $d_{A1}, d_{A2}, \ldots, d_{AJ}$ are not Gaussian independent r.v. but, for convenience, they can be considered as such in thinking of Q as it may be shown that this latter is asymptotically distributed as a chi-square r.v. with 1 d.f. for any arbitrary censoring process (Andersen *et al.*, 1982). According to the values given to the weights w_j in (4.11) one obtains different tests, namely:

$$w_j = 1 \qquad\qquad Q = Q_{M-H} \quad \text{Mantel and Haenzel (1959)}$$

$$w_j = n_j \qquad\qquad Q = Q_G \quad \text{Gehan (1965)}$$

$$w_j = \sqrt{n_j} \qquad\qquad Q = Q_{T-W} \quad \text{Tarone and Ware (1977)}$$

$$w_j = \prod_{t=1}^{j} \frac{n_l - d_l + 1}{n_l + 1} \quad Q = Q_P \quad \text{Prentice (1978)}$$

The Q_{M-H} statistic gives a constant weight to the differences between the two survival experiences at each of the J death times, while the Q_G statistic tends to weight the early differences more heavily than the late ones.

In commenting on the properties of Q_{M-H} and Q_G, Tarone and Ware (1977) argue that Gehan's weighting scheme may imply some loss of sensitivity when the distributions of censoring times widely differ in the two groups to be compared. In trying to overcome this drawback they suggested their own weights which appear to be a compromise between weights used by Q_{M-H} and Q_G. The same aim is pursued by Prentice's weighting scheme whose weights are very·similar to the Kaplan–Meier estimate of survival probability pooled over the two groups.

Harrington and Fleming (1982) suggested a class of weights generalizing those previously given and thoroughly investigated its theoretical properties. These weight are $w_j(\rho) = [\hat{P}(t_j)]^\rho$ where ρ is chosen to be equal to or greater than zero and $\hat{P}(t_j) = \prod_{l=1}^{j} (n_l - 1)/n_l$ is an estimate of the surviving probability of the whole set of patients. This latter estimate equals the K–M estimate in the absence of ties. For $\rho = 0$ the Q_{M-H} statistic is obtained, whilst for $\rho = 1$ the statistic behaves very similarly to Q_P.

To get a deeper insight into the properties of the previously reported tests, consider again the generic 2×2 contingency table by assuming that one death only is recorded at every time $t_{(j)}$. In the case of death in group A the table is as follows:

Treatment groups	Dead at time $t_{(j)}$	Alive at time $t_{(t)}$	Set at risk just before $t_{(j)}$
A	1	$n_{Aj} - 1$	n_{Aj}
B	0	n_{Bj}	n_{Bj}
Totals	1	$n_{Aj} + n_{Bj} - 1$	n_j

Under the generic alternative hypothesis $\lambda_A(t_j) \neq \lambda_B(t_j)$ for each j, the expected value of the difference in the numerator of (4.4), conditioned to the fixed marginals, may be written

$$E[d_{Aj} - E(d_{Aj})] = \frac{n_{Aj}\lambda_A(t_j)}{n_{Aj}\lambda_A(t_j) + n_{Bj}\lambda_B(t_j)} - \frac{n_{Aj}}{n_j}$$

After defining

$$\tau_j = \frac{n_{Aj}n_{Bj}}{n_j\left(\dfrac{n_{Aj}\lambda_A(t_j)}{\lambda_B(t_j)} + n_{Bj}\right)}$$

the previous expression may be rewritten as

$$E[d_{Aj} - E(d_{Aj})] = \tau_j\left(\frac{\lambda_A(t_j)}{\lambda_B(t_j)} - 1\right)$$

and the sum with respect to j becomes

$$\sum_{j=1}^{j} \tau_j \left(\frac{\lambda_A(t_j)}{\lambda_B(t_j)} - 1 \right) \tag{4.12}$$

This quantity is not the expected value of the sum $\sum_{j=1}^{j}(d_{Aj} - E(d_{Aj}))$; nevertheless Schoenfeld (1981) showed that it is, asymptotically, the non-centrality parameter of the Gaussian distribution of the sum of differences given by (4.5). Thus it is convenient to use (4.12) in order to investigate the behaviour of the non-parametric tests.

Referring now to Q_{M-H} we may suppose the following.

(1) The null hypothesis $\lambda_A(t_j) = \lambda_B(t_j)$, is true for each j. In this case (4.12) is null and the value of Q_{M-H} is expected to differ from zero only as a result of random fluctuation. This correctly implies not to reject the null hypothesis $(100 - \alpha)\%$ times, where α is the prefixed significance level of the test.

(2) The alternative hypothesis of proportional hazards is true, $\lambda_A(t_j)/(\lambda_B(t_j)) = \theta$ with $\theta < 1$ (or $\theta > 1$) as depicted in figure 4.1(a). The quantity (4.12) is the sum of J terms, all negative (or positive); thus Q_{M-H} is expected to be large and correctly to lead to the rejection of the null hypothesis $100(1 - \beta)\%$ times, where β is the second type error probability. In this case, in fact, the M–H test is known to be the most powerful test.

(3) The alternative hypothesis of proportionality between hazards is true, but the true value of the hazard ratio depends on time, i.e. $\lambda_A(t_j)/\lambda_B(t_j) = \theta_j$, with $\theta_j < 1$ (or $\theta_j > 1$) for each j (see figure 4.1(b). The J terms to be added in (4.12) are, also in this case, all negative (or positive) and Q_{M-H} is still expected to be large but not the most powerful. As a matter of fact, a test giving more weight to the differences corresponding to the larger θ_j, for example Gehan's test if the θ_j tend to be larger at the beginning of the follow-up period, would be more appropriate. Anyway it is worth noting that the M–H test has a remarkable power in this situation too.

(4) The alternative hypothesis of non-proportional hazards is true and the ratio $\lambda_A(t_j)/\lambda_B(t_j)$ is more that 1 for some values of t_j and less than 1 for others. This implies that the graphs of the curves $\lambda_A(t_j)$ and $\lambda_B(t_j)$ cross each other at at least one time point as shown in Fig. 4.1(c). In this case it may happen that the positive terms in (4.12), corresponding to $\theta_j > 1$, are balanced by the negative ones corresponding to $\theta_j < 1$; therefore Q_{M-H} may tend to zero without picking up any difference between the two curves. Gehan's and other tests, which give greater weight to differences in the initial part of the survival curves, could not be decreased so much by this cancelling of positive and negative terms. This would happen in the case depicted in figure 4.1(c) where the sum of negative terms is expected to be greater than the sum of positive terms and these terms are more likely to give a statistically significant result. However, relying on

a single test to assess the relative effect of the two treatments is deemed to be unsatisfactory. In this example, treatment A (dotted line) compares favourably to B (continuous line) in the early post-treatment period, while B is preferred later on. This behaviour may occur if treatment B causes early deaths related to high toxicity but shows a therapeutical impact on

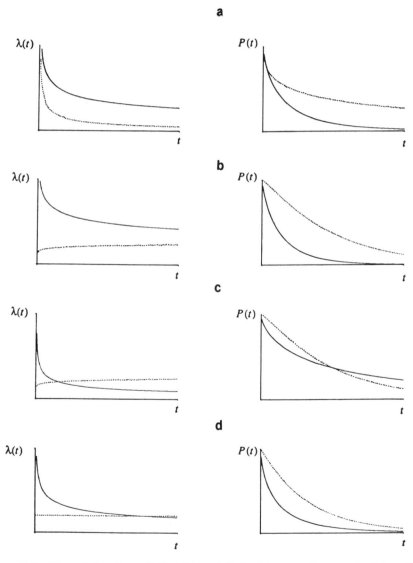

Figure 4.1 Hazard functions (left side) and their corresponding survival functions (right side) for four different hypothetical situations (see text). Treatment A . . . . ; treatment B ——.

those who survive the early period. This calls for a careful evaluation of timing and causes of death and eventually for making statements on mortality in groups A and B pertinent to follow-up periods.

It is worth noting that crossing hazard functions do not necessarily produce crossing survival functions, but these latter may result at in figure 4.1(d) where long-term survivorships are overlapping but treatment B has a higher early mortality rate. Gill (1980) showed that, when survival curves do not cross, the M–H test keeps its consistency properties. All these considerations emphasize that carefully studying the pattern of the survival and hazard curves as well as checking the assumption of proportional hazard are preliminary to a shrewd choice of the test to be adopted.

A graphical check of the assumption (2) above may be easily accomplished as follows: let $\hat{P}_A(t_j)$ and $\hat{P}_B(t_j)$ be the Kaplan–Meier or the life table survivorship estimates for groups A and B respectively. The values of $\log[-\log \hat{P}_A(t_j)]$ and $\log[-\log \hat{P}_B(t_j)]$ are the empirical estimates by (5.7) of the log cumulative failure rates in the two groups. When the values are plotted (y-axis) against each of the corresponding values of t_j (x-axis), and two broken lines connecting the points pertinent to each of the two treatments are drawn, they are expected to show parallel patterns if the proportional hazard assumption is fulfilled. Figure 4.2 gives an example of this graph for the Freireich *et al.* (1963) data; the curves do not contradict the assumption of proportional hazards.

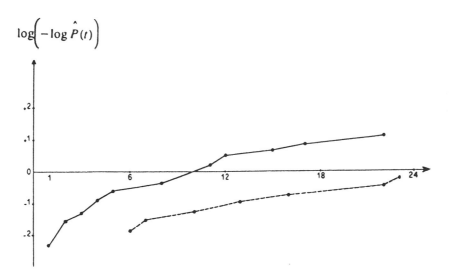

Figure 4.2 Graphical check of proportional hazard of the survival probability curves depicted in figure 3.1. Data from Freireich *et al.* (1963). Group treated with 6-mercaptopurine-----; placebo group ———. Abscissa: survival time (weeks).

4.5   THE INCLUSION OF STRATA IN THE MANTEL–HAENZEL TEST

One could anticipate that the appealing feature of the statistical procedure in the presence of strata should consist of comparing the two treatments within each stratum and in obtaining an average measure of their relative effect suitably weighted between strata. Mantel and Haenzel (1959) showed that the Q_{M-H} test has this feature and its computation in this context is a straightforward extension of the test given in the previous section.

4.5.1   Formulation of the test statistic

Let m ($m = 1, 2, \ldots, M$) indicate the strata; within each stratum the distinct failure times in the two groups together are ordered and the quantities $_m d_A, E_m(d_A)$ and $var_m(d_A)$ are computed. The overall test suitable for testing the hypotheses (4.1) is

$$Q_{M-H}(\text{overall}) = \frac{\left\{ \sum_{m=1}^{M} \left[\sum_{j=1}^{J_m} {}_m d_{Aj} - \sum_{j=1}^{J_m} E_m(d_{Aj}) \right] \right\}^2}{\sum_{m=1}^{M} \sum_{j=1}^{J_m} var_m(d_{Aj})} = \frac{\left(\sum_{m=1}^{M} U_m \right)^2}{\sum_{m=1}^{M} var(U_m)} \quad (4.13)$$

The number of failure times may vary in the M strata; coherently a suffix m has been added to J to indicate the maximum value of j in the mth stratum. The statistic (4.13) is asymptotically distributed as chi-square with 1 d.f.

Analogously:

$$\hat{\theta}_{M-H}(\text{overall}) = \frac{\sum_{m=1}^{M} \sum_{j=1}^{J_m} {}_m d_{Aj}({}_m n_{Bj} - {}_m d_{Bj})/n_j}{\sum_{m=1}^{M} \sum_{j=1}^{J_m} {}_m d_{Bj}({}_m n_{Aj} - {}_m d_{Aj})/n_j} = \frac{\sum_{m=1}^{M} {}_m R_+}{\sum_{m=1}^{M} {}_m S_+}$$

and the $100(1 - \alpha)\%$ confidence limits are obtained by direct extension of (4.10), namely:

$$(\hat{\theta}_L, \hat{\theta}_U) =$$

$$\frac{2 \sum_{m=1}^{M} {}_m R_+ \sum_{m=1}^{M} {}_m S_+ + z_{(1-\alpha/2)}^2 \sum_{m=1}^{M} {}_m W_+ \pm \sqrt{\left(4 \sum_{m=1}^{M} {}_m R_+ \sum_{m=1}^{M} {}_m S_+ + z_{(1-\alpha/2)}^2 \sum_{m=1}^{M} {}_m W_+ \right) z_{(1-\alpha/2)}^2 \sum_{m=1}^{M} {}_m W_+}}{2 \left(\sum_{m=1}^{M} {}_m S_+ \right)^2}$$

$$(4.14)$$

Equation (4.13) shows that the overall M–H test tests the effect of the two treatments within each stratum defined by different modalities of a prognostic variable or by combination of different levels of two or more prognostic variables. Thus, stratified analysis not only adjusts for imbalances on important

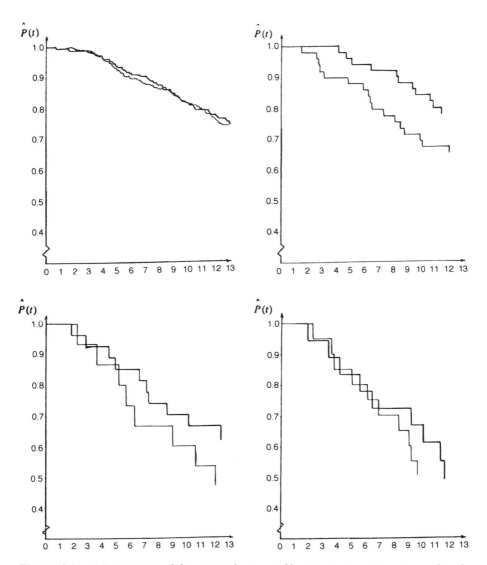

Figure 4.3   K–M estimate of the survival curves of breast cancer patients treated with mastectomy (light line) or quadrantectomy (bold line) for four strata defined according to the number (0, 1, 2–3, 4 or more) of metastatic lymph nodes (from Veronesi *et al.*, 1986). Ordinate: suvival probability; abscissa: survival time (years).Upper left: 0 nodes; upper right: 1 node; lower left: 2–3 nodes; lower right: 4 or more nodes.

Table 4.3 Computation procedure of the stratified M–H test (data from the trial shown in figure 1.3).

No. of metastatic nodes	No. of patients	No. of deaths from all causes		$E_m(d_A) =$ No. of deaths expected in the Quad. group	$var(U_m)$	$\hat{\theta}_{\text{M-H}}$	$exp(\hat{\xi})$
(1)	(2)	(3)	(4)	(5)	(6)	(7)	(8)
		Mast.	Quad.				
0	520	68	64	65.345 88	32.993 70	0.959	0.960
1	100	17	11	15.020 10	6.956 66	0.554	0.561
2–3	43	8	10	11.897 98	4.028 14	0.638	0.624
≥4	38	10	9	9.320 94	4.734 33	0.933	0.934
Totals	701	103	94	101.584 90	48.712 83		

prognostic factors, but resorting to the "compare like to like" principle, it increases the precision of the estimate the treatment effect.

As an example we consider the results of the conservative treatment trial to breast cancer described in figure 1.3. According to the number of axillary metastatic lymph nodes at surgery, four strata were identified retrospectively, namely 0, 1, 2–3 and 4 or more. The K–M estimates of survival experience (all causes of death) of patients treated with mastectomy or quadrantectomy for each stratum are given in figure 4.3; 13 years is a maximum length of follow-up. As is well known, the probability of surviving decreases dramatically as the number of metastatic nodes increases.

Table 4.3 reports the number of patients, the number of observed deaths, $E_m(d_A)$ and $var(U_m)$ for each stratum; moreover columns (6) and (7) give the Mantel–Haenzel and the one-step estimates of θ respectively.

From equation (4.13) and the total of columns (4), (5) and (6) one obtains

$$Q_{M-H}(\text{overall}) = \frac{(94 - 101.5849)^2}{48.712\,83} = 1.181$$

From the chi-square distribution with 1 d.f., $p = 0.2772$ implying not to reject the null hypothesis. By processing the data without retrospective stratification $E(d_A) = 100.330\,43$, $var(d_A) = 49.229\,50$ and $Q_{M-H} = 0.814$ with $p = 0.3669$. Though in this example the conclusions reached by the two approaches are overlapping, the results point out that stratification, as expected, tends to increase the power of the test both by increasing the difference at the numerator and by decreasing the variance (denominator). Column (7) of table 4.3 shows that the relative hazard rates $\hat{\theta}_{M-H}$ are consistently less than 1; they are concentrated around the estimate pooled over the four strata: $\hat{\theta}_{M-H}(\text{overall}) = 0.856$ (the 95% confidence interval computed according to (4.14) is (0.647, 1.132)).

It may happen that the relative failure rates vary widely among strata, and particularly in some strata they are greater and in some others less than 1. This may be evidence against pooling across strata to compute an overall test of difference in survivorship of the two treatment groups and a common estimate of θ. But how much variation is expected to occur just by random fluctuation and how much is adequate evidence of a true heterogeneity?

4.5.2   Testing homogeneity of treatment effects

The test suitable for testing treatment effect homogeneity among strata, i.e. the null hypothesis $H_0: \theta_1 = \theta_2 = \cdots = \theta_m = \cdots = \theta_M = \theta$ may be built up by resorting to the one-step estimates of $\ln \theta$ as given by (4.7) and computed for each of the M strata.

The θ common to all strata under H_0 may be estimated by the weighted average:

$$\bar{\hat{\xi}} = \frac{\sum\limits_{m=1}^{M} w_m \hat{\xi}_m}{\sum\limits_{m=1}^{M} w_m}$$

with weights inversely proportional to $var(\hat{\xi}_m)$. From (4.8) we obtain

$$\bar{\hat{\xi}} = \frac{\sum\limits_{m=1}^{M} var(U_m)\hat{\xi}_m}{\sum\limits_{m=1}^{M} var(U_m)} = \frac{\sum\limits_{m=1}^{M} U_m}{\sum\limits_{m=1}^{M} var(U_m)}$$

and therefore

$$var(\bar{\hat{\xi}}) = \frac{\sum\limits_{m=1}^{M} w_m^2 \, var(\hat{\xi}_m)}{\left(\sum\limits_{m=1}^{M} w_m\right)^2} = \left[\sum\limits_{m=1}^{M} var(U_m)\right]^{-1} \tag{4.15}$$

Furthermore, since each $\hat{\xi}_m$ is asymptotically normally distributed with variance $[var(U_m)]^{-1}$, it may be shown that under H_0 the statistic

$$Q_{\text{hom}} = \sum\limits_{m=1}^{M} var(U_m)(\hat{\xi}_m - \bar{\hat{\xi}})^2 \tag{4.16}$$

is asymptotically distributed as chi-square with $M-1$ d.f.

From the totals of columns (4), (5) and (6) in table 4.3 we obtain

$$\bar{\hat{\xi}}_m = \frac{94 - 101.584\,90}{48.712\,83} = -0.155\,71$$

which after exponentiating gives 0.856 as the overall estimate of θ common to all strata.

From columns (4), (5) and (6) the four stratum specific estimates of ξ are

$$\hat{\xi}_1 = -0.040\,79; \quad \hat{\xi}_2 = -0.577\,88; \quad \hat{\xi}_3 = -0.471\,18; \quad \hat{\xi}_4 = -0.067\,79$$

Moreover $\sum_{m=1}^{4} var(U_m)\hat{\xi}_m^2 = 3.294\,09$ and $\sum_{m=1}^{4} var(U_m)\hat{\xi}_m = -7.584\,85$; hence

$$Q_{\text{hom}} = 3.294\,09 - (-7.584\,85)^2/48.712\,83 = 2.113\,09 \quad \text{with} \quad 3 \text{ d.f.}$$

As expected, the null hypothesis is not rejected, being $0.60 > p > 0.50$, and the overall value 0.86 may be sensibly adopted as an estimate of the hazard rate ratio, quadrantectomy/mastectomy.

Unlike the conclusion reached in this example, suppose that the Q_{hom} results be statistically significant, suggesting that the stratum estimates $\hat{\theta}_m$ differ by more than they should under the assumption of homogeneity.

At this point the investigator might wish to explore the nature of a possible interaction between treatment and prognostic factors defining the strata. In doing this, he must be aware that quite striking-looking quantitative interactions can be produced by the play of chance (section 2.7). For example Collins *et al.* (1987) considered a randomized trial of over 16 000 patients with suspected acute myocardial infarction in which a highly significant $(2p < 0.002)$ overall benefit of treatment was found. They categorized patients according to prognostically utterly absurd criterion, i.e. their astrological birth sign, and found that the effect of treatment was confined to Scorpios (i.e to patients born between 24 October and 22 November) whilst it appeared small and non-significant in others. On the other hand, there might be plausible suggestions for looking for possible qualitative interactions. A suitable test to be applied in this context was proposed by Gail and Simon (1985).

4.5.3   Gail and Simon test of qualitative interaction

Consider figure 4.4: it shows the space of treatment effects with reference to two subsets of patients only; ξ_1 and ξ_2 are the true values of the log failure rate ratio of the treated to the control group in the first and second subset respectively. The origin represents the null hypothesis of equality or rates for either subset. The line $\xi_1 = \xi_2$ is the locus of points for which the two treatments have different effects (except for the origin), but the difference between the log failure rates of the two treatment groups is the same in the two subsets. Whenever the values ξ_1 and ξ_2 are not on the line $\xi_1 = \xi_2$ we say that treatments and subsets interact. The shaded regions, in the second $(\xi_1 < 0$ and $\xi_2 > 0)$ and fourth $(\xi_1 > 0)$ and $\xi_2 < 0)$ quadrants, indicate where qualitative (cross-over) interactions exist. Quantitative interactions consist of points in the first and third quadrants, which are not on the line $\xi_1 = \xi_2$. These definitions generalize to M subsets of individuals; there are 2^{M-1} orthants with cross-over and two without cross-over.

If the point estimates for the difference in treatment effects are in opposite directions, the important issue is to test whether these can be expressions of a true cross-over effect or whether they are due just to random fluctuations. This implies testing the null hypothesis that there are no cross-over interactions. This corresponds to testing the null hypothesis that the vector $\Xi = (\xi_1, \xi_2, \ldots, \xi_M)$ lies either in the orthant in which all components are non-negative (the positive orthant $\mathbf{0}^+$) or in the orthant in which all components are non-positive (the negative orthant $\mathbf{0}^-$).

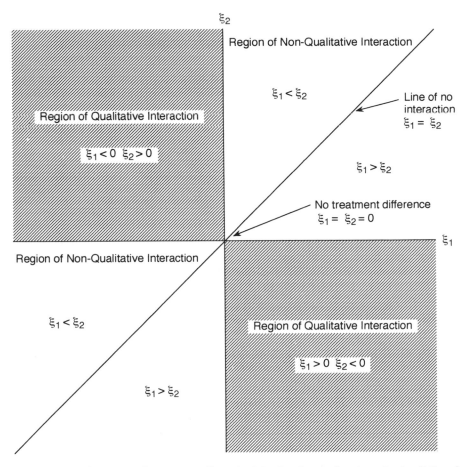

Figure 4.4   The space of treatment effects for $M = 2$ subsets of patients in the Gail and Simon (1985) test.

By resorting to the likelihood ratio principle, Gail and Simon (1985) developed a test for qualitative interaction based upon the two statistics Q^- and Q^+. The first one, computed by means of all $\hat{\xi}_m > 0$, is

$$Q^- = \sum_{m:\hat{\xi}_m > 0} \frac{\hat{\xi}_m^2}{var(\hat{\xi}_m)}$$

The second statistic, computed by means of all $\hat{\xi}_m < 0$, is

$$Q^+ = \sum_{m:\hat{\xi}_m < 0} \frac{\hat{\xi}_m^2}{var(\hat{\xi}_m)}$$

The test statistic turns out to be

$$Q_{G-S} = \min(Q^-, Q^+)$$

and the null hypothesis is rejected if $Q_{G-S} > c$, where c are the critical values of the test. They are such that, for all $\Xi \in \mathbf{0}^- \cup \mathbf{0}^+$ (i.e. for all vectors pertaining to the null hypothesis space: $\mathbf{0}^- \cup \mathbf{0}^+$) the probability that the inequalities $Q^- > c$ and $Q^+ > c$ are both satisfied is no greater than the significance level α. Selected critical values of c are given in Appendix A4.2.

Consider the following example on two surgical treatments, mastectomy vs quadrantectomy compared in the breast cancer trial shown in figure 1.3. Patients with metastatic lymph nodes were classified, according to the adjuvant treatment they received after surgery, in three subsets: no adjuvant therapy, radiotherapy and chemotherapy, as shown in table 4.4. The statistic $\hat{\xi}_m$ and its standard error, computed by means of (4.7) and (4.8), are reported in columns (2) and (3); the test statistic (4.16) is easily computed from the totals of table 4.4; from columns (4) and (5) we have

$$\bar{\bar{\xi}} = -\frac{4.792\,14}{18.935\,24} = -0.253\,08$$

This allows us to compute the terms reported in column (6); the total of this column is $Q_{\text{hom}} = 5.613$, asymptotically distributed as χ^2 with 2 d.f. ($p = 0.0604$). Note in table 4.4 that in the no adjuvant treatment group $\hat{\xi}_1$ has positive sign indicating a mortality rate in the quadrantectomy group higher than the corresponding one in the Halsted group, whilst the opposite happens in the other two subsets.

From column (7) of table 4.4 the test statistic is easily obtained:

$$Q^+ = 2.452\,60 \quad \text{and} \quad Q^- = 4.372\,79$$

For $\alpha = 0.05$ and three subsets, the critical value from Appendix A4.2 is $c = 4.23$. Since $Q_{G-S} = 2.453 < c$ the null hypothesis of no cross-over cannot be rejected. Thus the overall estimate $\hat{\theta} = \exp(-0.253) = 0.78$ may be taken as a measure of the rate ratio of the two treatments, with 95% confidence limits $(0.49, 1.22)$ computed by means of $var(\bar{\bar{\xi}})$ given by (4.15). Resorting to the M–H estimator we have $\hat{\theta}_{M-H}$ (overall) $= 0.80$ with 95% confidence interval $(0.52, 1.24)$ computed according to (4.14). In this example the agreement of the two appraoches is satisfactory.

Suppose that in this example one were to adopt the method of analysis commented upon in section 2.7, consisting of comparing the treatment effects within each subset, for instance by means of the M–H test. One would obtain:

Table 4.4 Test of interaction surgical treatments by subsets in the breast cancer trial presented in figure 1.3.

Subset (1)	$\hat{\xi}_m$ (2)	$SE(\hat{\xi}_m)$ (3)	$\hat{\xi}_m/var(\hat{\xi}_m)$ (4)	w_m (5)	$w_m(\hat{\xi}_m - \bar{\hat{\xi}})^2$ (6)	$\hat{\xi}_m^2/var(\hat{\xi}_m)$ (7)
No adjuvant treatment	0.817 95	0.522 29	2.998 50	3.665 87	4.205 11	2.452 60
Radiotherapy	−0.285 02	0.439 25	−1.477 24	5.182 94	0.005 29	0.421 06
Chemotherapy	−0.625 93	0.314 87	−6.313 40	10.086 43	1.402 21	3.951 73
Totals			−4.792 14	18.935 24	5.612 61	

1. no adjuvant therapy: $\hat{\theta} = 2.24$ $Q_{M-H} = 2.61$ $p = 0.1064$
2. radiotherapy: $\hat{\theta}_{M-H} = 0.75$ $Q_{M-H} = 0.42$ $p = 0.5150$
3. chemotherapy: $\hat{\theta}_{M-H} = 0.53$ $Q_{M-H} = 4.08$ $p = 0.0434$

(in computing θs, mastectomy was considered the reference treatment).

These results could lead to the wrong conclusion that in patients with metastatic lymph nodes, conservative surgery followed by adjuvant chemotherapy compares favourably with mastectomy followed by adjuvant chemotherapy in terms of survival for all death causes. The clinical implausibility of this statement appears to be evident.

4.6   TWO SAMPLE COMPARISON IN PRESENCE OF TIME-DEPENDENT COVARIATES: THE MANTEL–BYAR TEST

In perusing the oncological literature concerning the efficacy evaluation of new drugs or of drug combinations in patients with advanced measurable disease, one comes across papers comparing survivorship of subsets of the case series defined in terms of radiologically assessed tumor response.

The measure of treatment efficacy on the tumor mass is usually coded as complete remission (CR), partial remission (PR), no change (NC) and progression of disease (PD). Very frequently, the first two (CR + PR) are pooled in the group of *responders* and the last two subsets (NC + PD) are pooled in the group of *non-responders*; difference in survival between these two groups is tested by the M–H test. It is worth noting that in previous sections this test was used to compare the survivorship of subjects as classified according to a variable known at the origin of follow-up; in this case, on the contrary, patients are categorized by information emerging during follow-up on the response of the tumor to the administered treatment.

The problems underlying the analysis and interpretation of the comparison of survivorships by response category may be better understood by first considering the arguments supporting why such comparisons are done. These arguments concern two different aspects of clinical activity, the first being the methodology of clinical research and the second the strategy of patient management.

The analysis of survivorship by response category is adopted as a device to replace a controlled clinical trial with two arms: treatment versus no treatment. As responders benefited from treatment they are thought of as components of the treatment group. On the other hand, since non-responders did not benefit from the treatment, their survivorship is thought of as equivalent to that of untreated patients. As far as the second aspect is concerned, the finding that responders experience longer survival than non-responders is inappropriately employed as evidence to justify a more aggressive therapy by arguing that it will

Table 4.5 Results of a second phase clinical trial on patients with multiple myeloma categorized according to the tumor response. From Anderson *et al.* (1983).

Time from treatment (days) $t_{(j)}$ (1)	Non-responders (A)			Responders (B)			$E(d_{Aj})$ (8)	$var(d_{Aj})$ (9)
	No. at risk n_{Aj} (2)	No. of failures d_{Aj} (3)	Survival probability $\hat{P}_{Aj}$ (4)	No. at risk n_{Bj} (5)	No. of failures d_{Bj} (6)	Survival probability $\hat{P}_{Bj}$ (7)		
8	10	1	0.900	25	0	1.000	0.29	0.204
9	9	1	0.800	25	0	1.000	0.26	0.195
11	8	1	0.700	25	0	1.000	0.24	0.184
23	7	0	0.700	25	1	0.960	0.22	0.171
30	7	1	0.600	24	0	0.960	0.23	0.175
34	6	0	0.600	24	1	0.920	0.20	0.160
49	6	0	0.600	23	1	0.880	0.21	0.164
64	6	0	0.600	22	1	0.840	0.21	0.168
78	5	1	0.500	21	0	0.840	0.22	0.173
94	5	0	0.500	21	1	0.800	0.19	0.155
99	5	0	0.500	20	1	0.760	0.20	0.160
102	5	1	0.400	19	0	0.760	0.21	0.165
106	4	0	0.400	19	1	0.720	0.17	0.144
108	4	0	0.400	18	1	0.680	0.18	0.149
118	4	0	0.400	17	1	0.640	0.19	0.154
121	4	1	0.300	16	0	0.640	0.20	0.160
132	2	0	0.300	16	1	0.600	0.11	0.099
155	2	1	0.150	14	0	0.600	0.13	0.109
188	1	0	0.150	14	1	0.557	0.07	0.062
242	1	0	0.150	10	1	0.501	0.09	0.083
262	1	0	0.150	9	1	0.446	0.10	0.090
269	1	0	0.150	8	2	0.334	0.22	0.173
278	1	0	0.150	6	1	0.279	0.14	0.122
326	1	0	0.150	5	1	0.223	0.17	0.139
364	1	0	0.150	3	1	0.149	0.25	0.188
411	1	1	0.000	2	0	0.149	0.33	0.222
424	0	0		2	1	0.074	0.00	0.000
746	0	0		1	1	0.000	0.00	0.000
Totals		9			20		5.04	3.967

. yield a higher response percentage, which in turn will result in increased survivorship.

Data published by Anderson *et al.* (1983), and reported in table 4.5, will be used as a working example throughout this section along the lines proposed by these authors. A phase two trial was conducted on 35 patients to investigate the efficacy of a chemotherapy treatment in multiple myeloma. In the examinations performed during the follow-up, 25 patients showed CR or PR. Thus, according to the response of tumor to the chemotherapeutic agent, patients can be categorized in the two groups: A, non-responders, 10pts, and B, responders, 25 pts. For the two groups at each $t_{(j)}$, table 4.5 gives the number of patients at risk, the number of failures and the K–M estimate of the survival probability, calculated from the start of treatment, together with the two ingredients $E(d_{Aj})$ and $var(d_{Aj})$ needed for computing the M–H test. The two survival curves are drawn in figure 4.5 and appear to differ substantially. Suppose the M–H test is used for comparison; by the total number of deaths in group A (column (3)) and by the totals of $E(d_{Aj})$ and $var(d_{Aj})$ (columns (8) and (9)) of table 4.5, the M–H statistic is:

$$Q_{M-H} = \frac{(9 - 5.04)^2}{3.967} = 3.95$$

From the chi-square distribution with 1 d.f., $p = 0.047$ which allows the investigator to reject the null hypothesis that the hazard rate of responders equals that of non-responders. Unfortunately the conclusion a clinician might reach on the grounds of this analysis by response category are misleading because the approach is wrong and should never be adopted.

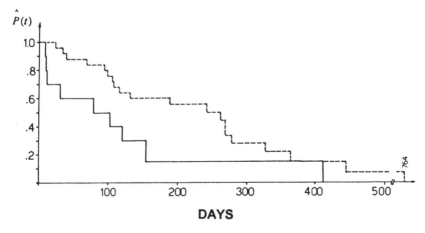

Figure 4.5 Survival curves of responders (----) and non-responders (——) to a chemo-therapeutic agent for treatment of multiple myeloma (see table 4.5). From Anderson *et al.* (1983).

The survival estimates are biased; in fact patients who are categorized as responders are required to live long enough for the response to be elicited whilst this requirement is not needed to categorize non-responders. Consider the waiting time to response, i.e. the period elapsing from the initiation of treatment and the response. It must be noted that the previously used approach ascribes this time to the response category and, as a consequence, the survival of the responders is inappropriately favoured whilst the survival of non-responders is inappropriately unfavoured. It appears that problems in survival estimation and comparison arise here because of the time dependent nature of the variable "response".

An alternative approach one can find in the literature consists in considering the time in which the response was observed as the starting point of the survival curve of responders and in comparing this survival experience to the non-responders' survivorship from the start of treatment. At first glance it seems that removing the waiting time to response from the responders' survival may avoid the above-mentioned bias. Moreover if the comparison of this curve with the whole survival curve of non-responders (i.e. the one drawn with a continuous line in figure 4.5) could intuitively appear to be biased in favour of these latter; paradoxically bias may be reversed because the mortality rate of non-responders at the beginning of follow-up may still be overestimated. In fact, though responders are exposed to the risk of death in the time period preceding the response and, as such, they are non-responders in all respects, they do not enter the set at risk to be used as a denominator to compute the mortality rate of the non-responders group. A further source of bias might be due to a possible sudden change of mortality rate. Assume that mortality rate for responders and non-responders is independent of response, but varies with time and, for example, that the failure rate function, $\lambda(t)$, common to either group, looks like the one depicted in figure 4.6. Moreover assume that the majority of responders respond after four months from the start of therapy. Since in the responders group survivorship is considered from the response time forwards, the portion of $\lambda(t)$ that is estimated is the one shown by a bold line in the figure. Thus the comparison of the survival of responders versus non-responders would reflect nothing but the clear reduction of the failure rate common to the two groups.

One could try to overcome the problem due to the time dependent nature of the variable "response' by selecting some fixed time, say x, from the start of therapy as a landmark for classifying all patients according to tumor response. The landmark, though arbitrary, should be a time point relevant from a clinical viewpoint, for example, the end of the induction period in the example here considered. Patients still alive and under follow-up at the landmark are categorized as responders or non-responders according to the tumor response assessed within the landmark time. From this point forward patients are followed up to investigate whether their survivorship depends on their response status at the landmark. The M–H test may be adopted to make a valid comparison of the

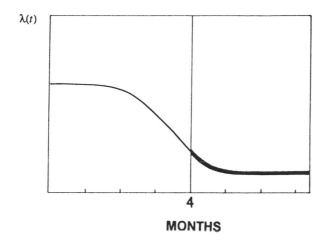

MONTHS

Figure 4.6 Hypothetical pattern of the hazard function for responders and non-responders in 7.5 months of treatment. The response is assumed to be elicited 4.0 months after the start of treatment; the bold line shows the hazard attributed erroneously to the responders group.

survival after the landmark time and the K–M approach to provide unbiased estimates of survivorship conditional on being at risk at the landmark time.

The role of a time dependent covariate can be better understood by means of the model shown in figure 4.7. It shows the state of the patient over time: at the start of treatment he is in state 1 and possibly moves to state 2 if death precedes response. Otherwise the patient moves to state 3 and possibly to state 4 if he dies during follow-up. Let $\lambda_{k,l}(t)$ denote the failure rate function for a patient moving from the kth state to the lth state at time t from the initiation of treatment. In this model the equation: $\lambda_{1,2}(t) = \lambda_{3,4}(t)$ implies that survival is independent of tumor response. This hypothesis can be true even if responders seem to survive longer than non-responders just as a result of the bias arising from the variation of the failure rate shown in figure 4.6.

The test suggested by Mantel and Byar (1974) (M–B), as an extension of the M–H test, is appropriate to test the null hypothesis $H_0 : \lambda_{1,2}(t) = \lambda_{3,4}(t)$ for all values of t. As in the computation of the M–H test, a 2×2 contingency table is constructed at each time of failure $t_{(j)}$ from the start of treatment. The main difference is that in the M–B test at the time of tumor response a patient moves from the non-responder to the responder group and in this group he remains up to death or to the cut-off date (censored observation). In this way, at each failure time, the number of patients at risk in states 1 and 3 for non-responders and responders respectively is updated. The quantities $E(d_{2j})$ and $var(d_{2j})$ are computed according to the usual formulas (4.2) and (4.3) after substituting suffix A with suffix 2. Table 4.6 is constructed by taking into account the response state of the 35 multiple myeloma patients at each of the 28 failure times. Consider, for

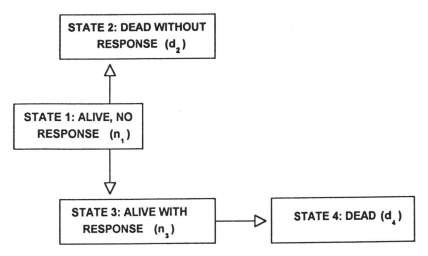

Figure 4.7   Treatment states of patients according to tumor response to the chemotherapeutic agent.

instance, the contingency table related to the first failure observed at the eighth day from treatment start: two patients among the 35 accrued are classified in state 3 as they were evaluated as responders within the eighth day. The first failure among responders is observed on the 23rd day and, by that time, 11 patients were at risk as responders. The M–B test is based upon the statistic:

$$Q_{M-B} = \frac{(9 - 8.56)^2}{4.028} = 0.05$$

This result, being statistically not significant, contradicts the one previously obtained by an improper use of the M–H test.

Alternatively the problem may be solved by resorting to the Cox (1972) model with time dependent covariates as shown in section 6.8.

The information yielded by the whole case series can be taken into account in estimating the survival curve from treatment start of non-responders by censoring responders at the response time. This approach allows for avoiding the risk of overestimating the non-responders' mortality rate. The pertinent estimates are headed $\hat{P}_{1,2}(t)$ in table 4.6 and the resulting curve is reported in figure 4.8 together with the survival curve of the whole case series. Under the previously specified null hypothesis, the two curves are expected to overlap. On the other hand, a figure reporting the $\hat{P}_{1,2}(t)$ curve and the survival curve of the responder group from the start of treatment would be utterly misleading because of the waiting time needed to obtain the response.

Table 4.6 Data from table 4.5; multiple myeloma patients classified according to the state at each time of failure together with the ingredients needed to compute the Mantel–Byar (1974) test (see text) (from Anderson *et al.*, 1983).

Time from treatment start (days) $t_{(j)}$	Patients in states 1–2 (non-responders)			Patients in states 3–4 (responders)		$E(d_{2j})$	$var(d_{2j})$
	No. at risk n_{1j}	No. of failures d_{2j}	Survival probability $\hat{P}_{1.2}(t_j)$	No. at risk n_{3j}	No. of failures d_{4j}		
8	33	1	0.9697	2	0	0.94	0.054
9	31	1	0.9384	3	0	0.91	0.080
11	30	1	0.9071	3	0	0.91	0.083
23	21	0		11	1	0.66	0.226
30	19	1	0.8594	12	0	0.61	0.237
34	13	0		17	1	0.43	0.246
49	12	0		17	1	0.41	0.243
64	9	0		19	1	0.32	0.218
78	8	1	0.7520	19	0	0.30	0.209
94	6	0		19	1	0.24	0.182
99	5	0		19	1	0.21	0.165
102	5	1	0.6016	18	0	0.22	0.170
106	4	0		18	1	0.18	0.149
108	4	0		17	1	0.19	0.154
118	4	0		16	1	0.20	0.160
121	4	1	0.4512	15	0	0.21	0.166
132	2	0		16	1	0.11	0.099
155	2	1	0.2256	14	0	0.13	0.109
188	1	0		14	1	0.07	0.062
242	1	0		10	1	0.09	0.083
262	1	0		9	1	0.10	0.090
269	1	0		8	2	0.22	0.173
278	1	0		6	1	0.14	0.122
326	1	0		5	1	0.17	0.139
364	1	0		3	1	0.25	0.188
411	1	1	0.0000	2	0	0.33	0.222
424	0	0		2	1	0.00	0.000
746	0	0		1	1	0.00	0.000
Totals		9			20	8.56	4.028

Being aware of the importance of the graphical presentation of the survivorship of responders and non-responders, Simon and Makuch (1984) proposed unbiased estimators alternative to those used to obtain figure 4.8 combining ideas from the landmark and Mantel–Byar procedure. To estimate unbiased survival curves from the landmark for both groups, Simon and Makuch suggest

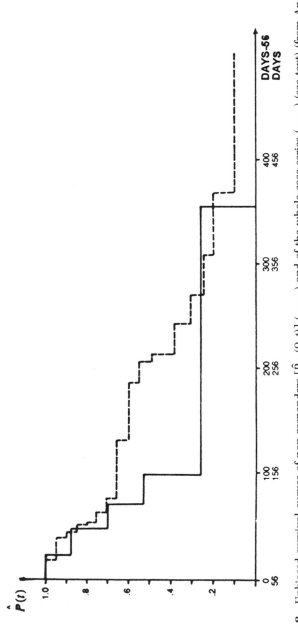

Figure 4.8 Unbiased survival curves of non-responders $[\hat{P}_{1,2}(0, t)]$ (———) and of the whole case series (----) (see text) (from Anderson *et al.* 1983).

the following statistic which is formally similar to the K–M estimate:

$$\hat{P}_{k,l}(x, t) = \prod_{u \in (t-x)} [1 - d_{k,l}(u)/n_k(u)] \tag{4.17}$$

where x is the landmark and $d_{k,l}(u)$ is the number of transitions from state k to state l at time u, for $x \leqslant u \leqslant t$.

$\hat{P}_{1,2}(x, t)$ estimates the probability of being alive at time $t\,(t \geqslant x)$ for a patient classified as non-responder at time x. This patient, however, could respond at any moment after x; if this happens his observation time is censored at that moment. $\hat{P}_{3,4}(x, t)$ estimates the probability of being alive at time $t\,(t \geqslant x)$ for a patient classified as responder at time x or whose response was elicited between x and t. This estimator does not allow for responders who die before time x, but it takes into account all patients responding after time x as they are included in the risk set $n_3(u)$ for any u greater than their response time. Due to these entries in $n_3(u)$ the size of the risk set may increase during follow-up. The curve $\hat{P}_{3,4}(x, t)$, cannot be interpreted straightforwardly as the usual K–M estimator of survival

Table 4.7 Estimate of $\hat{P}_{k,l}(x, t_j)$ according to (4.17) for multiple myeloma patients of table 4.5.

Time from treatment (days) t_j	Patients in states 1–2			Patients in states 3–4		
	Hazard rate	$\hat{P}_{1,2}(56, t_j)$	SE of $\hat{P}_{1,2}(56, t_j)$	Hazard rate	$\hat{P}_{3,4}(56, t_j)$	SE of $\hat{P}_{3,4}(56, t_j)$
64				0.0526	0.9474	0.512
78	0.1250	0.8750	0.1169			
94				0.0526	0.8975	0.0686
99				0.0526	0.8503	0.0796
102	0.2000	0.7000	0.1824			
106				0.0556	0.8030	0.0881
108				0.0588	0.7558	0.0948
118				0.0625	0.7086	0.0999
121	0.2500	0.5250	0.2041			
132				0.0625	0.6643	0.1030
155	0.5000	0.2625	0.2118			
188				0.0714	0.6168	0.1060
242				0.1000	0.5551	0.1119
262				0.1111	0.4935	0.1153
269				0.2500	0.3701	0.1148
278				0.1667	0.3084	0.1110
326				0.2000	0.2467	0.1046
362				0.3333	0.1645	0.0968
411	1.000	0.000				
424				0.5000	0.0823	0.0757
746				1.0000	0.0000	

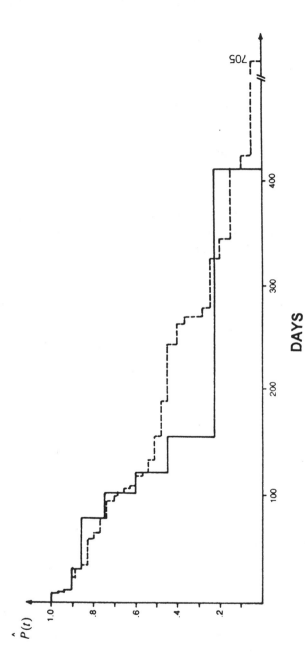

Figure 4.9 Survival curves $\hat{P}_{1,2}(56, t)$ (———) and $\hat{P}_{3,4}(56, t)$ (– – –). Time is given in days minus 56; this latter is the landmark time corresponding to the end of the induction phase (from Anderson *et al.* 1983).

in responders. In fact $\hat{P}_{3,4}(x,t)$ gives the probability of a patient being alive at time t conditional upon being in response at the landmark time or entering the response state at any moment between x and t.

By assuming $x = 56$ days (end of the induction period) as the landmark for multiple myeloma patients, and using (4.17), the estimates $\hat{P}_{1,2}(56,t)$ and $\hat{P}_{3,4}(56,t)$ reported in table 4.7 are obtained for the time interval from 56 to 746 days. The standard errors of these statistics are the square root of the variances given by (3.9a) where $\hat{P}(t)$ is replaced by $\hat{P}_{1,2}(x,t)$ and $\hat{P}_{3,4}(x,t)$ respectively.

The two corresponding curves are drawn in figure 4.9 which shows that the two curves cross at many points; the graphical message is similar to that emerging from figure 4.8 and the seeming difference between the curves displayed in figure 4.5 is an artifact due to the biased estimates computed in table 4.5.

It must be stressed here that even when proper methods of statistical analysis of survivorship by tumor response are used, neither of the arguments advanced by oncologists for justifying this analysis and reported at the beginning of this section can be supported. The conclusion that responders' survival is significantly ($p < 0.05$) longer than that of non-responders does not prove that the treatment prolongs survivorship or that a more aggressive therapy is justified. In fact these patients cannot be differentiated from those for whom tumor response is only a marker for a favourable subset of patients. This subset may be formed by patients with favourable prognosis because of unknown or known features of the patient (for example age and sex) and/or of the disease (for example extent of tumor, number of metastatic nodes, tumor markers). Therefore responders may survive more than non-responders not because of the effect of therapy on survivorship, but because tumor response picks up patients whose pretreatment features hold with longer survivorship. Similar considerations apply in case of comparison of groups retrospectively defined in terms of the amount of assumed treatment.

4.7* COMPARISON OF MORE THAN TWO SAMPLES

This section faces the problem of comparing the survival experience of $G\,(G > 2)$ samples, one sample from each of the G treatment populations. The overall null hypothesis to be tested is

$$H_0: \lambda_1(t) = \lambda_2(t) = \cdots = \lambda_g(t) = \cdots = \lambda_G(t) \qquad (4.18)$$

against the composite alternative hypothesis that there are at least two groups with different failure rates.

The test is computed by extending the M–H test developed in section 4.2 in order to be able to deal with $G \times 2$ contingency tables. Information pertinent to

Table 4.8 $G \times 2$ contingency table at failure time $t_{(j)}$.

Treatment	Dead at time $t_{(j)}$	Alive at time $t_{(j)}$	Set at risk just before $t_{(j)}$
1	d_{1j}	$n_{1j} - d_{1j}$	n_{1j}
2	d_{2j}	$n_{2j} - d_{2j}$	n_{2j}
$\vdots$	$\vdots$	$\vdots$	$\vdots$
g	d_{gj}	$n_{gj} - d_{gj}$	n_{gj}
$\vdots$	$\vdots$	$\vdots$	$\vdots$
G	d_{Gj}	$n_{Gj} - d_{Gj}$	n_{Gj}
Totals	d_j	$n_j - d_j$	n_j

each failure time may be conveniently arranged in a contingency table like that reported in table 4.8.

The extension necessarily involves matrix manipulation.

Under the null hypothesis (4.18) and conditioned to the fixed marginals, it may be shown that the number of failures in the G samples is distributed according to a multiple hypergeometric distribution of dimension $G - 1$. In presenting the computation procedure of the M–H test we assumed the B sample as the reference sample; on the analogy of this, we will assume now the Gth sample as the reference one.

Coherently with (4.2), the number of failures expected at time $t_{(j)}$ in the gth sample under the null hypothesis may be computed as follows:

$$E(d_{gj}) = n_{gj} \times \frac{d_j}{n_j}$$

The difference between the total number of observed failures and the number of failures expected under H_0 for the gth sample is:

$$U_g = \sum_{j=1}^{J} [d_{gj} - E(d_{gj})] \tag{4.19}$$

Therefore the scalar U of (4.5) has to be substituted now by the $G - 1$ dimensional column vector $\mathbf{U}$, with generic element U_g.

For each contingency table, a $(G-1) \times (G-1)$ dimensional covariance matrix $\mathbf{V}(t_j)$ must be computed, with generic component at time t_j

$$\mathbf{V}_{g,l} = cov(d_{gj}, d_{lj}) = \begin{cases} \dfrac{n_{gj}(n_j - n_{gj})d_j(n_j - d_j)}{n_j^2(n_j - 1)} & \text{for } g = l \\[3mm] -\dfrac{n_{gj} \times n_{lj} \times d_j(n_j - d_j)}{n_j^2(n_j - 1)} & \text{for } g \neq l \end{cases} \tag{4.20}$$

where $g, l = 1, 2, \ldots, G - 1$ are the sample suffixes. Then, adding these quantities over the J failure times we obtain

$$\mathbf{V} = \sum_{j=1}^{J} \mathbf{V}(t_j) \tag{4.21}$$

The test suitable for testing the null hypothesis (4.18) is based on the quadratic form

$$\mathbf{Q}_{\text{overall}} = \mathbf{U}' \mathbf{V}^{-1} \mathbf{U} \tag{4.22}$$

Under H_0, this statistic is asymptotically distributed as a chi-square r.v. with $G - 1$ d.f. For $G = 2$ this statistic reduces to $Q_{\text{M}-\text{H}}$ of (4.6).

Assume that the G samples may be classified with regard to a quantitative variable, for example the total dosage of an adjuvant chemotherapy, or an ordered variable. In this case the alternative hypothesis of major interest for the investigator concerns the possible decrease (or increase) of the failure rate as a function of the increase (or decrease) of the dosage. Test for *trend* provides a technique to investigate such ordering.

Suppose the researcher is interested in testing an ordered relationship among the G samples defined in terms of the weights $w_1, w_2, \ldots, w_g, \ldots, w_G$ so that

$$\lambda_1(t) = w_1 \lambda_G(t); \ldots; \lambda_g(t) = w_g \lambda_G(t); \ldots; \lambda_{G-1} = w_{G-1} \lambda_G(t)$$

where it has been assumed without loss of generality that $w_G = 1$.

A test for trend is given by

$$Q_{\text{trend}} = \frac{\left[\displaystyle\sum_{g=1}^{G} {}_g U_g \right]^2}{\mathbf{w}' \mathbf{V}^* \mathbf{w}} \tag{4.23}$$

where $\mathbf{w} = [w_1, w_2, \ldots, w_g, \ldots, w_G]$ is the row vector of weights defining the relationship of clinical interest. The matrix $\mathbf{V}^*$ is calculated as in (4.20) and (4.21) but considering all the G groups and thus has dimension $(G \times G)$. To make the calculation easier, Q_{trend} can be rewritten as

$$Q_{\text{trend}} = \frac{\left\{ \displaystyle\sum_{g=1}^{G} w_g \sum_{j=1}^{J} [d_{gj} - E(d_{gj})] \right\}^2}{\displaystyle\sum_{j=1}^{J} \frac{n_j - d_j}{n_j - 1} \left\{ \sum_{g=1}^{G} w_g^2 E(d_{gj}) - \frac{1}{d_j} \left[\sum_{g=1}^{G} w_g E(d_{gj}) \right]^2 \right\}} \tag{4.24}$$

This statistic is asymptotically distributed as a chi-square r.v. with 1 d.f.

If the weights are linearly spaced, say $w_g = g$, the so-called linear test for trend is obtained.

As an example let us utilize the data set reported in table 4.9 which may be considered the result of a carcinogenesis experiment: $G = 3$ levels of dose were tested and the time to event was considered as the time in days until the tumor was observed.

In analogy with the computation procedure of the M–H test, it is convenient to construct a 3×2 contingency table at each time in which an event occurs; correspondingly the ingredients needed to calculate (4.22) and (4.23) are computed. The first time at which a tumor was observed is $t_{(1)} = 47$ and the corresponding 3×2 contingency table is given in the upper part of table 4.10 together with the values of $E(d_{g1})$ and the $G - 1 = 2$ components of the vector U pertinent to the first failure time. In the central part of the table the computation of the components of the variance matrix (4.20) is shown. In the lower part the calculation of the ingredients of Q_{trend} is given in detail; note that the doses tested are used as weights w_g.

The same procedure is adopted for each of the following eight failure times. The following results are obtained:

$$\mathbf{U} = \begin{vmatrix} 3.208\,58 \\ -0.803\,36 \end{vmatrix} \quad \mathbf{V} = \begin{vmatrix} 1.318\,82 & -0.641\,32 \\ -0.641\,32 & 2.662\,69 \end{vmatrix} \quad \mathbf{V}^{-1} = \begin{vmatrix} 0.858\,84 & 0.206\,86 \\ 0.206\,86 & 0.425\,38 \end{vmatrix}$$

and $Q_{\text{overall}} = \mathbf{U}'\mathbf{V}^{-1}\mathbf{U} = 8.050$; this statistic is asymptotically distributed as chi-square with 2 d.f. ($p = 0.018$). The test for trend gives $Q_{\text{trend}} = 3.662$ which is asymptotically distributed as chi-square with 1 d.f. ($p = 0.056$). This result suggests that there is evidence for the ordering of the survival experiences according to the dose of carcinogen tested. This component may be removed from the overall statistic (4.22) leaving a residual:

$$Q_{\text{residual}} = Q_{\text{overall}} - Q_{\text{trend}}$$

which is asymptotically distributed as chi-square r.v. with $G - 2$ d.f. Both this statistic and Q_{trend} should be looked at in order to reach a sensible interpretation

Table 4.9 Set of data from a carcinogenesis experiment (from Thomas *et al.*, 1977, reproduced by permission of Academic Press, Inc.).

Group g	Dose	Initial group size n_{0g}	Times to event (t_j) (asterisk indicates censored times)
1	2.0	10	41*, 41*, 47, 47*, 47*, 58, 58, 58, 100*, 117
2	1.5	10	43*, 44*, 45*, 67, 68*, 136, 136, 150, 150, 150
3	0.0	9	73*, 74*, 75*, 76, 76, 76*, 99, 166, 246*

Table 4.10 Computation procedure for ingredients of tests (4.22) and (4.23) for data from table 4.9.

Dose	Dead at time $t_{(1)}$	Alive at time $t_{(1)}$	Set at risk just before $t_{(1)}$	$E(d_{g1})$	U_1	
2	1	7	8	0.333 33	$1 - 0.333\,33 =$	0.666 67
1.5	0	7	7	0.291 67	$0 - 0.291\,67 =$	$-0.291\,67$
0	0	9	9	0.375 00		
Totals	1	23	24			

Components of $V(t_1)$

$$\begin{bmatrix} \dfrac{8(24-8)1(24-1)}{24^2(24-1)} = 0.222\,22 & -\dfrac{8 \times 7 \times 1 \times 23}{24^2 \times 23} = -0.097\,22 \\[3mm] -0.097\,22 & \dfrac{7(24-7) \times 1 \times 23}{24^2 \times 23} = 0.206\,60 \end{bmatrix}$$

Test for trend

Numerator $= 2 \times 0.666\,67 + 1.5(-0.291\,67) = 0.895\,835$

Denominator $= \dfrac{24-1}{24-1}[(2^2 \times 0.333\,33 + 1.5^2 \times 0.291\,67) - (2 \times 0.333\,33 + 1.5$

$\times 0.291\,67)^2] = 1.989\,58 - 1.219\,18 = 0.770\,40$

of how appropriate the weights are for describing the dose–response relationship. In the previous example $Q_{\text{residual}} = 8.050 - 3.662 = 4.388$ with 1 d.f. and $p = 0.036$; this could mean that a set of weights different from the one we adopted could better fit the ordering of the three survivorships elicited in this experiment.

4.8 SAMPLE SIZE

In previous sections, dealing with results of statistical tests, our attention was focused on α the probability of type I error (false positive): rejecting H_0 when H_0 is true. The test procedure, however, runs another risk: to make the error of *not* rejecting H_0 when H_0 is false. This false-negative error is the so-called type II error and the probability of it occurring is named β. The quantity $1 - \beta$ is called the power of the test.

The probability of the type I error is computed by assuming that H_0 is correct, i.e., with reference to the hypothesis (4.1b), when $P_A(t) - P_B(t) = 0$; therefore α has one single value. On the contrary β is computed in situations in which H_0 is false, namely for difference $P_A(t) - P_B(t) = \delta \neq 0$ and δ can have an infinite number of values. Correspondingly, the probability of type II error assumes distinct values for each δ; the function relating β to δ is called the *operating characteristic curve*. In our context it allows the investigator to know the probability the trial has to miss a difference in survival between the two treatments at a prefixed time point of clinical relevance. There are three major factors that determine the profile of the operating characteristic curve: outcome level in the reference group (standard treatment or placebo), significance level of the test and sample size.

As an example, consider figure 4.10 where the operating characteristic curve pertinent to the M–H test for comparing two survival experiences is drawn by assuming $\alpha = 0.05$; $n_1 = n_2 = 350$, $t = 5$ years and $P(5) = 0.90$ for the standard treatment. It is the operating characteristic curve of the surgical trial described in figure 1.3 in which Halsted mastectomy was the standard treatment. The probability of the II type error is reported on the ordinate while the horizontal axis reports both the ratio θ of failure rates and the difference δ of survival probabilities in the two groups. The curve shows that the greater the difference between the two treatment effects, the smaller is β. In particular this trial has probabilities 20% and 10% of missing a difference of 5% and 5.5% respectively.

Freiman *et al.* (1978) drew the attention of clinical trialists to the size of II type error and to the trial size. A set of 71 trials drawn from three prestigious medical

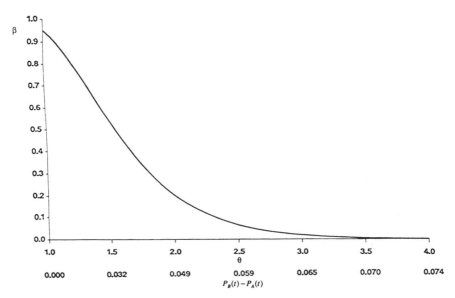

Figure 4.10 Operative characteristic curve for the M–H test to compare two survivorship curves (see text).

journals (*New England Journal of Medicine, Lancet* and *Journal of the American Medical Association*) were studied. In all of them the authors stated that therapies were not different from control ($p > 0.05$). Freiman *et al.* investigated the operating characteristic curve of these "negative" trials in terms of relative frequency of failure, say $\hat{F}_A$ and $\hat{F}_B$ for the control and experimental group respectively. They found that 50% of the trials had $\beta > 0.74$ for a relative reduction $(\hat{F}_A - \hat{F}_B)/\hat{F}_A$ equal to 25%, in other words half of the trials had a greater than 74% risk of missing a 25% reduction on the control group values. For a 50% reduction the figure is, of course, better but even so half of the trials had a 40% risk of missing such an important relative difference.

The lesson is clear: in planning a trial investigators should make a reliable estimate of $P_A(t)$, specify what they believe are clinically relevant values of δ, and fix α in order to compute a sample size that has a low β of missing these clinically relevant values of δ. If no significant difference between groups is found with these criteria, then investigators can conclude that even if there is a difference between the two groups, it is not clinically relevant. On the other hand, if a significant result at the prefixed α level is found, statistical significance coincides with clinical relevance. Moreover the properly computed trial size, when compared to the probable number of patients who will be accrued in the planned time period, enables the investigator to critically reconsider the whole project. Then he will be able to decide to activate the trial as planned or to modify the original design, for example involving other centres in the project in order to reach the planned accrual in a reasonable time period. In some cases it will be wiser and ethically correct not to start the trial at all as it would be inconclusive.

Several handbooks giving tables for sample size calculation in comparative trials are now available in the statistical literature (e.g. Machin and Campbell, 1987; Lemeshow *et al.*, 1990); they enable investigators to answer the question "how big a sample do I need" in a great number of situations one can face in the clinical setting.

This section and Appendix A4.3 aim at introducing the rationale of the procedure suitable for computing the sample size for the comparison of two survival curves by means of the M–H test as it is now widely used in several branches of medicine; Freedman's (1982) approach will be followed. The following calculations will be in terms of θ.

The choice of the three ingredients θ, α and β must be done on the basis of a compromise between the scientific goal and the trial feasibility, recalling that:

(1) for α and β fixed, the number of patients N tends to increase for values of θ approaching 1;
(2) for α and N fixed, β tends to increase for values of θ approaching 1.

From equation (4.1b) it is easy to see that

$$\theta = \frac{\log P_{A}(t)}{\log P_{B}(t)} \tag{4.25}$$

Freedman's procedure entails the assumption that analysis occurs at a fixed time t^* after the *last* patient was accrued; information on patient follow-up extending beyond t^* is neglected . This assumption frequently accords with practice and, thus, it appears to be partly motivated. Since the statistical analysis does not utilize all information, the required number of patients tends to be over-estimated. However, it is sensible to accomplish a definitive analysis only when the great majority of information has been gathered. This suggests choosing a minimum follow-up time t^* beyond which the frequency of occurrence of responses is low. Such a choice of t^* tends to limit any overestimation of the trial size. The assumption has the positive consequence that the number of required subjects does not depend on the rates of accrual and occurrence of deaths but only on the proportion of living patients in the two treatments after minimum followup time t^*. From previous studies one can reasonably foresee the survival probability at time t^*, $P_{A}(t^*)$, for the group of patients treated with the standard treatment A and then, on pharmacological or radiological or clinical bases, to guess the minimum gain in survival he is expecting by applying treatment B; with reference to expression (4.25) this corresponds with making a reasonable guess at θ. By assuming distinct times of event occurrence, no withdrawals, and an equal number exposed to risk in the two groups at each response time, it may be shown (Appendix A4.3) that the total number of events needed for the trial (one-tailed test) is

$$d = \left[z_{1-\alpha} + 2z_{1-\beta}\frac{\theta^{1/2}}{(1+\theta)} \right]^{2}\left(\frac{1+\theta}{1-\theta}\right)^{2}$$

By further assuming that $\theta^{1/2}/(1+\theta) \simeq \frac{1}{2}$ this equation may be written

$$d = (z_{1-\alpha} + z_{1-\beta})^{2}\left(\frac{1+\theta}{1-\theta}\right)^{2} \tag{4.26}$$

where z indicates the standardized Gaussian deviate. The two suffixes α and β specify the I and II type error probabilities respectively. In the case of a two-tailed test, $z_{1-\alpha/2}$ replaces $z_{1-\alpha}$. The total number N of patients to be entered into the study can be computed by arguing that if d events have been recorded at time t^*, they have occurred in the two groups, of $N/2$ size each, according to the pertinent probabilities, that is:

$$d = \frac{N}{2}[1 - P_{A}(t^*)] + \frac{N}{2}[1 - P_{B}(t^*)]$$

from which it is easy to obtain

$$N = \frac{2d}{2 - P_A(t^*) - P_B(t^*)}$$

(4.27)

Since some withdrawals almost always happen, the researcher has to make some allowance for this. Suppose that the anticipated percentage of patients to withdraw is x; N being the required number of patients given by (4.27), $100N/(100 - x)$ subjects should be recruited.

As pointed out by Freedman, when patient accrual lasts several years, the use of the minimum follow-up time t^* to compute θ via (4.25) and then d via (4.26) would be *wrong* as it overestimates the number of patients needed, often seriously. Instead, as an approximate device, the *average* follow-up time should be adopted. The trial size estimated in this way is approximate but adequate for practical aims.

One of the research protocols supported by the Italian Consiglio Nazionale delle Ricerche in the project Oncologia, concerned a multicentre R.C.T. on ovarian cancer of stage I. The purpose of the applicants was to evaluate the effect of a single chemotherapeutic agent as adjuvant post-surgery treatment compared to surgery alone (standard treatment). By perusing the literature and on the basis of their own experience, the investigators assumed that in surgically treated patients the survival probability at five years was 60%; furthermore they argued that a survival increment δ of at least 25% was clinically relevant. The analysis according to M–H was anticipated with a two-tailed test at a significance level of $\alpha = 0.05$ and $\beta = 0.20$. For $t^* = $ five years, $P_A(5) = 0.60$ and $P_B(5) = 0.85$; from (4.25) one calculates that $\theta = \ln 0.60/\ln 0.85 = 3.14$. With $z_{1-\alpha/2} = 1.96$ and $z_{1-\beta} = 0.841$, (4.26) and (4.27) give $d = 2.801^2(4.14/2.14)^2 = 29.36 \simeq 30$ and $N = 2 \times 30/(2 - 0.60 - 0.85) = 109.09 \simeq 110$. Allowance for a possible 20% withdrawal rate increases this number to $100 \times 110/(100 - 20) = 137.5 \simeq 150$. Thus the investigators decided to accrue at least 75 patients per treatment.

Tables A and B of Appendix A4.4 reproduce the original tables from Freedman (1982) for the one-tailed and two-tailed M–H test respectively. Each cell in the tables contains six numbers. The numbers without brackets specify the study size N whilst the numbers in parentheses are the total number of deaths d required. The three sets of N and d correspond to three combinations of significance levels and powers of the test; $\alpha = 0.05$ and $\beta = 0.20$; $\alpha = 0.05$ and $\beta = 0.10$; $\alpha = 0.01$ and $\beta = 0.05$. It is easy to check that for $P_A = 0.60$ and $P_B - P_A = 0.25$ the first two numbers of the pertinent cell coincide with those previously computed.

The investigator, particularly in chemoprevention studies, may wish to make allowance for other important features of the trial such as time lag in the effectiveness of medication, staggered entry, drop-in rate and non-compliance.

To face these situations the interested reader is referred to the paper by Lakatos (1988) for which the theoretical background is above the level of this book.

These authors' experience in dealing with clinical research in the determination of the number of patients to be accrued suggests that it is a matter of bargaining between the researcher and the biostatistician. The trial size N, computed by the statistician according to the ingredients α, β, θ and $P_A(t^*)$ discussed with the clinician, rather frequently shocks the latter and leads to subsequent modification of the trial design. The modification may concern expansion from a single centre to several centres to increase the number of patients, changes in the eligibility criteria to make recruitment easier, or increase of the recruitment and/or the follow-up periods.

In trying to reach a compromise between the desired degree of protection and what the clinician thinks is a feasible trial size, the computation of the operating characteristic curve of the test appears to be a helpful tool.

APPENDIX A4.1   HYPERGEOMETRIC DISTRIBUTION

Consider the marginals of table 4.1 as fixed quantities; under the null hypothesis (4.1) the n_j patients in study can be thought of as a finite population, the units of which are classified into two mutually exclusive categories of d_j patients dead at $t_{(j)}$ and $n_j - d_j$ still alive at $t_{(j)}$. A sample of size n_{Aj} is assumed to be randomly selected from this population and the investigator wishes to compute the probability that exactly d_{Aj} dead be in the sample. It is convenient to divide the problem into two parts: in the first part the number of ordered arrangements of d_{Aj} units and $n_{Aj} - d_{Aj}$ units is computed, while in the second the probability of obtaining any of these arrangements is determined; the product of these two results is the solution of the problem.

We think of n_{Aj} blank spaces to be filled in, with d_{Aj} units of the first category (dead) and with $n_{Aj} - d_{Aj}$ units of the second category (alive). The number of ways of selecting the d_{Aj} spaces for units of the first category is $\binom{n_{Aj}}{d_{Aj}}$ and, after this has been done, the units can be arranged in the selected spaces in just one way. Next the units of the second category can be arranged in the remaining $n_{Aj} - d_{Aj}$ spaces in just $\binom{n_{Aj} - d_{Aj}}{n_{Aj} - d_{Aj}} \times 1$, or 1, way. Hence the total number of arrangements is:

$$\binom{n_{Aj}}{d_{Aj}} = \frac{n_{Aj}!}{d_{Aj}!(n_{Aj} - d_{Aj})!} \tag{A4.1}$$

i.e. the number of combinations of n_{Aj} different objects (the n_{Aj} blank spaces), taken d_{Aj} at a time. As regards the second part we may think of one of the possible arrangements, namely the one in which the first category units are arranged in the first d_{Aj} spaces and the second category units in the remaining $n_{Aj} - d_{Aj}$ spaces. Owing to the fact that we are sampling from a finite population, we must resort to the product of n_{Aj} terms of conditional probabilities:

$$\frac{d_j}{n_j} \times \frac{d_j - 1}{n_j - 1} \times \cdots \times \frac{d_j - d_{Aj} + 1}{n_j - d_{Aj} + 1} \times \frac{n_j - d_j}{n_j - d_{Aj}} \times \frac{n_j - d_j - 1}{n_j - d_{Aj} - 1} \times \cdots \times \frac{n_j - d_j - (n_{Aj} - 1 - d_{Aj})}{n_j - n_{Aj} + 1}$$

By using factorials this product may be written

$$\frac{d_j!}{(d_j - d_{Aj})!} \times \frac{(n_j - d_j)!}{(n_j - d_j - n_{Aj} + d_{Aj})!} \times \frac{(n_j - n_{Aj})!}{n_j!} \tag{A4.2}$$

The product of (A4.1) times (A4.2) gives us the probability we were looking for:

$$p(d_{Aj} \mid d_j, n_j) = \frac{\binom{d_j}{d_{Aj}} \binom{n_j - d_j}{n_j - d_j - n_{Aj} + d_{Aj}}}{\binom{n_j}{n_{Aj}}} \tag{A4.3}$$

For fixed values of n_{Aj} and d_j, the r.v. d_{Aj} may range from $\max(0, n_{Aj} - n_j + d_j)$ to $\min(n_{Aj}, d_j)$. In a simple and elegant way Feller (1968, pp. 232–233) proves that

$$E(d_{Aj}) = n_{Aj} \frac{d_j}{n_j} = n_{Aj} q_j$$

$$var(d_{Aj}) = n_{Aj} q_j (1 - q_j) \frac{n_j - n_{Aj}}{n_j - 1}$$

Note that the variance of the hypergeometric distribution equals that of the binomial distribution times the finite population correction. The effect of this latter consists in reducing the variance of the binomial distribution as the sampling fraction n_{Aj}/n_j tends to increase.

APPENDIX A4.2   CRITICAL VALUES FOR THE GAIL AND SIMON TEST

Critical values (c) for the likelihood ratio test $\min(Q^+, Q^-) > c$ are given in the following table, from Gail and Simon (1985).

Number of groups	Significance level				
	0.20	0.10	0.05	0.025	0.001
2	0.71	1.64	2.71	3.84	9.55
3	1.73	2.95	4.23	5.54	11.76
4	2.59	4.01	5.43	6.86	13.47
5	3.39	4.96	6.50	8.02	14.95
6	4.14	5.84	7.48	9.09	16.32
7	4.86	6.67	8.41	10.09	17.58
8	5.56	7.48	9.29	11.05	18.78
9	6.75	8.26	10.15	11.98	19.93
10	6.92	9.02	10.99	12.87	21.03
12	8.24	10.50	12.60	14.57	23.13
14	9.53	11.93	14.15	16.25	25.17
16	10.79	13.33	15.66	17.85	27.10
18	12.03	14.70	17.13	19.41	28.96
20	13.26	16.04	18.57	20.94	30.79
25	16.28	19.34	22.09	24.65	35.15
30	19.25	22.55	25.50	28.23	39.33

APPENDIX A4.3   FREEDMAN'S PROCEDURE FOR SAMPLE SIZE

With reference to the generic 2×2 contingency table 4.1, define $\phi_j = n_{Aj}/n_{Bj}$, the ratio of the number of patients exposed to risk of death in the two treatment groups just before the jth failure time. Conditional on the observed marginal totals, in Appendix A4.1 it was shown that the distribution of d_{Aj} is hypergeometric. In order to test the null hypothesis (4.1) we resort to the test statistic (4.6) or equivalently to

$$z_{1-\alpha} = \frac{\sum\limits_{j=1}^{J} d_{Aj} - \sum\limits_{j=1}^{J} E(d_{Aj})}{\left[\sum\limits_{j=1}^{J} var(d_{Aj}) \right]^{1/2}} \tag{A4.4}$$

In section 4.8 it was stressed that in computing the sample size investigators must take into account the failure rate ratio θ and the type II error probability β. This implies computing the critical ratio $z_{1-\beta}$ under the alternative hypothesis and this, in turn, implies computing expected value and variance of d_{Aj} under the alternative hypothesis, say, $E_a(d_{Aj})$ and $var_a(d_{Aj})$.

Let us think of the first and second row of the jth 2×2 contingency table as empirical observations of two distinct binomial r.v. having parameters $[n_{Aj}, q_{Aj}]$ and $[n_{Bj}, q_{Bj}]$ respectively. We recall that $q_{Aj} = \lambda_A(t_j)dt$ and $q_{Bj} = \lambda_B(t_j)dt$ (section 3.6.3*) and that, under the alternative hypothesis, $\lambda_A(t_j) = \theta \lambda_B(t_j)$, with $\theta \neq 1$.

The total d_j is expected to be the sum of two components, namely

$$d_j = E(d_{Aj}) + E(d_{Bj}) = n_{Aj}\theta \lambda_B(t_j)dt + n_{Bj}\lambda_B(t_j)dt$$

Hence we obtain

$$\lambda_B(t_j)dt = \frac{d_j}{\theta n_{Aj} + n_{Bj}}$$

As a consequence the expected value of d_{Aj} under the alternative hypothesis is

$$E_a(d_{Aj}) = n_{Aj}\lambda_A(t_j)\,dt = n_{Aj}\frac{\theta d_j}{\theta n_{Aj} + n_{Bj}} \tag{A4.5}$$

The variance of d_{Aj} can be obtained by multiplying the variance of the corresponding binomial r.v. by the finite population correction factor which, under H_a, is

$$\frac{(\theta n_{Aj} + n_{Bj}) - \theta n_{Aj}}{(\theta n_{Aj} + n_{Bj}) - \theta}$$

Thus

$$var_a(d_{Aj}) = n_{Aj}\frac{\theta d_j}{\theta n_{Aj} + n_{Bj}}\left(1 - \frac{\theta d_j}{\theta n_{Aj} + n_{Bj}}\right)\frac{n_{Bj}}{\theta n_{Aj} + n_{Bj} - \theta} \tag{A4.6}$$

Now we can write

$$z_\beta = \frac{\sum\limits_{j=1}^{J} d_{Aj} - \sum\limits_{j=1}^{J} E_a(d_{Aj})}{\left[\sum\limits_{j=1}^{J} var_a(d_{Aj})\right]^{1/2}} \tag{A4.7}$$

If one assumes $d_j = 1$ for $j = 1, 2, \ldots, J$, expectation and variance of d_{Aj} under H_0 can be rewritten in terms of θ and ϕ_j, where

$$\phi_j = \frac{n_{Aj} + n_{Bj}}{n_{Bj}} - 1$$

We have

$$E(d_{Aj}) = 1 - \frac{n_{Bj}}{n_{Aj} + n_{Bj}} = \frac{\phi_j}{\phi_j + 1}$$

$$var(d_{Aj}) = \frac{n_{Aj} \times n_{Bj} \times (n_{Aj} + n_{Bj} - 1)}{(n_{Aj} + n_{Bj})^2 \times (n_{Aj} + n_{Bj} - 1)} = \frac{\phi_j}{(\phi_j + 1)^2}$$

Analogously, under the H_a, $E_a(d_{Aj})$ and $var_a(d_{Aj})$ can be written in terms of θ and ϕ_j, namely

$$E_A(d_{Aj}) = \frac{\theta\phi_j}{\theta\phi_j + 1}$$

$$var_a(d_{Aj}) = \frac{\theta\phi_j}{(\theta\phi_j + 1)^2}$$

From equation (A4.4) we see that the total number of events, $\sum_{j=1}^{J} d_{Aj}$, observed in the treatment A group is

$$\sum_{j=1}^{J} d_{Aj} = \sum_{j=1}^{J} E(d_{Aj}) + z_{1-\alpha} \left[\sum_{j=1}^{J} var(d_{Aj}) \right]^{1/2}$$

and similarly from (A4.7):

$$\sum_{j=1}^{J} d_{Aj} = \sum_{j=1}^{J} E_a(d_{Aj}) + z_\beta \left[\sum_{j=1}^{J} var_a(d_{Aj}) \right]^{1/2}$$

By equating these two expressions we obtain

$$\sum_{j=1}^{J} E_a(d_{Aj}) - \sum_{j=1}^{J} E(d_{Aj}) = z_{1-\alpha} \left[\sum_{j=1}^{J} var(d_{Aj}) \right]^{1/2} - z_\beta \left[\sum_{j=1}^{J} var_a(d_{Aj}) \right]^{1/2}$$

which in terms of θ and ϕ_j becomes

$$\sum_{j=1}^{J} \left(\frac{\theta\phi_j}{\theta\phi_j + 1} - \frac{\phi_j}{\phi_j + 1} \right) = z_{1-\alpha} \left[\sum_{j=1}^{J} \frac{\phi_j}{(\phi_j + 1)^2} \right]^{1/2} + z_{1-\beta} \left[\sum_{j=1}^{J} \left(\frac{\theta\phi_j}{(\theta\phi_j + 1)^2} \right) \right]^{1/2}$$

Under the already mentioned assumption that $d_j = 1$ for $j = 1, 2, \ldots, J$, J equals the total number of events, d, observed in the two groups. Therefore for $\phi_j = 1$, the previous equation becomes

$$\frac{\theta d}{\theta + 1} - \frac{d}{2} = z_{1-\alpha} \frac{d^{1/2}}{2} + z_{1-\beta} \frac{(\theta d)^{1/2}}{\theta + 1}$$

From this equation the total number of events is

$$d = \left(z_{1-\alpha} + 2z_{1-\beta}\frac{\theta^{1/2}}{\theta+1}\right)^2 \left(\frac{\theta+1}{\theta-1}\right)^2$$

This computation can be made simpler by assuming $var(d_{Aj}) = var_a(d_{Aj})$; in this case we have

$$d\left(\frac{\theta}{\theta+1} - \frac{1}{2}\right)^2 \simeq \left(\frac{z_{1-\alpha}}{2} + \frac{z_{1-\beta}}{2}\right)^2$$

and then

$$d \simeq (z_{1-\alpha} + z_{1-\beta})^2 \left(\frac{\theta+1}{\theta-1}\right)^2$$

which corresponds to (4.26).

APPENDIX A4.4   SAMPLE SIZE ACCORDING TO FREEDMAN

Table A Number of patients required to detect an improvement of $(P_B - P_A)$ in survival probability over a baseline survival probability (P_A), when $\alpha = 5\%$, $1 - \beta = 80\%$; $\alpha = 5\%$, $1 - \beta = 90\%$; and $\alpha = 1\%$, $1 - \beta = 95\%$; one-tailed test (from Freedman, 1982).

P_A	$P_B - P_A$ 0.05	0.10	0.15	0.20	0.25	0.30	0.35	0.40	0.45	0.50
0.05	391 (362) 542 (501) 997 (922)	137 (123) 189 (170) 348 (313)	79 (69) 109 (95) 200 (175)	55 (46) 75 (64) 136 (117)	42 (34) 58 (48) 106 ((87)	34 (27) 47 (38) 86 (69)	29 (22) 40 (31) 73 (56)	25 (19) 35 (26) 64 (48)	23 (16) 31 (22) 57 (41)	21 (14) 28 (20) 52 (36)
0.10	759 (664) 1051 (920) 1934 (1692)	232 (196) 322 (273) 592 (503)	122 (101) 169 (139) 311 (256)	80 (64) 110 (88) 202 (161)	58 (45) 80 (62) 146 (113)	45 (34) 62 (47) 114 (86)	37 (27) 51 (37) 93 (68)	32 (22) 43 (30) 79 (55)	28 (18) 38 (25) 69 (46)	24 (16) 34 (22) 61 (39)
0.15	1115 (919) 1544 (1273) 2840 (2343)	320 (256) 443 (354) 815 (652)	161 (124) 222 (172) 406 (316)	101 (75) 139 (104) 255 (191)	71 (51) 98 (71) 180 (130)	54 (38) 75 (52) 137 (95)	43 (29) 60 (40) 109 (73)	36 (23) 50 (32) 90 (59)	31 (19) 42 (26) 77 (48)	27 (16) 37 (22) 67 (40)
0.20	1439 (1115) 1993 (1545) 3668 (2843)	396 (296) 551 (413) 1013 (760)	194 (140) 268 (194) 492 (357)	118 (83) 163 (114) 300 (210)	82 (55) 113 (76) 207 (140)	61 (40) 84 (55) 154 (100)	48 (30) 66 (41) 121 (76)	40 (24) 54 (32) 99 (59)	33 (19) 46 (26) 83 (48)	29 (16) 40 (22) 72 (39)
0.25	1723 (1249) 2386 (1730) 4391 (3183)	464 (325) 643 (450) 1182 (827)	221 (149) 306 (206) 562 (379)	133 (86) 183 (119) 336 (218)	90 (56) 124 (78) 228 (142)	66 (40) 92 (55) 168 (100)	52 (30) 71 (41) 130 (75)	42 (23) 58 (31) 105 (58)	35 (18) 48 (25) 87 (46)	30 (15) 41 (20) 75 (37)
0.30	1960 (1323) 2714 (1832) 4995 (3371)	518 (336) 717 (466) 1319 (857)	242 (151) 335 (209) 616 (385)	143 (86) 198 (119) 364 (218)	96 (55) 133 (76) 244 (140)	70 (38) 97 (53) 177 (97)	54 (28) 74 (39) 136 (71)	43 (21) 60 (30) 108 (54)	36 (17) 49 (23) 90 (42)	31 (14) 42 (19) 76 (34)
0.35	2147 (1341) 2975 (1858) 5471 (3419)	559 (335) 774 (464) 1423 (853)	258 (148) 357 (205) 656 (377)	151 (83) 208 (114) 382 (210)	100 (52) 138 (72) 253 (133)	72 (36) 100 (50) 182 (91)	55 (26) 76 (36) 136 (65)	44 (20) 60 (27) 110 (49)	36 (15) 50 (21) 90 (38)	31 (12) 42 (16) 76 (30)
0.40	2281 (1312) 3159 (1816) 5814 (3343)	586 (322) 812 (446) 1493 (821)	268 (140) 370 (194) 680 (357)	155 (77) 214 (107) 392 (196)	102 (48) 140 (66) 257 (122)	73 (33) 100 (45) 183 (82)	55 (23) 76 (32) 136 (58)	44 (17) 60 (24) 109 (43)	36 (13) 49 (18) 88 (33)	30 (10) 41 (14) 74 (26)

0.45	2363(1240) 3272(1718) 6022(3161)	600(300) 831(415) 1529(764)	271(129) 375(178) 689(327)	155(70) 214(96) 393(177)	101(43) 140(59) 255((106)	72(28) 99(39) 180(72)	54(20) 74(28) 135(50)	43(15) 58(20) 105(37)	35(11) 47(15) 85(27)	30(9) 40(12) 71(21)
0.50	2391(1135) 3311(1572) 6092(2894)	601(270) 832(374) 1530(686)	269(114) 372(158) 683(290)	153(61) 211(84) 386(154)	99(37) 136(51) 248(93)	70(24) 95(33) 174(80)	52(17) 71(23) 129(42)	41(12) 56(16) 100(30)	33(9) 45(12) 80(22)	
0.55	2365(1005) 3275(1391) 6025(2561)	588(235) 814(325) 1497(598)	261(97) 360(135) 681(248)	146(51) 202(70) 370(129)	94(30) 129(42) 236(76)	66(19) 90(27) 163(49)	49(13) 67(18) 120(33)	38(9) 52(13) 92(23)		
0.60	2285(856) 3164(1186) 5821(2183)	562(196) 778(272) 1429(500)	246(80) 340(110) 624(202)	137(41) 189(56) 345(103)	87(24) 120(33) 217(59)	61(15) 83(20) 149(37)	45(10) 61(13) 109(24)			
0.65	2151(699) 2979(968) 5481(1781)	523(156) 723(216) 1328(398)	226(62) 312(85) 572(157)	125(31) 171(42) 312(78)	79(17) 107(24) 194(43)	54(10) 74(14) 132(26)				
0.70	1965(540) 2721(748) 5005(1376)	470(117) 650(162) 1193(296)	201(45) 276(62) 505(113)	109(21) 149(29) 271(54)	68(12) 93(16) 166(29)					
0.75	1726(368) 2389(537) 4395(968)	405(81) 559(111) 1025(205)	169(29) 233(40) 424(74)	91(13) 124(18) 222(33)						
0.80	1436(251) 1986(347) 3652(639)	327(49) 450(67) 824(123)	134(16) 182(22) 329(41)							
0.85	1094(136) 1513(189) 2778(347)	238(23) 326(32) 592(59)								
190	705(52) 972(72) 1778(133)									

Table B Number of patients required to detect an improvement of $(P_B - P_A)$ in survival probability over a baseline survival probability (P_A), when $\alpha = 5\%$, $1-\beta = 80\%$; $\alpha = 5\%$, $1-\beta = 90\%$; $\alpha = 1\%$, $1-\beta = 95\%$, two-tailed test (from Freedman, 1982).

P_A	$P_B - P_A$ 0.05	0.10	0.15	0.20	0.25	0.30	0.36	0.40	0.45	0.50
0.05	497 (459)	174 (156)	100 (87)	69 (59)	53 (44)	43 (34)	37 (28)	32 (24)	29 (21)	26 (18)
	664 (615)	232 (209)	133 (116)	92 (78)	71 ((58)	58 (46)	49 (38)	43 (32)	38 (27)	35 (24)
	1126 (1042)	393 (354)	225 (197)	156 (133)	119 (96)	97 (78)	82 (64)	72 (54)	64 (46)	58 (41)
0.10	963 (843)	295 (251)	155 (128)	101 (80)	73 (57)	57 ((43)	47 (34)	40 (28)	35 (23)	31 (20)
	1289 (1128)	395 (335)	207 (171)	135 (108)	96 (76)	76 (57)	63 (45)	53 (37)	46 (31)	41 (26)
	2185 (1912)	668 (568)	351 (289)	228 (182)	165 (123)	129 (97)	105 (78)	89 (62)	77 (52)	69 (44)
0.15	1415 (1167)	406 (325)	204 (158)	127 (95)	90 (65)	68 (48)	55 (37)	46 (29)	39 (24)	34 ((20)
	1894 (1562)	544 (435)	272 (211)	170 (128)	120 (87)	91 (64)	73 (49)	61 (39)	52 (32)	45 (27)
	3209 (2648)	921 (737)	461 (357)	288 (216)	203 (147)	154 (108)	124 (83)	102 (66)	87 (54)	76 (45)
0.20	1827 (1416)	505 (379)	245 (178)	150 (105)	104 (70)	77 (50)	61 (38)	50 (30)	42 (24)	36 (20)
	2445 (1895)	676 (507)	328 (238)	200 (140)	138 (93)	103 (67)	81 (51)	66 (40)	56 (32)	48 (26)
	4144 (3212)	1145 (858)	568 (403)	339 ((237)	234 (158)	174 (113)	137 (85)	112 (67)	94 (54)	81 (44)
0.25	2187 (1585)	589 (412)	280 (189)	188 (109)	114 (71)	84 (50)	65 (37)	53 (29)	44 (23)	38 (19)
	2927 (2122)	788 (551)	375 (253)	224 (146)	152 (95)	112 (67)	87 (50)	70 (39)	59 (31)	50 (25)
	4961 (3597)	1335 (934)	634 (426)	380 (247)	258 (161)	189 (113)	147 (84)	118 (65)	99 (52)	84 (42)
0.30	2487 (1679)	657 (427)	307 (192)	182 (109)	122 (70)	89 (49)	68 (36)	55 (27)	45 (21)	39 (17)
	3330 (2247)	879 (571)	411 (257)	243 (145)	163 (93)	118 (65)	91 (47)	73 (36)	60 (28)	51 (23)
	5643 (3809)	1490 (968)	696 (435)	411 (246)	275 (158)	200 (110)	153 (80)	122 (61)	101 (48)	86 (38)
0.35	2724 (1703)	709 (425)	327 (188)	191 (105)	127 (66)	91 (45)	70 (33)	55 (25)	46 (19)	39 (15)
	3647 (2279)	949 (569)	438 (251)	255 (140)	169 (89)	122 (61)	93 (44)	74 (33)	60 (25)	51 (20)
	6181 (3663)	1607 (964)	741 (426)	432 (237)	286 (150)	206 (102)	156 (74)	124 (55)	101 (43)	85 (34)
0.40	2695 (1665)	744 (409)	338 (178)	196 (98)	129 (41)	92 (41)	70 (29)	55 (22)	45 (17)	38 (13)
	3876 (2228)	995 (547)	454 (238)	262 (131)	172 (81)	123 (55)	93 (39)	73 (29)	60 (22)	50 (17)
	6569 (3777)	1687 (927)	769 (403)	443 (221)	290 (138)	207 (93)	156 (66)	122 (49)	99 (37)	83 (29)

0.45	2999 (1574) 4014 (2107) 6804 (3572)	762 (381) 1019 (509) 1727 (863)	344 (163) 480 (218) 779 (370)	197 (88) 263 (118) 444 (200)	128 (54) 171 (72) 288 (122)	91 (36) 121 (48) 203 (81)	68 (25) 91 (34) 152 (57)	54 (18) 71 (24) 119 (41)	44 (14) 58 (18) 96 (31)
									37 (11) 48 (14) 80 (24)
0.50	3034 (1441) 4061 (1929) 6883 (3269)	763 (343) 1020 (459) 1728 (777)	341 (145) 456 (193) 771 (327)	193 (77) 258 (103) 436 (174)	125 (46) 166 (62) 280 (106)	88 (30) 116 (40) 196 (68)	65 (21) 87 (28) 145 (47)	51 (15) 67 (20) 112 (33)	42 (11) 55 (15) 90 (24)
0.55	3001 (1275) 4017 (1707) 6808 (2893)	746 (298) 998 (399) 1691 (676)	330 (123) 441 (165) 746 (280)	185 (64) 247 (86) 417 (146)	119 (38) 158 (51) 266 (86)	83 (24) 110 (33) 184 (55)	61 (17) 81 (22) 135 (37)	48 (12) 63 (15) 104 (26)	
0.60	2900 (1087) 3881 (1455) 6577 (2466)	713 (249) 953 (333) 1615 (586)	312 (101) 417 (135) 705 (229)	173 (52) 231 (69) 389 (116)	110 (30) 146 (40) 245 (67)	78 (19) 101 (25) 168 (42)	56 (12) 74 (16) 122 (27)		
0.65	2730 (667) 3654 (1187) 6192 (2012)	663 (198) 886 (265) 1500 (450)	286 (78) 382 (105) 646 (177)	157 (39) 209 (52) 352 (88)	99 (22) 131 (29) 219 (49)	68 (13) 89 (17) 148 (29)			
0.70	2493 (685) 3337 (917) 5654 (1555)	596 (149) 796 (199) 1347 (336)	253 (57) 336 (76) 570 (128)	137 (27) 182 (36) 306 (61)	85 (15) 112 (19) 187 (32)				
0.75	2190 (492) 2930 (659) 4965 (1117)	512 (102) 684 (136) 1157 (231)	214 (37) 284 (49) 478 (83)	114 (17) 150 (22) 250 (37)					
0.80	1821 (318) 2436 (426) 4125 (722)	413 (62) 551 (82) 930 (139)	168 (21) 222 (27) 371 (46)						
0.85	1367 (173) 1854 (231) 3138 (392)	300 (30) 396 (39) 668 (66)							
0.90	892 (66) 1189 (89) 2007 (150)								

5

Distribution Functions for Failure Time T

5.1 INTRODUCTION

In previous chapters the analysis of failure times was faced by means of non-parametric procedures which enable investigators to estimate survival curves and mortality rates and to compare survivorship of two or more patient groups. Statistical techniques were introduced by a heuristic approach; in a more formal way this chapter will deal with equivalent functions describing survival, introduce the simplest parametric model adopted for fitting failure data and finally present introductory issues to survival regression analysis.

5.2 *T* CONTINUOUS

All data considered in previous examples measured the time elapsed between some specified event (diagnosis, surgical treatment, randomization etc.) and the time of failure (relapse, death, etc.).

As such, these failure times are empirical realizations of the random variable (r.v.) T defined in section 1.2.1. Let us consider a *homogeneous* population; the function $F(t)$ gives the probability that the failure time is less than or equal to t:

$$F(t) = \Pr\{T \leqslant t\}$$

and is called the *distribution* function of the r.v. T. This is a continuous non-negative variable, i.e. with sample space $0 \leqslant t < \infty$. Moreover, since

$$F(0) = 0 \quad \text{and} \quad \lim_{t \to \infty} F(t) = 1$$

$F(\cdot)$ is called a *proper* distribution function.

Also, $F(t)$ is a monotone non-decreasing function of t since, for $t^* < t^{**}$, we have

$$F(t^{**}) - F(t^*) = \Pr\{T \leqslant t^{**}\} - \Pr\{T \leqslant t^*\} = \Pr\{t^* < T \leqslant t^{**}\}$$

and this is non-negative.

Let us now consider distribution functions $F(t)$ which have a derivative at all points t. The function

$$\frac{\mathrm{d}}{\mathrm{d}t} F(t) = f(t)$$

is the *probability density function* (p.d.f.) at a point t. The differential

$$\mathrm{d}F(t) = f(t)\mathrm{d}t \sim [F(t + \mathrm{d}t) - F(t)]$$

is called the *probability element* and gives the *unconditional* probability of failure in the infinitesimal interval $(t, t + \mathrm{d}t)$.

The *survival function* $S(t)$ is defined to be the complementary function of $F(t)$, i.e.

$$S(t) = 1 - F(t) = \Pr\{T > t\} \tag{5.1}$$

and gives the probability of surviving at t, for any non-negative $t (0 \leqslant t < \infty)$.

Thus, the probability element $f(t)\mathrm{d}t$ is

$$f(t)\mathrm{d}t = \mathrm{d}F(t) = -\mathrm{d}S(t) \sim [S(t) - S(t + \mathrm{d}t)] \tag{5.2}$$

and

$$S(0) = 1 \quad \text{and} \quad \lim_{t \to \infty} S(t) = 0$$

Consider now the probability of dying at t *conditional* on having survived at a given time t^* with $t \geqslant t^*$. The *conditional* density function is defined as

$$f(t|t^*) = \lim_{\Delta t \to 0^+} \frac{\Pr\{t \leqslant T < t + \Delta t | T \geqslant t^*\}}{\Delta t} \tag{5.3}$$

hence the probability of dying in any interval $(t, t + \mathrm{d}t)$, given survival at t^*, is

$$f(t|t^*)\mathrm{d}t = \frac{f(t)\mathrm{d}t}{S(t^*)}$$

and $\int_{t^*}^{\infty} f(u|t^*)\mathrm{d}u = 1$.

The hazard function is defined as

$$\lambda(t) = \lim_{\Delta t \to 0^+} \frac{\Pr\{t \leqslant T < t + \Delta t \mid T \geqslant t\}}{\Delta t}$$

so that $\lambda(t)dt$ is the probability of dying in the infinitesimal interval $(t, t + dt)$, given survival at time t. This function is not a conditional density function as will be shown later, although, at each instant t, as t varies, it is

$$\lambda(t) = f(t \mid t) = \frac{f(t)}{S(t)} \tag{5.4}$$

The function (5.4) is called the *hazard rate function, intensity rate* or *force of mortality*. The function

$$\Lambda(t) = \int_0^t \lambda(u)du \tag{5.5}$$

is called the *cumulative hazard* function.

5.3   RELATIONSHIP AMONG *f(t)*, *S(t)* AND *λ(t)*

Any of the three functions $f(t)$, $S(t)$ and $\lambda(t)$ defines uniquely a specific survival distribution; since each of them provides the investigator with a different view of the data, it comes in handy to known the relationship among these functions in order to deduce one from the other. The hazard function has been defined in terms of $f(t)$ and $S(t)$ in (5.4) and, given the relationship (5.2), it follows that

$$\lambda(t) = \frac{f(t)}{S(t)} = -\frac{d}{dt} \log S(t) \tag{5.6}$$

By integrating both members of this equation we have

$$-\int_0^t \lambda(u)du = \log S(u) \Big|_0^t$$

so that, because $S(0) = 1$,

$$\Lambda(t) = -\log S(t) \tag{5.7}$$

Exponentiating we obtain an expression for $S(t)$:

$$S(t) = \exp\left[-\int_0^t \lambda(u)du \right] = \exp[-\Lambda(t)] \qquad (5.8)$$

Owing to the fact that $S(\infty) = 0$ and $\lambda(t) \geqslant 0$, from (5.7) we can see that

$$\lim_{t \to \infty} \int_0^t \lambda(u)du = \infty$$

Since the cumulative hazard function diverges, it appears that $\lambda(t)$ is *not* a probability density function as was stated in section 5.2. Based on (5.4), it appears that the *unconditional* probability of failure in $(t, t + dt)$ is approximately equal to the product of the probability of surviving beyond t times the conditional probability of failure in $(t, t + dt)$:

$$f(t)dt = S(t)\lambda(t)dt \qquad (5.9)$$

From this latter we can define the survival function as

$$S(t) = \int_t^\infty f(u)du = \int_t^\infty S(u)\lambda(u)du \qquad (5.10)$$

This formula expresses the survival function in terms of future lifetime; on the contrary (5.8) expresses the survival function in terms of past lifetime.

5.4 *T* DISCRETE

Consider now a discrete distribution of the r.v. T at points $\{t_{(j)}\}$, $t_{(1)} < t_{(2)} < \dots$, and the corresponding probability function made up of atoms:

$$f_j = \Pr\{T = t_{(j)}\} > 0, \quad j = 1, 2, \dots$$

each representing the probability of dying at the point $t_{(j)}$. The probability of surviving beyond t is given as

$$S(t) = \Pr\{T > t\} = \sum_{t_{(j)} > t} f_j$$

For example, $S(t_{(j-1)}) = f_j + f_{j+1} + \dots$ and $S(t_{(j)}) = f_{j+1} + f_{j+2} + \dots$ and therefore

$$f_j = S(t_{(j-1)}) - S(t_{(j)}) \qquad (5.11)$$

The value of $S(t)$ varies at each time $t_{(j)}$ only and, just before any $t_{(j)}$, is greater than the value at $t_{(j)}$. This means that the function $S(t)$ is a monotonically decreasing step function, continuous to the right, i.e.

$$S(t_{(j)}^-) > S(t_{(j)}) \quad \text{and} \quad S(t_{(j)}^-) = S(t_{(j-1)}), \quad t_{(j-1)} \leqslant t_{(j)}^- < t_{(j)}$$

where $S(t_{(j)}^-)$ indicates the probability of surviving up to $t_{(j)}$.

By resorting to definition (5.3) and taking $t^* = t_{(j)}^-$, the hazard function is defined as having atoms λ_j at points $\{t_{(j)}\}$ which express the conditional probability of failure at $t_{(j)}$:

$$\lambda_j = \Pr\{T = t_{(j)} | T > t_{(j)}^-\}$$

Analogously to (5.4),

$$\lambda_j = \frac{f_j}{S(t_{(j)}^-)} = \frac{f_j}{f_j + f_{j+1} + \cdots}$$

Thus by (5.11),

$$\lambda_j = \frac{S(t_{(j-1)}) - S(t_{(j)})}{S(t_{(j-1)})}$$

from which

$$S(t_{(j)}) = S(t_{(j-1)})(1 - \lambda_j)$$

By applying this formula recursively for $t_{(1)} < t_{(2)} < \ldots$, and having defined that $S(t_0) = S(0) = 1$, we obtain

$$S(t) = \prod_{j|t_{(j)} \leqslant t} (1 - \lambda_j) \tag{5.12}$$

Expression (5.12) shows that survival beyond $t_{(j)}(T > t)$ requires survival at any time point $t_{(j)}$ in the range between 0 and t, the extremes included. Therefore (5.12) may be seen as the result of the application of the product law of conditional probabilities which founded the product limit and the actuarial estimators. In this formal presentation of the survival function the notation $S(t)$ has been used; this corresponds to the notations $P(t)$ and P_j adopted in sections 3.2 and 3.3 respectively for an introduction to survivorship estimation. Furthermore the conditional probability of survival at $t_{(j)}$, named p_j, corresponds to the atom λ_j introduced in this section.

In order that the relation (5.8) still holds in the discrete setting, the cumulative hazard function $\Lambda(t)$ follows from (5.12):

$$-\Lambda(t) = \sum_{t_{(j)} \leqslant t} \log(1 - \lambda_j)$$

However, if the λ_j are small, $\log(1 - \lambda_j)$ is approximately equal to $-\lambda_j$ and an alternative definition of $\Lambda(t)$ is

$$\Lambda(t) = \sum_{t_{(j)} \leqslant t} \lambda_j$$

which justifies the Aalen estimator introduced in section 3.6.3.

5.5   MODELS IN SURVIVAL ANALYSIS

Typically, survival data tent to have a positively skewed distribution and consequently the Gaussian distribution cannot be taken as a reasonable standard for analysis. The simplest distribution which plays a central role in the analysis of survival and epidemiological data is the exponential distribution, which will be introduced here.

In survival analysis the distributional model is commonly specified by defining a functional form of the hazard $\lambda(t)$, from which the functions $S(t)$ and $f(t)$ can be derived by using the relationships (5.8) and (5.4). Once a distributional model has been fitted to the data, the estimates of the model parameters convey succinctly the information on the distribution formulated for T. If the assumed parametric form is plausible, these estimates can be used for inference, for describing the hazard and survival for current and future patients.

5.5.1   Specifying a parametric model

Suppose we wish to model the survival experience of individuals in a given age class of 10-year width, from a population whose risk of death is mainly due to a rare disease. It is sensible to argue that a distribution of T based on a constant failure rate is a plausible model. Then the hazard function is specified as

$$\lambda(t) = \lambda \tag{5.13}$$

and depends on one parameter only, the rate λ, with $\lambda > 0$.

From (5.8) we obtain

$$S(t) = \exp\left[-\lambda \int_0^t \mathrm{d}u\right] = \exp(-\lambda t) \tag{5.14}$$

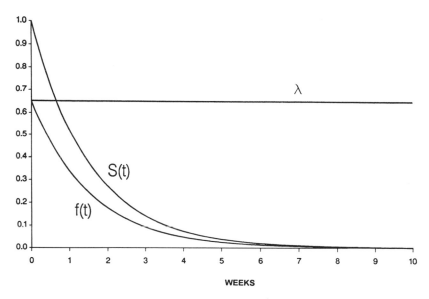

Figure 5.1 Exponential survival function [$S(t)$], density function [$f(t)$] and hazard function (λ) for $\lambda = 0.65$.

and from (5.9):

$$f(t) = \lambda \exp(-\lambda t) \tag{5.15}$$

This is called the exponential model.

As an example the graphs of the three functions (5.13), (5.14) and (5.15) are drawn in figure 5.1 for a specific value of the rate, $\lambda = 0.65$. The mean of the exponential distribution is the reciprocal of the constant rate as obtained through integrating by parts:

$$\mu = \int_0^\infty u\lambda \exp(-\lambda u)du = \frac{1}{\lambda} \tag{5.16}$$

The variance is

$$\sigma^2 = \mu_2' - \mu^2 = \int_0^\infty u^2 \lambda \exp(-\lambda u)du - \left(\frac{1}{\lambda}\right)^2 = \left(\frac{1}{\lambda}\right)^2 \tag{5.17}$$

5.5.2 Estimation of the parameter

In the absence of censoring, let **t** be the observation vector whose components, $t_1, t_2, \ldots, t_i, \ldots, t_N$ are times to failure. If we consider these times as realizations of a random variate T exponentially distributed, we wish to estimate the parameter λ. To this end we can resort to the *maximum likelihood (ML) method*. The likelihood function is proportional to the joint probability (or density) of the N outcomes, regarded as a function of the parameters of the statistical model; in the present case we have

$$L(\lambda; t_1, t_2, \ldots, t_i, \ldots, t_N) = \prod_{i=1}^{N} f(t_i; \lambda) = \prod_{i=1}^{N} \lambda \exp(-\lambda t_i) \qquad (5.18)$$

We wish to find the value of λ which maximizes (5.18) or, equivalently, its logarithm:

$$\log L(\lambda; \mathbf{t}) = LL(\lambda; \mathbf{t}) = \sum_{i=1}^{N} \log[\lambda \exp(-\lambda t_i)] = N \log \lambda - \lambda \sum_{i=1}^{N} t_i$$

In order to write the likelihood function in the presence of censoring it is convenient to use the pair of variables (t_i, δ_i), introduced in section 1.3.2 to identify the outcome of each individual. The contribution to the sample likelihood of an individual who fails at $t_i (\delta_i = 1)$ is the probability of failing as in (5.18). For an individual whose survival time is censored at $t_i (\delta_i = 0)$ the contribution to the likelihood is given by the probability of surviving beyond that point in time, $S(t_i)$.

Thus, in general, the likelihood function in the presence of censoring is

$$L(\lambda; \mathbf{t}) = \prod_{i=1}^{N} [f(t_i)]^{\delta_i} [S(t_i)]^{1-\delta_i} \qquad (5.19)$$

In particular, under the exponential model, the contributions to the function are

$$\lambda \exp(-\lambda t_i) \quad \text{if} \quad \delta_i = 1$$
$$\exp(-\lambda t_i) \quad \text{if} \quad \delta_i = 0$$

and the function is then

$$L(\lambda; \mathbf{t}) = \prod_{i=1}^{N} [\lambda \exp(-\lambda t_i)]^{\delta_i} [\exp(-\lambda t_i)]^{1-\delta_i}$$

Taking logarithms, we have

$$LL(\lambda; \mathbf{t}) = \sum_{i=1}^{N} [\delta_i(\log \lambda - \lambda t_i) + (1 - \delta_i)(-\lambda t_i)] \qquad (5.20)$$

$$= d \log \lambda - \lambda \sum_{i=1}^{N} t_i$$

where the number of individuals who fail, $d = \sum_{i=1}^{N} \delta_i$, is assumed to be greater than zero.

Differentiating with respect to λ we obtain

$$\frac{d}{d\lambda} LL(\lambda; \mathbf{t}) = \frac{d}{\lambda} - \sum_{i=1}^{N} t_i \qquad (5.21)$$

Equating to zero and solving the equation we have the estimate

$$\hat{\lambda} = \frac{d}{\sum\limits_{i=1}^{N} t_i} \qquad (5.22)$$

corresponding to the one suggested in section 3.6.1. There the model was not made explicit, but it was implicitly assumed that the failure rate did not vary in time. Hence, the estimate heuristically introduced for the rate is proved here to be the ML estimate under the exponential model. Furthermore, the function $S(t)$ in (5.14) corresponds to the cumulative probability of survival beyond t obtained in (3.18). Note that in the absence of censoring $d = N$ and thus $\hat{\lambda} = N/\sum_{i=1}^{N} t_i$ and from (5.16) the reciprocal of $\hat{\lambda}$ gives the ML estimator of the mean failure time, given that ML estimators are invariant to one-to-one transformations. In the presence of censoring, the ML estimator of μ is the reciprocal of the estimate (5.22) of λ, namely the sum of all the observed failure and censored times divided by *the number of failures*. However, ML estimates of μ are of little practical use, and even less so the more the number of censored observations increases. As the distribution is positively skewed, the preferred summary measure of location is the median survival time, t_{Me}, where $S(t_{Me}) = 0.50$. The estimate of t_{Me} is:

$$t_{Me} = \lambda^{-1} \log 2$$

5.5.3   Variance of the estimator

For each observation t_i, the contribution to the sum in (5.21) is called the *efficient score*:

$$U_i(\lambda) = \frac{d}{d\lambda} LL(\lambda; t_i) = \frac{\delta_i}{\lambda} - t_i \qquad (5.23)$$

It has expectation zero:

$$E(U_i) = 0$$

and variance (see Cox and Hinkley, 1974, pp. 107–113):

$$\mathfrak{J}_i(\lambda) = E[U_i(\lambda)]^2 = E\left[\frac{d}{d\lambda} LL(\lambda; t_i)\right]^2 = -E\left[\frac{d^2}{d\lambda^2} LL(\lambda; t_i)\right]$$

Since the t_i are independent, the total score statistic is the sum of N independent contributions:

$$U(\lambda) = \sum_{i=1}^{N} U_i(\lambda) \qquad (5.24)$$

A central limit theorem will apply and, consequently, $U(\lambda)$ is asymptotically normally distributed with mean 0 and variance

$$\Im(\lambda) = \sum_{i=1}^{N} \Im_i(\lambda)$$

This quantity is known as Fisher information.

Under mild conditions a consistent estimator of $\Im_i(\lambda)$ is given by

$$I_i(\hat{\lambda}) = -\frac{d^2}{d\lambda^2} LL(\lambda; t_i)\Big|_{\lambda = \hat{\lambda}}$$

and therefore the variance estimate of the total score is the observed information:

$$I(\hat{\lambda}) = \sum_{i=1}^{N} I_i(\hat{\lambda}) \qquad (5.25)$$

Statistical theory shows that under very general conditions, maximum likelihood estimators are consistent, asymptotically efficient and can be dealt with as asymptotically normally distributed about the true value as mean and with variance:

$$var(\hat{\lambda}) = \left[-E\left(\frac{d^2}{d\lambda^2} LL(\lambda; \mathbf{t})\right) \right]^{-1} = \Im^{-1}(\lambda)$$

This latter can be estimated by resorting to be observed information (5.25) as

$$var(\hat{\lambda}) = I^{-1}(\hat{\lambda}) \qquad (5.26)$$

Under the exponential model, the total score is given in (5.21) and the observed information is

$$I(\hat{\lambda}) = \frac{d}{\hat{\lambda}^2} = \frac{\left(\sum_{i=1}^{N} t_i\right)^2}{d}$$

5.5.4 An example

Consider the 6-MP group of the Freireich *et al.* (1963) trial (section 3.1): $d = 9$ and $\sum_{i=1}^{21} t_i = 359$. Thus the estimate of the relapse rate is $\hat{\lambda}_{6-MP} = 9/359 = 0.025\,07$.

The statistic $U(\lambda)/\sqrt{I(\hat{\lambda})}$ has an asymptotic standard normal distribution and enables us to compute the $100(1 - \alpha)\%$ confidence interval of λ. We can write

$$\left| \frac{\left(\dfrac{d}{\lambda} - \sum_{i=1}^{N} t_i \right)\lambda}{d^{1/2}} \right| = \left| d^{1/2} - \lambda d^{-1/2} \sum_{i=1}^{N} t_i \right| \leqslant z_{1-\alpha/2}$$

and therefore the 95% confidence interval is

$$\left| 3 - \frac{359}{3}\lambda \right| \leqslant 1.96, \text{i.e. } 0.008\,69 \leqslant \lambda \leqslant 0.041\,45$$

However, the confidence interval of λ can be computed directly accounting for the $var(\hat{\lambda})$ given by (5.26). Thus the usual expression

$$\hat{\lambda} - z_{1-\alpha/2}\sqrt{I^{-1}(\hat{\lambda})} \leqslant \lambda \leqslant \hat{\lambda} + z_{1-\alpha/2}\sqrt{I^{-1}(\hat{\lambda})}$$

applies. Substituting, the limits for λ are

$$0.025\,07 \pm 1.96 \times \frac{3}{359}$$

and therefore

$$0.008\,69 \leqslant \lambda \leqslant 0.041\,45$$

In this case the interval is precisely the same as the one we computed by means of the score procedures.

5.6 REGRESSION MODELS IN SURVIVAL ANALYSIS

In the previous section a homogeneous population was assumed with failure time distributed according to an exponential model, the unique parameter of which was estimated by the maximum likelihood method. Such a situation, however, is not realistic in the medical field where the populations are made heterogeneous by risk factors in epidemiology or treatments and prognostic factors in clinical research. Thus survival models are extended to include regression variables, also called covariates or factors, measured on the individuals in the study. These models enable us to investigate the role of these covariates in modifying the response and to take into account confounding factors in the estimate or treatment or exposure effects. An introduction to the use of regression models is given here with reference to the exponential distribution.

5.6.1 The exponential regression model

By analogy with the regression analysis with quantitative dependent variables, one could argue in favour of modelling the expectation of time to failure $E(T)$ by factors of clinical concern. Indicate with $\mathbf{x}'_i = [x_{1i}, \ldots, x_{ki}, \ldots, x_{Ki}]$ the vector of K covariates observed on the ith individual who is now characterized by the set of data $(t_i, \delta_i, \mathbf{x}_i)$. If we assume an exponential distribution for T, modelling $E(T)$ in terms of $\mathbf{x}_i$ implies modelling λ given that $E(T) = 1/\lambda$ by (5.16). Thus, adopting a linear predictor in $\mathbf{x}$, we have the model suggested by Feigl and Zelen (1965):

$$\lambda_i^{-1} = \beta_0 + \boldsymbol{\beta}'\mathbf{x}_i = \mu_i$$

Here β_0 indicates the intercept and may be seen as the coefficient of an additional regression covariate x_{0i} taken to be identically 1 for all individuals. The linear predictor:

$$\mu_i = \beta_0 x_{0i} + \beta_1 x_{1i} + \cdots + \beta_k x_{ki} + \cdots + \beta_K x_{Ki}$$

includes $K + 1$ regressors.

The log-likelihood (5.20) is rewritten accounting for the fact that each individual i is related to a different λ_i. Substituting $\mu_i^{-1} = (\beta_0 + \boldsymbol{\beta}'\mathbf{x}_i)^{-1}$ for λ in (5.20) we obtain the log-likelihood of the sample:

$$LL(\beta_0, \boldsymbol{\beta}; \mathbf{t}, \mathbf{x}) = \sum_{i=1}^{N} [\delta_i(\log \mu_i^{-1} - \mu_i^{-1} t_i) + (1 - \delta_i)(-\mu_i^{-1} t_i)]$$

The $K + 1$ parameters β_0 and $\boldsymbol{\beta}$ can be estimated by maximizing this function.

An alternative model (Glasser, 1967), adopts the following definition for λ_i:

$$\lambda_i = \lambda_0 \exp(\boldsymbol{\beta}'\mathbf{x}_i) = \exp \mu_i \tag{5.27}$$

with $\beta_0 = \log \lambda_0$.

The quantity $\exp(\beta_0)$ may be seen as the failure rate in the reference category identified by $\mathbf{x}_i = \mathbf{0}$. This quantity, multiplied by a factor depending on the covariate values, gives the hazard rate in any other category identified by a pattern in the covariates. The model assumes that there is proportional hazard among different groups.

5.6.2 Estimation of the regression parameters

Let us fit model (5.27) to the Freireich *et al.* (1963) data with one independent variable x having values 0 and 1; 0 will be used to denote the placebo group and 1 the 6-MP group. From (5.20) the log-likelihood of the sample is

$$LL(\beta_0, \beta_1) = \sum_{i=1}^{N} \{\delta_i(\beta_0 + \beta_1 x_i) - \exp(\beta_0 + \beta_1 x_i)t_i\} \tag{5.28}$$

The first derivatives with respect to β_0 and β_1 are

$$\frac{\partial}{\partial \beta_0} LL(\beta_0, \beta_1) = \sum_{i=1}^{N} \{\delta_i - \exp(\beta_0 + \beta_1 x_i)t_i\}$$

$$= \sum_{i=1}^{n_1} \{\delta_i - \exp(\beta_0)t_i\} + \sum_{i=n_1+1}^{N} \{\delta_i - \exp(\beta_0 + \beta_1)t_i\}$$

$$= d_1 + d_2 - \exp(\beta_0) \sum_{i=1}^{n_1} t_i - \exp(\beta_0 + \beta_1) \sum_{i=n_1+1}^{N} t_i \tag{5.29}$$

$$\frac{\partial}{\partial \beta_1} LL(\beta_0, \beta_1) = \sum_{i=1}^{N} \{\delta_i x_i - \exp(\beta_0 + \beta_1 x_i)t_i x_i\}$$

$$= \sum_{i=n_1+1}^{N} \{\delta_i - \exp(\beta_0 + \beta_1)t_i\}$$

$$= d_2 - \exp(\beta_0 + \beta_1) \sum_{i=n_1+1}^{N} t_i \tag{5.30}$$

where d_1 and d_2 are the observed deaths and n_1 and $N - n_1$ indicate the number of patients in the placebo and 6-MP group respectively.

Since two parameters must be estimated, two scores are computed for each individual and the total scores $U(\beta_0) = (5.29)$ and $U(\beta_1) = (5.30)$ share the same properties we mentioned for $U(\lambda)$ in the previous section.

The function $LL(\beta_0, \beta_1)$ in (5.28) is a three-dimensional surface (figure 5.2) which may be conveniently represented by the contours in the space of the parameters β_0 and β_1 as in figure 5.3. Each contour is traced by connecting points whose coordinates correspond to the same value of the log-likelihood. With the aid of a computer, the surface may be studied directly in order to determine $\hat{\beta}_0$ and $\hat{\beta}_1$ which maximize $LL(\beta_0, \beta_1)$. Since from the figures the surface looks like a hill, this means finding the coordinates of the top of the hill, marked with a cross in figure 5.3, as the point with $\hat{\beta}_0 = -2.15948$ and $\hat{\beta}_1 = -1.52661$. Alternatively, β_0 and β_1 may be estimated by an iterative procedure after setting (5.29) and (5.30) equal to zero (see Morabito and Marubini, 1976). In the example, $\exp(\hat{\beta}_0) = 0.11539$ is the estimate of the

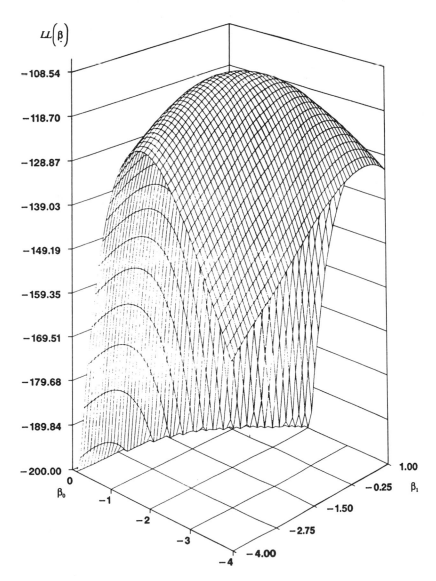

Figure 5.2 $LL\,(\boldsymbol{\beta})$ (vertical axis) for the Freireich *et al.* (1963) data as a function of β_0 and β_1.

hazard rate in the placebo group ($x = 0$) and $\exp(\hat{\beta}_0 + \hat{\beta}_1) = 0.02507$ is that in the 6-MP group ($x = 1$). The treatment effect is estimated in terms of the relative failure rate in the 6-MP group with respect to the placebo group as

$$\hat{\theta} = \frac{\exp(\hat{\beta}_0 + \hat{\beta}_1)}{\exp(\hat{\beta}_0)} = \exp(\hat{\beta}_1) = 0.217$$

Note, however, that the assumptions of constant hazard and of multiplicative effect of the covariate x on the hazard itself adopted in (5.27) make the parameter estimation procedure so straightforward as to be accomplished on a desk calculator. The basic concept is that the hazard rate is constant in each of the two treatment groups. Therefore the estimates of the hazard rates we

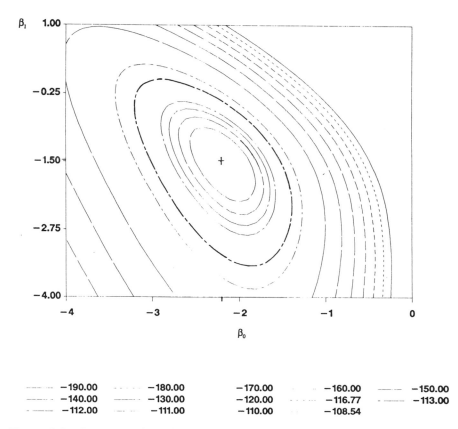

−190.00	−180.00	−170.00	−160.00	−150.00
−140.00	−130.00	−120.00	−116.77	−113.00
−112.00	−111.00	−110.00	−108.54	

Figure 5.3 Contour surface of the log-likelihood (5.28) for the Freireich *et al.* (1963) data. The point + corresponds to the top of the hill ($LL = -108.52405$); the bold contour corresponds to $LL(\boldsymbol{\beta}) = -116.76666$.

obtained by exponentiating $\hat{\beta}_0$ and $(\hat{\beta}_0 + \hat{\beta}_1)$ equal those we would obtain by means of the estimator (5.22) applied to the data of the two treatment groups separately. In the placebo group there are 21 relapses and the total amount of remission person–weeks is 182, while the corresponding figures in the 6-MP group are nine and 359; thus $\hat{\lambda}_p = 21/182 = 0.115\,39$ and $\hat{\lambda}_{6-MP} = 9/359 = 0.025\,07$. Hence, under the assumption of proportional hazards it is straightforward to estimate $\hat{\theta} = \hat{\lambda}_{6-MP}/\hat{\lambda}_p = 0.217$. This estimate is slightly higher than the two values of $\hat{\theta}$ computed in section 4.3 ($\hat{\theta} = 0.194$ and $\hat{\theta}_{M-H} = 0.192$) without making any assumption on the underlying hazard. However, both the model (5.27) and the non-parametric approach of Chapter 4 make the assumption of proportional hazards.

5.6.3 Statistical tests on regression coefficients

The statistical inference on $\boldsymbol{\beta}' = (\beta_1, \ldots, \beta_K)$ relies on three tests which are based on the properties of the likelihood function. Suppose the hypothesis to be tested is

$$H_0: \beta_1 = \beta_{10}, \ldots, \beta_g = \beta_{g0}$$

The null hypothesis specifies a particular value, denoted with the second suffix 0, for g of the K parameters. Without loss of generality, we take the parameters in H_0 to be the first g components of the vector $\boldsymbol{\beta}$ ($g \leqslant K$). Since the model contains the unknown intercept β_0, we deal in fact with $K + 1$ parameters $(\boldsymbol{\beta}, \beta_0)$; these may be split into two sets of components $(\boldsymbol{\beta}_g, \boldsymbol{\beta}_r)$ so that the null hypothesis is more simply written as

$$H_0: \boldsymbol{\beta}_g = \boldsymbol{\beta}_{g0} \tag{5.31}$$

The test imposes g restrictions on the parameters to be estimated. The notation $\boldsymbol{\beta}_r$ indicates the set of the remaining $r = K + 1 - g$ parameters which are left unspecified by the hypothesis tested and may be regarded as "nuisance" parameters. The vector $\boldsymbol{\beta}_g$ may contain just one or more parameters; a typical hypothesis we shall test in our examples is $H_0: \boldsymbol{\beta}_g = \mathbf{0}$.

Consistently, we denote the log-likelihood function as $LL(\boldsymbol{\beta}_g, \boldsymbol{\beta}_r)$. Since we deal here with a vector of $K + 1$ parameters, the Fisher information $\mathfrak{J}$ is a $(K + 1) \times (K + 1)$ matrix with elements

$$\mathfrak{J}_{lj} = -E\left[\frac{\partial^2 LL(\boldsymbol{\beta}_g, \boldsymbol{\beta}_r)}{\partial \beta_l \partial \beta_j}\right]$$

In analogy to the one-dimensional case (section 5.5.3), this is the variance–covariance matrix of the score vector with $(K + 1)$ components $U_l = \partial LL(\boldsymbol{\beta}_g, \boldsymbol{\beta}_r)/\partial \beta_l$.

The inverse matrix $\mathfrak{J}^{-1}(\boldsymbol{\beta}_g, \boldsymbol{\beta}_r)$ is the variance–convariance matrix of the set of parameters $(\boldsymbol{\beta}_g, \boldsymbol{\beta}_r)$. Coherently with (5.25), the variance estimate of the score vector is the observed information matrix $\mathbf{I}$. This is the matrix of minus the second derivatives of LL with respect to $(\boldsymbol{\beta}_g, \boldsymbol{\beta}_r)$ evaluated at $(\hat{\boldsymbol{\beta}}_g, \hat{\boldsymbol{\beta}}_r)$. The inverse matrix $\mathbf{I}^{-1}(\hat{\boldsymbol{\beta}}_g, \hat{\boldsymbol{\beta}}_r)$ is the estimate of the variance–covariance matrix of the vector of the ML estimates $(\hat{\boldsymbol{\beta}}_g, \hat{\boldsymbol{\beta}}_r)$.

The three tests on the hypothesis (5.31) against the alternative $H_1 : \boldsymbol{\beta}_g \neq \boldsymbol{\beta}_{g0}$ are as follows.

(1) *The likelihood ratio (LR) test*: to use this test we need to fit both the unrestricted and the restricted models. We shall obtain the value of the log-likelihood function $LL(\hat{\boldsymbol{\beta}}_g, \hat{\boldsymbol{\beta}}_r)$ where $(\hat{\boldsymbol{\beta}}_g, \hat{\boldsymbol{\beta}}_r)$ is the joint ML estimate of $\boldsymbol{\beta}_g$ and $\boldsymbol{\beta}_r$ in the unrestricted model, and of $LL(\boldsymbol{\beta}_{g0}, \tilde{\boldsymbol{\beta}}_r)$ where $\tilde{\boldsymbol{\beta}}_r$ is the ML estimate of $\boldsymbol{\beta}_r$ when the model imposes the g restrictions in H_0. This estimator of $\boldsymbol{\beta}_r$ has been given a tilde to highlight that $\tilde{\boldsymbol{\beta}}_r$ does not generally coincide with the $\hat{\boldsymbol{\beta}}_r$ obtained in the unrestricted model. The test statistic for H_0 is based on the difference of the log-likelihood values. Under H_0, the statistic

$$Q_{\mathrm{LR}} = 2\,[LL(\hat{\boldsymbol{\beta}}_g, \hat{\boldsymbol{\beta}}_r) - LL(\boldsymbol{\beta}_{g0}, \tilde{\boldsymbol{\beta}}_r)] \tag{5.32}$$

is asymptotically distributed as χ^2 with a number g of degrees of freedom equal to the number of restrictions imposed by the null hypothesis on the linear predictor (see Cox and Hinkley, 1974, pp. 314–317).

(2) *The Wald test*: this requires fitting the unrestricted model, and is based on the ML estimator $\hat{\boldsymbol{\beta}}_g$. The test statistic is

$$Q_{\mathrm{W}} = (\hat{\boldsymbol{\beta}}_g - \boldsymbol{\beta}_{g0})'\,[I_{g \times g}^{-1}(\hat{\boldsymbol{\beta}}_g, \hat{\boldsymbol{\beta}}_r)]^{-1}(\hat{\boldsymbol{\beta}}_g - \boldsymbol{\beta}_{g0}) \tag{5.33}$$

Note that $I_{g \times g}^{-1}(\hat{\boldsymbol{\beta}}_g, \hat{\boldsymbol{\beta}}_r)$ is a submatrix of dimension $(g \times g)$ of the entire $((K + 1) \times (K + 1))$ variance–covariance matrix, of $(\hat{\boldsymbol{\beta}}_g, \hat{\boldsymbol{\beta}}_r)$ estimated under the unrestricted model. The submatrix contains the variance–covariance estimates corresponding to the g parameters in H_0. The quadratic form in (5.33) requires the inverse of the submatrix $I_{g \times g}^{-1}$ and, under H_0 is asymptotically distributed as χ^2 with g degrees of freedom. If H_0 imposes one restriction only, $g = 1$, on the parameter β_k, $H_0 : \beta_k = \beta_{k0}$, then the statistic (5.33) reduces to the square of a Gaussian standardized deviate:

$$Q_{\mathrm{W}} = \frac{(\hat{\beta}_k - \beta_{k0})^2}{var(\hat{\beta}_k)} \tag{5.34}$$

Computer programs performing survival regression analysis report the result of this test on the null hypothesis $H_0 : \beta_k = 0$ for each regression coefficient β_k $(k = 1, 2, \ldots, K)$.

(3) *Rao's score test*: this test is performed after fitting the restricted model and obtaining $LL(\boldsymbol{\beta}_{g0}, \tilde{\boldsymbol{\beta}}_r)$. It is based on the gradient of this function at $\boldsymbol{\beta}_{g0}$ i.e. on the score vector of g components:

$$\mathbf{U}'_{Ho} = \mathbf{U}'(\boldsymbol{\beta}_{g0}, \tilde{\boldsymbol{\beta}}_r) = \left[\frac{\partial LL(\boldsymbol{\beta}_g, \boldsymbol{\beta}_r)}{\partial \beta_1}, \dots, \frac{\partial LL(\boldsymbol{\beta}_g, \boldsymbol{\beta}_r)}{\partial \beta_g} \right]_{\boldsymbol{\beta}_g = \boldsymbol{\beta}_{g0}, \boldsymbol{\beta}_r = \tilde{\boldsymbol{\beta}}_r}$$

evaluated at $\boldsymbol{\beta}_g = \boldsymbol{\beta}_{g0}$ and $\boldsymbol{\beta}_r = \tilde{\boldsymbol{\beta}}_r$. The Rao score test considers the statistic

$$Q_R = \mathbf{U}'_{Ho} I^{-1}_{g \times g}(\boldsymbol{\beta}_{g0}, \tilde{\boldsymbol{\beta}}_r) \mathbf{U}_{Ho} \qquad (5.35)$$

where $I^{-1}_{g \times g}(\cdot)$ indicates the submatrix, of dimension $(g \times g)$, extracted from the inverse of the $(K + 1) \times (K + 1)$ observed information matrix evaluated at $\boldsymbol{\beta}_g = \boldsymbol{\beta}_{g0}$ and $\boldsymbol{\beta}_r = \tilde{\boldsymbol{\beta}}_r$. The quadratic form (5.35) has approximately a χ^2 distribution with g degrees of freedom. From the computational viewpoint, this test compares favourably with the two previous ones because it does not need to estimate the regression coefficients included in the null hypothesis.

The three tests often lead to identical conclusions on the regression parameters. However, the use of the likelihood ratio test is recommended, on the basis of qualitative arguments (obtaining the variance of the test statistics under different parametrizations) and of results concerning the asymptotic theory of ML estimation (Cox and Oakes, 1984, p. 36).

5.6.4   An example

In order to apply the previous tests to the Freireich *et al.* (1963) data, it is useful to consider the following hierarchical predictors:

$$a: \mu_i = \beta_0 x_{0i} + \beta_1 x_{1i}; \qquad b: \mu_i = \beta_0 x_{0i}; \qquad c: \mu_i = 0$$

in model (5.27). We recall that $x_{0i} = 1$ for each individual, and the dummy variable x_{1i} assumes values 0 for the placebo and 1 for the 6-MP groups respectively. Since no other covariate is reported in table 4.2, the predictor *a* makes allowance for all the available sources of variability; it is the *saturated* predictor. The predictor *b* is suitable for fitting the 42 outcomes when they are assumed to be drawn from the same homogeneous population whose hazard rate of failure is $\lambda_0 = \exp(\beta_0)$. Besides homogeneity, the third predictor assumes that the hazard rate is $\lambda_0 = 1$; from a clinical point of view this appears to be meaningless and it is used here only for didactical purposes.

The three statistics Q_{LR}, Q_W and Q_R are now computed to test the hypothesis of no treatment effect, $H_0: \beta_1 = 0$. With the notation adopted in section 5.6.3, the vector $\boldsymbol{\beta}_g$ to be tested contains β_1 only and β_0 remains the only nuisance parameter in $\boldsymbol{\beta}_r$; thus $LL(\boldsymbol{\beta}_g, \boldsymbol{\beta}_r)$ is explicitly indicated here with $LL(\beta_1, \beta_0)$.

(1) *The LR test*: the unrestricted model contains the predictor a and has log-likelihood function (5.28). The value assumed at the ML estimates $\hat{\beta}_1$ and $\hat{\beta}_0$ is

$$LL(\hat{\beta}_1, \hat{\beta}_0) = d_1\hat{\beta}_0 - \exp(\hat{\beta}_0) \sum_{i=1}^{m_1} t_i + d_2(\hat{\beta}_0 + \hat{\beta}_1) - \exp(\hat{\beta}_0 + \hat{\beta}_1) \sum_{i=m_1+1}^{N} t_i$$

$$= 21 \times (-2.159\,48) - 0.115\,39 \times 182 + 9(-3.686\,09)$$
$$- 0.025\,07 \times 359$$
$$= -108.524\,05$$

Owing to the fact that predictor b assumes a homogeneous population, the estimate of β_0 is easily obtained from (5.22), namely $\bar{\beta}_0 = \log(30/541) = -2.892\,22$. Therefore:

$$LL(\bar{\beta}_0) = (d_1 + d_2)\bar{\beta}_0 - \exp(\bar{\beta}_0) \sum_{i=1}^{N} t_i$$

$$= 30 \times (-2.892\,22) - 0.055\,45 \times 541 = -116.766\,66$$

According to (5.32) we obtain

$$Q_{LR} = 2(-108.524\,05 + 116.766\,66) = 16.4852$$

(2) *The Wald test*: the prerequisite of the calculation of the Wald statistic is the computation of the observed information matrix for the unrestricted model. By substituting the ML estimates of the parameters in the predictor a into the second derivatives of the log-likelihood (5.28) (see Appendix A5.1), the observed information is

$$\mathbf{I}(\hat{\beta}_1, \hat{\beta}_0) = \begin{bmatrix} d_2 & d_2 \\ d_2 & d_1 + d_2 \end{bmatrix} = \begin{bmatrix} 9 & 9 \\ 9 & 30 \end{bmatrix}$$

The inverse is

$$\mathbf{I}^{-1}(\hat{\beta}_1, \hat{\beta}_0) = \frac{1}{d_1 d_2} \begin{bmatrix} d_1 + d_2 & -d_2 \\ -d_2 & d_2 \end{bmatrix} = \begin{bmatrix} 0.158\,73 & -0.047\,62 \\ -0.047\,62 & 0.047\,62 \end{bmatrix}$$

In this case $I_{g \times g}^{-1}$ of (5.33) corresponds to the upper left element of the inverse matrix (since $g = 1$) and is the estimate of the $var(\hat{\beta}_1)$. Thus (5.33) becomes

$$Q_W = (-1.526\,61 - 0) \times (0.158\,73)^{-1} \times (-1.526\,61 - 0)$$
$$= 14.6824$$

Since there is only one restriction in H_0, Q_W could have been straightforwardly calculated by means of (5.34).

(3) *The Rao score test*: in this case the restricted model, assuming the predictor b, is fitted to the data to obtain the ML estimate $\tilde{\beta}_0$; the score vector reduces to one component, that is the first derivative of the log-likelihood (5.28) with respect to β_1, evaluated in $\beta_1 = 0$ and $\beta_0 = \tilde{\beta}_0$:

$$U(0) = -10.9076$$

The observed information matrix (and its inverse) under the restricted model have elements

$$\mathbf{I}(0, \tilde{\beta}_0) = \begin{bmatrix} 19.9076 & 19.9076 \\ 19.9076 & 30 \end{bmatrix} \qquad \mathbf{I}^{-1}(0, \tilde{\beta}_0) = \begin{bmatrix} 0.149\,32 & -0.099\,08 \\ -0.099\,08 & 0.099\,08 \end{bmatrix}$$

The upper left element of $\mathbf{I}^{-1}(0, \tilde{\beta}_0)$ corresponds in this case to $I_{g \times g}^{-1}(0, \tilde{\beta}_0)$ in (5.35), given that $g = 1$ and is taken as the estimate of $1/var\ U(0)$. The test statistic is

$$Q_R = U(0) \times \frac{1}{var\ U(0)} \times U(0) = (-10.9076)^2 \times 0.149\,32 = 17.7655$$

Though all three tests allow us to reject the null hypothesis, their numerical values are rather different. A heuristic justification of this difference will be given in the next section.

We wish now to test the null hypothesis: $H_0': \beta_0 = 0$, $\beta_1 = 0$. Under this hypothesis the score vector is

$$\mathbf{U}_{H_0} = \begin{bmatrix} U(\beta_0)|_{\beta_0=0,\beta_1=0} \\ U(\beta_1)|_{\beta_0=0,\beta_1=0} \end{bmatrix} = \begin{bmatrix} 30 - 541 \\ 9 - 359 \end{bmatrix}$$

and the obtained observed information matrix and its inverse are

$$\mathbf{I}(0, 0) = \begin{bmatrix} 541 & 359 \\ 359 & 359 \end{bmatrix} \qquad \mathbf{I}^{-1}(0, 0) = \begin{bmatrix} 0.005\,49 & -0.005\,49 \\ -0.005\,49 & 0.008\,28 \end{bmatrix}$$

Therefore $Q_R = [-511, -350]\mathbf{I}^-(0, 0) \begin{bmatrix} -511 \\ -350 \end{bmatrix} = 483.6487$ which is asymptotically distributed as χ^2 with 2 d.f.

The computational advantage of the score test appears evident: it demands only the estimate of the restricted model. On the contrary the Wald test uses only the unrestricted model and the log-likelihood ratio test requires both them.

The three information matrices used in previous tests deserve some considerations. When the saturated predictor a is used, the number of failures predicted by the model on the grounds of failure times must equal the number of observed failures as it appears from the fact that (5.29) and (5.30) are equal to zero. This implies that the matrix $\mathbf{I}(\hat{\beta}_1, \hat{\beta}_0)$ is based upon the number of failures occurring in the two groups which convey the entire available information. In the example, the matrix elements are the number of failures in the group with $x_1 = 1$ (nine relapses in the 6-MP group) and the total number of failures. The matrix $\mathbf{I}(0, \tilde{\beta}_0)$ is computed under the assumption that we are dealing with a homogeneous population, with individuals exhibiting the same failure rate. This latter, as previously assessed, is estimated by (5.22): $\hat{\lambda}_0 = 30/541 = 0.055\,45$. Since 359 remission person–weeks are observed in the 6-MP group, $0.055\,45 \times 359 = 19.9076$ are the corresponding expected failures. Finally the matrix $\mathbf{I}(0, 0)$ is based on the remission person–weeks in the two groups. In fact, under the assumption of a homogeneous population with $\lambda_0 = 1$, the number of expected deaths coincides with the number of remission person–weeks.

Note that the square root of (5.34) is asymptotically a normal standard deviate; thus, with reference to the coefficient β_1, we can compute its $100(1 - \alpha)\%$ confidence interval with the usual expression: $\hat{\beta}_1 \pm z_{1-\alpha/2}\mathrm{ES}(\hat{\beta}_1)$. By exponentiating these limits, the confidence interval of the hazard propor-

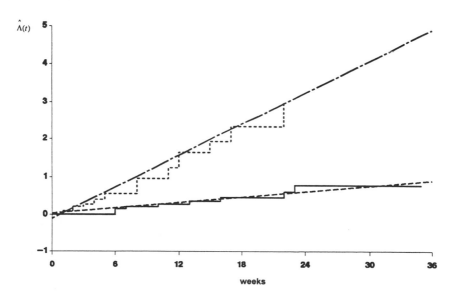

Figure 5.4 Graphical check of the assumption of exponential distribution for failure times of the Freireich *et al.* (1963) trial. Placebo group: --- 6-MP group: ——.

tionality constant (θ_1) is obtained:

$$\exp[\hat{\beta}_1 - z_{1-\alpha/2}\,ES(\hat{\beta}_1)] \leqslant \theta_1 \leqslant \exp[\hat{\beta}_1 + z_{1-\alpha/2}\,ES(\hat{\beta}_1)] \qquad (5.36)$$

In our example $\hat{\theta}_1 = 0.217$ and the 95% confidence interval is $\exp(-1.52661 \pm 1.96\sqrt{0.15873})$ so that $0.099 \leqslant \theta_1 \leqslant 0.474$.

In performing the computation of Q_{LR}, Q_W, Q_R and the confidence interval of the regression coefficient on the Freireich *et al.* (1963) data, we assumed an exponential model for the time of relapse. Given (5.8) and (5.14), it is $\Lambda(t) = \lambda t$. This suggests an easy graphical way to check the assumption of exponential distribution. The cumulative hazard estimates, $\hat{\Lambda}(t) = -\log \hat{S}(t)$, can be obtained from the K–M estimates of $S(t)$; if the failure times have an exponential distribution, the graph of $\hat{\Lambda}(t)$ against t is expected to be a straight line with null intercept and slope equal to $\hat{\lambda}$. These graphs, separately for the two treatment groups, are reported in figure 5.4 and suggest that the assumption of exponential hazard is questionable. Note, however, that on the grounds of the graphical check in section 4.4, the assumption of proportional hazards is tenable; thus more general parametric models or Cox's (1972) semi-parametric model could be more suitable for processing these data.

5.6.5   Relationship among LR, Wald and Rao score tests

Let us consider the linear predictors b and c of the previous section so that the model includes only one parameter β_0; moreover let $H_0 : \beta_0 = 0$ be the null hypothesis and $H_1 : \beta_0 \neq 0$ be the alternative hypothesis. In figure 5.5, the $LL(\beta_0)$ of a hypothetical sample (broken line, A) together with that of the Freireich *et al.* (1963) data (continuous line, B) are drawn against the possible values β_0 can assume. They reach their maximum $(-116.766\,66)$ for the same value of $\hat{\beta}_0 = -2.892\,22$; call this point P. However, the two log-likelihoods differ in their *curvature* $C(\hat{\beta}_0)$, that is, in the rate at which the tangent line is rotating as the point P of tangency travels along the curve. The value of $\frac{1}{2}Q_{LR}$ for sample B can be read directly on the ordinate as the difference of the two values of $LL(\beta_0)$, for $\beta_0 = \hat{\beta}_0$ and $\beta_0 = 0$, given by the continuous curve. We find that $\frac{1}{2}LR = -116.77 + 541.0 = 424.23$. Note that the distance $\frac{1}{2}Q_{LR}$ is a function of the distance $(\hat{\beta}_0 - 0)$ and of the curvature. In fact, for a given distance $(\hat{\beta}_0 - 0)$, $\frac{1}{2}Q_{LR}$ for curve A $(-116.77 + 710.00 = 593.23)$ is larger than the corresponding value for curve B because $C(\hat{\beta}_0)$ of curve A is greater than that of curve B. Conversely, given $C(\hat{\beta}_0)$, the larger the distance $(\hat{\beta}_0 - 0)$, the further will $LL(0)$ be from the maximum $LL(\hat{\beta}_0)$, and the larger will be $\frac{1}{2}Q_{LR}$.

From calculus we known that, for a generic function $y = f(x)$, the curvature is given by the positive value of

$$\frac{d^2 y}{dx^2}\left[1 + \left(\frac{dy}{dx}\right)^2\right]^{-3/2}$$

As $dLL(\beta_0)/d\beta_0$ evaluated at $\hat{\beta}_0$ is null, the term within brackets vanishes and the curvature $C(\hat{\beta}_0)$ of $LL(\beta_0)$ equals the observed information $I(\hat{\beta}_0)$. This is the key-point for a diagrammatic derivation of the Wald test (5.34). Instead of the difference in log-likelihood, this test considers the square of the distance $(\hat{\beta}_0-0)$. As a matter of fact, on intuitive grounds, large differences can be taken as evidence that the data do not support the null hypothesis. Figure 5.5 shows that the two samples provide the same squared distance $(\hat{\beta}_0-0)^2$; nonetheless, from the perspective of the LR test, it appears that sample B favours the null hypothesis more than sample A because the curvatures of the two $LL(\beta_0)$ differ. This suggests weighting the squared distance with weights directly proportional to the curvature; thus

$$Q_W = (\hat{\beta}_0-0)^2 I(\hat{\beta}_0) = \frac{(\hat{\beta}_0-0)^2}{var(\hat{\beta}_0)}$$

In considering the Rao score test we shall refer to figure 5.6; again two log-likelihood estimate functions have been drawn, for a hypothetical sample A and for set B of the Freireich *et al.* (1963) data. The maximum likelihood under the restriction of the null hypothesis is the pivot of this test and if H_0 is true this estimate will be close to the unrestricted one. We recall that this latter, that maximizes $LL(\beta_0)$, is obtained by solving the equation $U(\beta_0) = 0$ where $U(\beta_0) = dLL(\beta_0)/d\beta_0$. Therefore one can evaluate how the estimate under the restricted model fails to reach the maximum of $LL(\beta_0)$ by determining how much

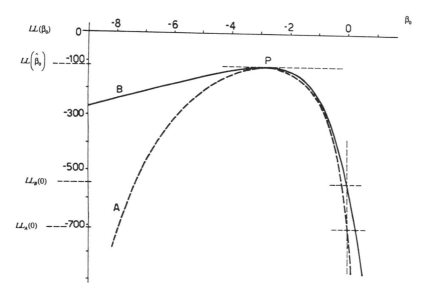

Figure 5.5　　The Wald test and its relationship with the LR test.

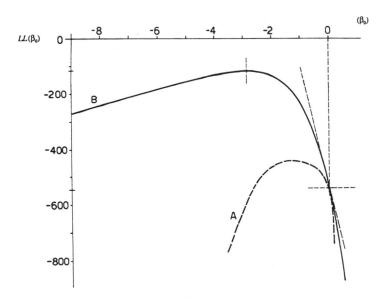

Figure 5.6   The Rao score test.

$U(0)$ differs from zero. As the sign of the slope $U(0)$ does not matter, $[U(0)]^2$ could be suggested as test statistic. However, figure 5.6 shows that two samples may produce the sample slope of $LL(\beta)$ at point $\beta_0 = 0$, but for one sample $\hat{\beta}_0$ is nearer to 0 than of the other. As in the case of the Wald test, we must make allowance for the curvature of $LL(\beta_0)$. Note that $\hat{\beta}_0$ maximizing the log-likelihood of sample A, the curvature of which is greater than that of sample B, is closer to $\beta_0 = 0$ than the corresponding value for sample B. Therefore we must weight $U(0)^2$ by the inverse of the curvature, $I^{-1}(0)$, so that the test statistic assumes small values when 0 is close to the ML estimate $\hat{\beta}_0$. Hence $U(0)^2 I^{-1}(0)$ is the statistic (5.37) in the case of a predictor with one regression coefficient.

When the likelihood function is quadratic the Wald and Rao score tests must give the same numerical values as the LR test which measures the vertical distance, $LL(\hat{\beta}_0) - LL(0)$, directly. This is because three parameters fully specify a quadratic function. In computing the Wald and Rao score tests three quantities are utilized to determine, implicity, the parameters of the quadratic and, consequently, the vertical distance. The Wald test uses the height of the hill at $\hat{\beta}_0$, $[LL(\hat{\beta}_0)]$, the value of the score $U(\hat{\beta}_0) = 0$ and the value of the second derivative $C(\hat{\beta}_0)$, whilst the Rao score test uses the height of the hill at 0, $[LL(0)]$, the value of the score $U(0)$ and the value of the second derivative $C(0)$. On the contrary, the non-quadratic form of $LL(\beta)$ accounts for different values of Q_{LR}, Q_W and Q_R.

The relationship among Q_{LR}, Q_W and Q_R computed in section 5.6.4 to test the null hypothesis $H_0: \beta_1 = 0$ is now investigated graphically. The predictor *a* implies the estimation of two parameters: our attention focuses here on β_1 and β_0 is thought of as a nuisance parameter. A way to deal with this latter consists in estimating its value. For each value of β_1, the value of β_0 which maximizes the likelihood can be obtained and substituted into the joint log-likelihood (5.28). The resulting curve is called the *profile* log-likelihood; coherently, when subtracted from its maximum value, we have the *profile log-likelihood ratio* (*LLR*). The profile log-likelihood is not a true log-likelihood because it cannot be obtained directly from the sample data; nevertheless in most situations it behaves in the same way as a log-likelihood.

As regards the example processed in section 5.6.4 the profile log-likelihood ratio against β_1 is drawn in figure 5.7 by a continuous line; the value at $\beta_1 = 0$ is

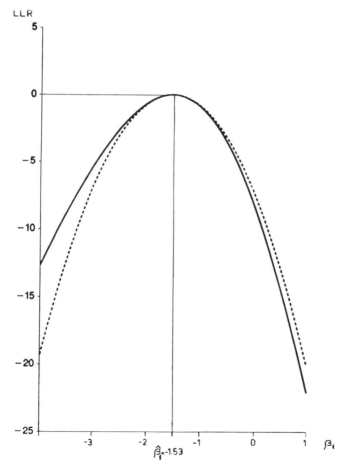

Figure 5.7 Profile log-likelihood ratio (*LLR*) (continuous line) and its approximation by the Wald test (dashed line).

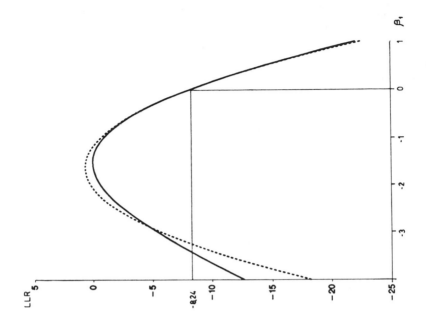

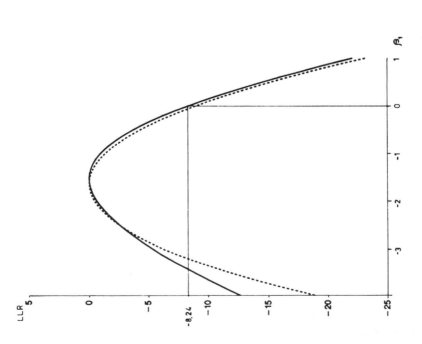

Figure 5.8 Profile log-likelihood ratio (*LLR*) (continuous line) and its approximation by the rao score test (dashed line). For explanation of left- and right-hand figures see text.

-8.247, i.e. $-\frac{1}{2}Q_{LR}$. Consider the expression (5.34); as Q_W is approximately the square of a Gaussian standardized deviate, $-\frac{1}{2}Q_W$ is the equation of the quadratic curve which approximates the log-likelihood ratio. This curve is drawn as a dashed line in figure 5.7 and one can see that it provides the closest approximation in the region of the most likely value, $\hat{\beta}_1 = -1.526\,61$.

As regards the score test, the approximation to the log-likelihood ratio may be obtained by the formula

$$-\frac{var\ U(0)}{2}\left(\beta_1 - \frac{U(0)}{var\ U(0)}\right)^2$$

where both the score and its variance are computed at the null value of the parameter. In our example $U(0)/var(0) = -10.9076/6.6970$ and the pertinent curve is drawn as a dashed line in figure 5.8 together with the profile log-likelihood curve (left-hand figure). The two curves show the same shape in the region of the null value, as may be better appreciated in the right-hand figure where the curve obtained by the Gaussian approximation has been displaced upwards. As a matter of fact this curve is the most accurate quadratic approximation to the profile log-likelihood ratio curve in the region of the null value.

APPENDIX A5.1 OBSERVED INFORMATION MATRIX FOR THE EXPONENTIAL REGRESSION MODEL

Consider one regression covariate x_1:

$$\lambda_i = \exp(\beta_0 + \beta_1 x_{1i})$$

with log-likelihood function as in (5.28). As for λ, in section 5.5.3, statistical inference on β_0 and β_1 relies on Fisher information; since in this case two parameters are estimated, we will deal with a 2×2 matrix based upon the partial second derivatives. On the principal diagonal there are:

$$\frac{\partial^2}{\partial\beta_0^2}LL(\beta_0, \beta_1) = \sum_{i=1}^{N}\{-\exp(\beta_0 + \beta_1 x_i)t_i\}$$

$$\frac{\partial^2}{\partial\beta_1^2}LL(\beta_0, \beta_1) = \sum_{i=1}^{N}\{-\exp(\beta_0 + \beta_1 x_i)t_i x_i^2\}$$

and out of the diagonal:

$$\frac{\partial^2}{\partial\beta_0\partial\beta_1}LL(\beta_0, \beta_1) = \sum_{i=1}^{N}\{-\exp(\beta_0 + \beta_1 x_i)t_i x_i\}$$

The entries of the Fisher's information matrix $\mathfrak{J}(\beta_0, \beta_1)$ are the negative expected values of these elements. Coherently with (5.25) the matrix $\mathbf{I}(\hat{\beta}_0, \hat{\beta}_1)$ with entries

$$I_{l,k} = -\left.\frac{\partial^2 LL(\beta)}{\partial \beta_l \partial \beta_k}\right|_{\beta_0 = \hat{\beta}_0, \beta_1 = \hat{\beta}_1}$$

evaluated at $\hat{\beta}_0$ and $\hat{\beta}_1$ is used as a consistent estimator of the information $\mathfrak{J}(\beta_0, \beta_1)$ under the unrestricted model. When a restricted model is considered, for example that under $H_0 : \beta_1 = 0$, the elements of the observed information matrix are evaluated at $\beta_0 = \tilde{\beta}_0$ and $\beta_1 = 0$, where $\tilde{\beta}_0$ is the ML estimate under the restricted model.

6

The Cox Regression Model

6.1 THE BASIC COX MODEL

In 1972 Cox introduced a regression model which is now widely used in the analysis of censored survival data for identifying differences in survival due to treatment and prognostic factors in clinical trials, and for studying the effect of exposure allowing for confounders in cohort studies.

Let N be the number of individuals in the study, each with the observed vector $(t_i, \delta_i, \mathbf{x}_i)$. The basic model assumes that the hazard function for failure time T for an individual i with covariate vector $\mathbf{x}_i' = (x_{1i}, x_{2i}, \ldots, x_{ki}, \ldots, x_{Ki})$ is

$$\lambda(t, \mathbf{x}_i) = \lambda_0(t) \exp(\boldsymbol{\beta}'\mathbf{x}_i) \qquad (6.1)$$

for $i = 1, \ldots, N$.

The covariates are assumed to be constant in time, as typically occurs when treatment, sex, age and clinical or biochemical features at study entry are considered.

The hazard (6.1) depends on both time and covariates, but through two separate factors: the first, $\lambda_0(t)$, is a function of time only, which is left arbitrary but is assumed to be the same for all subjects; the second is a quantity which depends on the individual covariates only through the $K \times 1$ vector $\boldsymbol{\beta}'$ of regression coefficients.

The Cox model is not a fully parametric model since it does not specify the form of $\lambda_0(t)$. It does, however specify the hazard ratio for any two individuals with covariate vectors $\mathbf{x}_1$ and $\mathbf{x}_2$, and for this reason it is defined as a semi-parametric model. In fact the hazard ratio turns out not to depend on $\lambda_0(t)$:

$$\frac{\lambda(t, \mathbf{x}_1)}{\lambda(t, \mathbf{x}_2)} = \frac{\lambda_0(t) \exp(\boldsymbol{\beta}'\mathbf{x}_1)}{\lambda_0(t) \exp(\boldsymbol{\beta}'\mathbf{x}_2)} = \exp[\boldsymbol{\beta}'(\mathbf{x}_1 - \mathbf{x}_2)] \qquad (6.2)$$

Thus model (6.1) is a *proportional hazard* (PH) regression model since it assumes that the failure rates of any two individuals are proportional, given that the ratio in (6.2) does not depend on time. If a logarithmic scale is adopted, (6.2) becomes:

$$\ln \lambda(t, \mathbf{x}_1) - \ln \lambda(t, \mathbf{x}_2) = \boldsymbol{\beta}'(\mathbf{x}_1 - \mathbf{x}_2) \tag{6.3}$$

showing that the model assumes a constant difference between the logarithm of the hazards. In particular, if in (6.2) the two individuals are taken to have covariate vectors $\mathbf{x}$ and $\mathbf{0}$, the ratio of their hazards is

$$\frac{\lambda(t, \mathbf{x})}{\lambda(t, \mathbf{0})} = \frac{\lambda_0(t) \exp(\boldsymbol{\beta}'\mathbf{x})}{\lambda_0(t)} = \exp(\boldsymbol{\beta}'\mathbf{x}) \tag{6.4}$$

This shows that $\lambda_0(t)$ may be regarded as the hazard function of an individual with all covariates of value zero, and for this reason $\lambda_0(t)$ is often termed the *baseline hazard*.

The Cox model (6.1) makes a second assumption which is implied by the use of the exponential function for linking the covariates to the hazard. The assumption is that independent covariates affect the hazard in a *multiplicative* way, as shown by (6.2) and (6.4), or equivalently the logarithm of the hazard in an additive way, as shown by (6.3). The two assumptions made by the Cox model are illustrated here with some examples, in which, for simplicity, we omit the index i for individual.

In clinical studies, the regression vector $\mathbf{x}$ contains treatment indicators and other variables representing characteristics which are thought to influence survival. Suppose, for the moment, that vector $\mathbf{x}$ includes only an indicator variable x_1 for treatment, with $x_1 = 0$ indicating the standard treatment A and $x_1 = 1$ the experimental treatment B. The Cox model assumes that the hazard functions are $\lambda_0(t)$ and $\lambda_0(t)e^{\beta_1}$ for individuals treated with A and B, respectively: at any time point, the failure rate related to B is that of treatment A *multiplied* by a *constant* factor e^{β_1}, no matter which form $\lambda_0(t)$ has, as in figure 6.1(a). Equivalently, the logarithm of these hazard functions has a constant distance β_1 (figure 6.1(b)). A negative value of β_1 indicates that treatment B is associated with a lower failure rate or, equivalently, is related to a higher survival probability.

Suppose now that another dichotomous covariate x_2 is included in the model to represent a characteristic of the patient, say gender, by specifying $\mathbf{x} = (x_1, x_2)$ and $\lambda(t, \mathbf{x}) = \lambda_0(t) \exp(\beta_1 x_1 + \beta_2 x_2)$. In each of the "strata" defined by the characteristic x_2, the hazard functions by treatment x_1, on a logarithmic scale, are assumed to be at a constant distance β_1 and meanwhile, in each treatment group, the hazards of patients of different genders are assumed to be at a constant distance β_2. An example of this model is represented in figure 6.2, where the hazard functions for the four groups identified by combining the possible values of x_1 and x_2 are plotted and specified. The model assumes that the two covariates have *independent* effects on the hazard rate. In other words, there is no interaction between x_1 and x_2 since treatment is assumed to have the same

Group	Values of covariate x_1	$\lambda(t,x_1) = \lambda_0(t)e^{\beta_1 x_1}$
A	0	$\lambda_0(t)$
B	1	$\lambda_0(t)e^{\beta_1}$

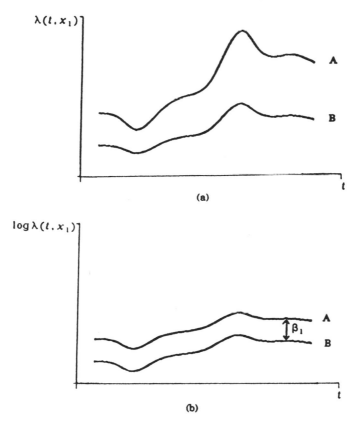

(a)

(b)

Figure 6.1 Plot of (a) the hazard functions and (b) the logarithm of the hazard functions according to the Cox model with one binary covariate ($x_1 = 0$ for standard treatment A and $x_1 = 1$ for experimental treatment B) and $\beta_1 < 0$.

effect on males and females. Here, the positive value of β_2 is related to a higher risk for males ($x_2 = 1$) than for females ($x_2 = 0$).

The introduction of an interaction term in model 6.1 would allow the description of a situation in which the hazard functions were, for example, as in figure 6.3. This figure results from a Cox model in which a covariate z is added to the

Group	Values of covariates x_1, x_2	$\lambda(t, \mathbf{x}) = \lambda_0(t)e^{\beta_1 x_1 + \beta_2 x_2}$
A, F	0, 0	$\lambda_0(t)$
B, F	1, 0	$\lambda_0(t)e^{\beta_1}$
A, M	0, 1	$\lambda_0(t)e^{\beta_2}$
B, M	1, 1	$\lambda_0(t)e^{\beta_1 + \beta_2}$

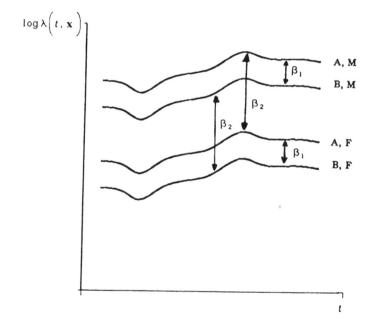

Figure 6.2 Plot of the logarithm of the hazard functions according to the Cox model with two binary covariates x_1 ($x_1 = 0$ for treatment A and $x_1 = 1$ for treatment B) and x_2 ($x_2 = 0$ for female (F) and $x_2 = 1$ for male (M)), and with $\beta_1 < 0$ and $\beta_2 > 0$.

vector of regressors $\mathbf{x}$ and it is

$$\lambda(t, \mathbf{x}, z) = \lambda_0(t) \exp(\beta_1 x_1 + \beta_2 x_2 + \gamma z) \tag{6.5}$$

where z is the product $x_1 \cdot x_2$ and has value $z = 1$ when $x_1 = 1$ and $x_2 = 1$ and $z = 0$ otherwise, and γ is less than zero. Figure 6.3 reflects the presence of a quantitative interaction since treatment B is beneficial in both males and females but is more so in males where a negative term γ is added to the negative coefficient β_1.

Group	Values of covariates x_1, x_2, z	$\lambda\left(t, \mathbf{x}, z\right) =$ $\lambda_0(t)e^{\beta_1 x_1 + \beta_2 x_2 + \gamma z}$
A, F	0, 0, 0	$\lambda_0(t)$
B, F	1, 0, 0	$\lambda_0(t)e^{\beta_1}$
A, M	0, 1, 0	$\lambda_0(t)e^{\beta_2}$
B, M	1, 1, 1	$\lambda_0(t)e^{\beta_1 + \beta_2 + \gamma}$

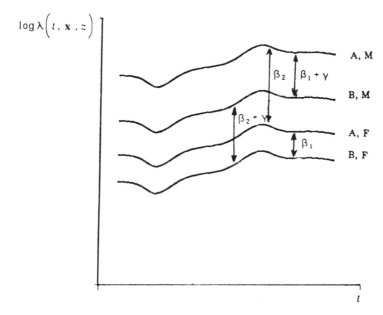

Figure 6.3 Plot of the logarithm of the hazard functions according to the Cox model with two binary covariates x_1, x_2 and an interaction term $z = x_1 \cdot x_2$, when $\beta_1 < 0$, $\beta_2 > 0$ and $\gamma < 0$, where γ is the coefficient of the z term.

In the hypothesis that γ is positive, two different situations are possible with model (6.5), which are represented in figure 6.4. If, for example, the regression coefficients were respectively $\beta_1 = -0.7$ and $\gamma = +0.5$, the resulting log hazards would be related as in figure 6.4(a). This still shows a quantitative interaction where treatment B is beneficial in both females and males, but less so in the latter. If the regression coefficients were instead $\beta_1 = -0.7$ and $\gamma = +2$, the corresponding plot would be that in figure 6.4(b). This shows a qualitative interaction, where treatment B is beneficial for females but harmful for males.

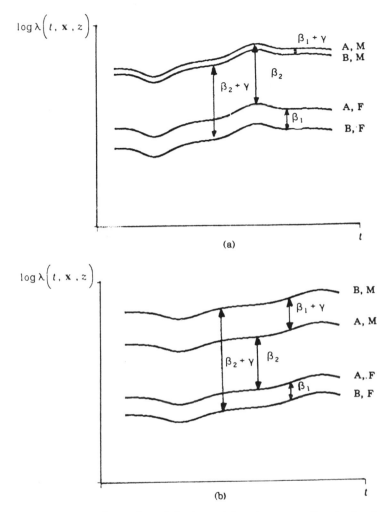

Figure 6.4 Plot of the logarithm of the hazard functions according to the Cox model with two binary covariates x_1, x_2 and an interaction term $z = x_1 \cdot x_2$, when $\beta_1 < 0$, $\beta_2 > 0$ and $0 < \gamma < -\beta_1$ (plot (a), quantitative interaction) or $\gamma > -\beta_1$ (plot (b), qualitative interaction).

If two covariates x_1 and x_2 are included in a Cox model without an interaction term, as in figure 6.2, they are supposed to act independently and multiplicatively on the hazard. The logarithm of the baseline hazard increases by a constant value (or decreases, if the regression coefficient is negative) at each increase of one unit of any covariate. From figures 6.3 and 6.4 it is clear that the introduction of an interaction term relaxes this assumption.

Group	Values of covariate x_1	$\lambda(t, x_1)$ $= \lambda_0(t)e^{\beta_1 x_1}$
0	0	$\lambda_0(t)$
1	1	$\lambda_0(t)e^{\beta_1}$
2	2	$\lambda_0(t)e^{2\beta_1}$
3	3	$\lambda_0(t)e^{3\beta_1}$
4	4	$\lambda_0(t)e^{4\beta_1}$

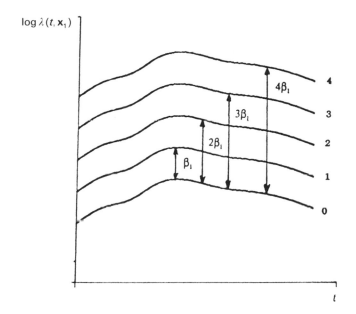

Figure 6.5 Plot of the logarithm of the hazard functions according to a Cox model with one variable x_1 taking values from 0 to 4 and with $\beta_1 > 0$.

For simplicity, we have considered here dichotomous covariates only, but the same concepts extend to covariates with three or more categories or to continuous covariates. An example of the Cox model $\lambda(t, x_1) = \lambda_0(t)e^{\beta_1 x_1}$ with the variable x_1 taking values from 0 to 4 is shown in figure 6.5, where the logarithm of the hazard functions is drawn. The model assumes that the baseline log hazard is added to the same constant factor β_1 for every increase of one unit in the value of x. If this feature is undesirable, alternative coding systems should be adopted for the variable, as discussed in section 6.5.

The Cox model specifies a form for the hazard ratio (6.2) which includes the regression coefficients β only as unknown parameters. Coherently, Cox introduced a method for estimating β, and hence the hazard ratio, allowing for an arbitrary $\lambda_0(t)$ in (6.1). This method is based on the formulation of a "partial likelihood" which will be shown in the following section.

The estimate of β allows us to quantify, by (6.2), the relative rate of failure for an individual with covariate vector x_1 with respect to an individual with vector x_2, and not the absolute rate for each individual. The former is usually the main point of the analysis in a comparative study on treatment or in a study on the effect of exposure to a risk factor. For example, in the treatment comparison represented by figure 6.1, an estimated value of $\hat{\beta}_1 = -0.7$ means that the failure rate for subjects treated with B is nearly half as much ($e^{\hat{\beta}_1} = e^{-0.7} = 0.497$) as that for treatment A, at all time points. In the example of figure 6.2, the estimated treatment effect would be adjusted by gender: $e^{\hat{\beta}_1}$ would estimate the hazard ratio between subjects treated with A and subjects treated with B, the other covariate being identical.

The likelihood approach to inference, introduced in section 5.6.3, can be extended to the partial likelihood. For instance, the partial likelihood ratio test for the hypothesis $\beta_1 = 0$ (against $\beta_1 \neq 0$ allows the statistical evaluation of the strength of association of the covariate x_1 to prognosis. A test on the hypothesis $\gamma = 0$ (against the alternative $\gamma \neq 0$) aims at evaluating whether an interaction term is needed in the model to account for departures from an independent multiplicative effect of each covariate, in the presence of the other covariates.

However, if estimation is restricted to the regression parameters, it is not possible to describe the hazard or the survival functions of groups of subjects according to the model. Suppose we wish to plot the survival curves predicted by the model for four subgroups of patients identified by two binary variables x_1 and x_2, say treatment and gender, as in figure 6.2. In particular, the survival function of individuals with covariate vector $\mathbf{0}$ (female patients treated with A) is, by (5.8), $S(t, \mathbf{0}) = \exp[-\int_0^t \lambda_0(u)du] = \exp[-\Lambda_0(t)]$ and we indicate it with $S_0(t)$. The group of female patients treated with B, identified by the covariate vector $\mathbf{x} = (1, 0)$, has survival function $S(t, \mathbf{x}) = \exp[-\Lambda_0(t)e^{\beta_1}] = \{\exp[-\Lambda_0(t)]\}^{\exp \beta_1} = [S_0(t)]^{\exp \beta_1}$.

The survival function for males, obtained in an analogous way, is a power of $S_0(t)$ with exponent e^{β_2} in treatment group A and $e^{\beta_1 + \beta_2}$ in treatment group B. Thus, in order to estimate the survival functions associated with model (6.1), an estimate of the baseline hazard is needed in addition to the values $\hat{\beta}$. This estimate will be introduced in section 6.3 and, on the basis of it, the survival function of any individual i can be estimated according to

$$\hat{S}(t, \mathbf{x}_i) = [\hat{S}_0(t)]^{\exp(\beta' \mathbf{x}_i)} \tag{6.6}$$

The remainder part of this chapter, while going into the details of estimation and application, will attempt to clarify the relationship of the Cox model with methods introduced in the first part of the book. Like any model, (6.1) makes assumptions that must be clearly understood and considered before its application.

The way in which the assumptions of proportional hazard and multiplicative effect of covariates may be relaxed will be explored later in this chapter, while the problem of model validation will be dealt with in Chapter 7. The flexibility of the Cox model as a tool for regressing prognosis on various factors lies in the non-parametric specification of the baseline hazard $\lambda_0(t)$. If its form were specified, model 6.1 would become a parametric regression model, introducing additional assumptions on the distribution of failure time T. For example, by restricting $\lambda_0(t)$ to be a constant λ_0, the Cox model would reduce to the exponential model (section 5.6.1).

6.2* ESTIMATION OF THE REGRESSION PARAMETERS

Since $\lambda_0(t)$ is not specified parametrically, it is not possible to use an ordinary likelihood such as that in (5.19) to estimate the regression coefficients $\boldsymbol{\beta}$. The arbitrary function $\lambda_0(t)$ is a nuisance function, and the aim is to estimate $\boldsymbol{\beta}$ on the basis of the information conveyed by the observed data without having to involve $\lambda_0(t)$. Cox (1972) argued conditionally on the set of observed failures and described the data with a function depending on $\boldsymbol{\beta}$ only. Consider a sample of N subjects and suppose a total of J failures occur, with J generally smaller than N, due to the presence of censoring. Let $t_{(1)} < t_{(2)} < \cdots < t_{(J)}$ be the J distinct ordered failure times observed and let $R(t)$ be the set of subjects, at risk at time t, who are alive and under observation just before t. With a slight abuse of notation, we indicate with j the label of the subject who fails at $t_{(j)}$ so that its vector of covariates is $\mathbf{x}_j$. In general, $\mathbf{x}_i$ is the vector of covariates for the ith subject and the covariates have a constant value in time. The probability that an individual with covariates $\mathbf{x}$ fails in a small interval $(t, t \mid dt)$, given the set at risk at t, is $\lambda(t, \mathbf{x})dt$ (section 3.6.3). Thus, conditional on the fact that one individual is observed to fail at $t_{(j)}$, the probability that it is an individual with covariates $\mathbf{x}_j$ is

$$\lambda(t_{(j)}, \mathbf{x}_j)dt \left/ \sum_{i \in R(t_{(j)})} \lambda(t_{(j)}, \mathbf{x}_i)dt \right.$$

It follows (Cox and Oakes, 1984, pp. 92–93) that the function describing the failure pattern is the product of J terms, one for each observed failure time:

$$L(\lambda_0(t), \boldsymbol{\beta}) = \prod_{j=1}^{J} \frac{\lambda(t_{(j)}, \mathbf{x}_j)dt}{\sum_{i \in R_j} \lambda(t_{(j)}, \mathbf{x}_i)dt}$$

where the hazard function is defined by (6.1) and $R_j = R(t_{(j)})$. Given expression (6.1), the baseline function $\lambda_0(t)dt$ cancels out and the product above simplifies to

$$L = L(\boldsymbol{\beta}) = \prod_{j=1}^{J} \frac{\exp(\boldsymbol{\beta}'\mathbf{x}_j)}{\sum_{i \in R_j} \exp(\boldsymbol{\beta}'\mathbf{x}_i)} \tag{6.7}$$

where $L(\boldsymbol{\beta})$ indicates that function (6.7) depends on the unknown parameters $\boldsymbol{\beta}$, the values of $\mathbf{x}$ being known.

It is clear that there is no contribution to the estimate of $\boldsymbol{\beta}$ from the gaps between successive failures. This is because, in the absence of knowledge of $\lambda_0(t)$, the values $t_{(j)}$ cannot provide information about $\boldsymbol{\beta}$, for their distribution depends heavily on $\lambda_0(t)$ (as an extreme example, $\lambda_0(t)$ might be zero except in the neighbourhood of each $t_{(j)}$). Furthermore, the observed censoring times are assumed not to contribute information on $\boldsymbol{\beta}$, but occurrence of censoring must be considered to identify the correct set at risk at each $t_{(j)}$. The usual convention (section 3.2.2) is adopted: if individual i has censored time c_i, with $c_i = t_{(j)}$ for some j, than he is considered at risk in $R(t_{(j)})$ and dropped thereafter. While the conditional probability at $t_{(j)}$ depends on the entire history up to $t_{(j)}$ through the identification of the set of subjects at risk, it does not depend on the values and on the distribution of the ordered failure times $t_{(j)}, j = 1, \dots, J$. In fact, function (6.7) does not vary under monotonic transformations of the time scale.

It was proposed by Cox that function (6.7), although it describes only part of the data, could be regarded as a likelihood function allowing the estimation of $\boldsymbol{\beta}$ with standard procedures. In his paper (1972), the author highlighted those issues concerning the model which needed further work to be clarified as follows:

"(i) It is assumed without proof in the paper that the usual asymptotic procedures and properties associated with maximum likelihood estimates and tests hold.
(ii) Is it possible and worthwhile to try to recover information which for any specific $\lambda_0(t)$ is contained in the gaps between failures?
(iii) What is the loss of information about the regression coefficients involved in using the procedures of the paper when some parametric representation of $\lambda_0(t)$ is in fact appropriate?"

These issues have in fact stimulated a lot of work since the 1972 paper. To summarize, the large sample properties of the maximum likelihood estimators of $\boldsymbol{\beta}$ based on (6.7) have been shown to be the same as those of any estimator from complete likelihood (Cox, 1975; Tsiatis, 1981b; Andersen and Gill, 1982). It is worth mentioning that (6.7) was given the name "partial likelihood" by Cox. In a paper published in 1975, he derived the full likelihood $L(\lambda_0(t), \boldsymbol{\beta})$ based on (6.1) and showed that inference on $\boldsymbol{\beta}$ could be made using $L_1(\boldsymbol{\beta})$, which coincides with (6.7), and depends on $\boldsymbol{\beta}$ only, while neglecting the factor $L_2(\lambda_0(t), \boldsymbol{\beta}) = L(\lambda_0(t), \boldsymbol{\beta})/L_1(\boldsymbol{\beta})$ which depends both on $\boldsymbol{\beta}$ and on the unknown

$\lambda_0(t)$. In large samples, the distribution of β can be approximated by a normal distribution with the true value as mean, which is estimated by maximizing the likelihood, and a variance–covariance matrix which is estimated from the second derivative of the likelihood function.

The efficiency of the likelihood (6.7) relative to a full likelihood based on a parametric form of $\lambda_0(t)$ was studied in various papers (Efron, 1977; Oakes, 1981). In brief, it was shown that if parameters β are not far from zero and censoring does not depend on covariates, the asymptotic efficiency of (6.7) is high relative to the full likelihood. In contrast, the ML estimates of β based on a parametric model, thus on a complete likelihood, are fully efficient whatever the value of β is, provided the model assumptions are satisfied, but may be inconsistent if the assumptions do not hold. The partial likelihood has been shown to be the profile likelihood for the regression parameters β given a set of (infinite) nuisance parameters corresponding to $\lambda_0(t)$ (Johansen, 1983; Clayton and Hills, 1993, pp. 299–302).

The partial likelihood can be easily modified to allow for the presence of left truncation, as illustrated by Cnaan and Ryan (1989) and Breslow and Day (1987, Chapter 5).

The asymptotic theory of maximum likelihood estimation requires that the likelihood function satisfies some "regularity conditions" which are met in most applications. The theoretical properties of the Cox model are well documented in several textbooks on survival data analysis and will not be discussed further here (Kalbfleisch and Prentice, 1980; Lawless, 1982; Cox and Oakes, 1984). The regression coefficients β are estimated by the values $\hat{\beta}$ which maximize the partial likelihood $L(\beta)$ or equivalently its logarithm $LL(\beta)$:

$$LL(\beta) = \sum_{j=1}^{J} \left\{ \beta'x_j - \log\left[\sum_{i \in R_j} \exp(\beta'x_i) \right] \right\} = \sum_{j=1}^{J} l_j \quad \cdot \tag{6.8}$$

where l_j is the contribution to the log-likelihood corresponding to the failure time $t_{(j)}$. The values $\hat{\beta} = (\hat{\beta}_1, \ldots, \hat{\beta}_K)$ are obtained by equating to zero the K first derivatives of $LL(\beta)$ with respect to $\beta_k (k = 1, \ldots, K)$. An iterative process such as the Newton–Raphson one has to be adopted to solve this system of equations for β. If the kth component of vector x_i is x_{ki}, the kth derivative of the contribution l_j to (6.8) is

$$\frac{\partial l_j}{\partial \beta_k} = x_{kj} - \frac{\sum\limits_{i \in R_j} x_{ki} \exp(\beta'x_i)}{\sum\limits_{i \in R_j} \exp(\beta'x_i)} \tag{6.9}$$

This derivative is the difference between the value of the kth covariate on the subject who fails at $t_{(j)}$ and the weighted average of the covariate over the risk set R_j, with exponential weights $\exp(\beta'x_i)$. This expression suggests, on an intuitive basis, how the likelihood "works": if, for example, patients who fail tend to have higher values of the kth covariate, the value of β_k will have to be large enough to reduce the derivative to zero. Summing (6.9) over all failure times, we have the kth component of the score $U(\boldsymbol{\beta})$:

$$U_k(\boldsymbol{\beta}) = \sum_{j=1}^{J} \frac{\partial l_j}{\partial \beta_k} \tag{6.10}$$

By taking the second derivative of l_j, an expression is obtained which has the form of a variance. For example, the derivative of (6.9) with respect to β_k is

$$\frac{\partial^2 l_j}{\partial \beta_k^2} = -\left[\frac{\sum\limits_{i \in R_j} x_{ki}^2 \exp(\boldsymbol{\beta}'\mathbf{x}_i)}{\sum\limits_{i \in R_j} \exp(\boldsymbol{\beta}'\mathbf{x}_i)} - \left(\frac{\sum\limits_{i \in R_j} x_{ki} \exp(\boldsymbol{\beta}'\mathbf{x}_i)}{\sum\limits_{i \in R_j} \exp(\boldsymbol{\beta}'\mathbf{x}_i)} \right)^2 \right] \tag{6.11}$$

In fact (6.11) may be written as $\partial^2 l_j / \partial \beta_k^2 = -[E_j(x_{ki}^2) - [E_j(x_{ki})]^2]$ where the probability of selecting a subject i from the risk set R_j is not simply $1/R_j$ but is proportional to $\exp(\boldsymbol{\beta}'\mathbf{x}_i)$ and the subscript j indicates that the expected value is taken conditional on R_j.

As in section (5.5.3), the inverse of the information matrix, evaluated at $\hat{\boldsymbol{\beta}}$, that is $\mathbf{I}^{-1}(\hat{\boldsymbol{\beta}})$, is the estimated covariance matrix of $\hat{\boldsymbol{\beta}}$.

The partial likelihood (6.7) has been written for T as a continuous variable. However, even when T is continuous in theory, tied observations may occur in practice because of the measurement units of time. Provided that ties are few in number, it is possible to deal satisfactorily with the problem by modifying the likelihood (6.7). The approach which is routinely used is the one proposed by Peto in the discussion of Cox's paper (1972) and Breslow (1974). Let d_j be the number of failures observed at time $t_{(j)}$ and $\mathbf{s}_j$ be the sum of the covariate vectors of the d_j subjects who fail. The logarithm of the partial likelihood is written as if the d_j failures occurred in any order and ignoring the fact that the risk set should successively be diminished by one subject at each failure. It is

$$LL(\boldsymbol{\beta}) = \sum_{j=1}^{J} \left\{ \boldsymbol{\beta}'\mathbf{s}_j - d_j \log\left[\sum_{i \in R_j} \exp(\boldsymbol{\beta}'\mathbf{x}_i) \right] \right\} = \sum_{j=1}^{J} l_j \tag{6.8a}$$

and $LL(\boldsymbol{\beta})$ reduces to (6.8) when no ties are present ($d_j = 1$ and $\mathbf{s}_j = \mathbf{x}_j$). Previous expressions (6.9) and (6.11) for the first and second derivative of l_j become

$$\frac{\partial l_j}{\partial \beta_k} = s_{kj} - d_j \frac{\sum\limits_{i \in R_j} x_{ki} \exp(\boldsymbol{\beta}'\mathbf{x}_i)}{\sum\limits_{i \in R_j} \exp(\boldsymbol{\beta}'\mathbf{x}_i)} \tag{6.9a}$$

$$\frac{\partial^2 l_j}{\partial \beta_k^2} = -d_j \left[\frac{\sum_{i \in R_j} x_{ki}^2 \exp(\boldsymbol{\beta}' \mathbf{x}_i)}{\sum_{i \in R_j} \exp(\boldsymbol{\beta}' \mathbf{x}_i)} - \left(\frac{\sum_{i \in R_j} x_{ki} \exp(\boldsymbol{\beta}' \mathbf{x}_i)}{\sum_{i \in R_j} \exp(\boldsymbol{\beta}' \mathbf{x}_i)} \right)^2 \right] \tag{6.11a}$$

where s_{kj} is the kth component of vector $\mathbf{s}_j$.

Other approaches to the problem of ties have been discussed by Cox (1972), Peto, in the discussion of Cox's paper, and Efron (1977). When survival time is subject to interval grouping, a discrete model was proposed by Cox (1972) as a more appropriate approach.

A simple example of calculation of partial likelihood and estimation of $\boldsymbol{\beta}$ is given in section 6.4.

6.3* ESTIMATION OF THE BASELINE HAZARD AND SURVIVAL PROBABILITY

The description of the estimated survival probability is often used to represent the results of a study. When the Cox model is applied, this requires a second phase in the estimation process, which follows the estimation of the regression parameters. As discussed by Oakes (1981), different approaches have been adopted to estimate the arbitrary baseline function $\lambda_0(t)$ or equivalently the cumulative function $\Lambda_0(t)$. A simple estimator of $\Lambda_0(t)$ proposed by Breslow (1974), the properties of which were discussed by Tsiatis (1981b), is presented here.

Breslow derived a maximum likelihood estimator of $\Lambda_0(t)$ after assuming that the failure time distribution has a hazard function which is constant between each pair of successive observed failure times. Also, censored observations which occur between $t_{(j)}$ and $t_{(j+1)}$ are assumed to be censored at $t_{(j)}$ and $t_0 = 0$ is taken as the origin of the observation. The estimate of $\lambda_0(t)$ in the interval $(t_{(j-1)}, t_{(j)}]$ is given by

$$\hat{\lambda}_j = \frac{d_j}{h_j \sum_{i \in R_j} \exp(\boldsymbol{\beta}' \mathbf{x}_i)}$$

where $h_j = t_{(j)} - t_{(j-1)}$ is the time interval between two consecutive failure times. The expression for $\hat{\lambda}_j$ is, as in section 3.6.1, a ratio between the number of events and the weighted number of person–time units at risk, where each individual in the rest set $\mathbf{R}_j$ contributes weight $\exp(\boldsymbol{\beta} \mathbf{x}_i)$ for time h_j. A rough estimate of $\Lambda_0(t_{(j)}) - \Lambda_0(t_{(j-1)})$ is $\hat{\lambda}_j h_j$. Summing such terms over all $t_{(j)} \leq t$ we obtain Breslow's estimator of the cumulative failure rate at time t

$$\hat{\Lambda}_0(t) = \sum_{t_{(j)} \leq t} \frac{d_j}{\sum_{i \in R_j} \exp(\boldsymbol{\beta}' \mathbf{x}_i)}$$

Thus the estimate of the cumulative baseline hazard is a step function which, for $\beta = 0$, is identical to the Aalen estimator introduced in section 3.6.3.

The baseline survivor function can be estimated by (5.8) as

$$\hat{S}_0(t) = \prod_{t_{(j)} \leqslant t} \left[\exp \frac{-d_j}{\sum_{i \in R_j} \exp(\beta' x_i)} \right] \tag{6.13}$$

Alternatively, an analogue of the Kaplan–Meier estimator can be derived for $\hat{S}_0(t)$ by thinking of $\hat{\lambda}_0(t)$ in terms of a discrete hazard having mass $d_j / \sum_{i \in R_j} \exp(\beta' x_i)$ at each failure time $t_{(j)}$. Intuitively, this gives approximately the probability that d_j subjects with covariates $x = 0$ fail at $t_{(j)}$ conditional on the set of covariates x_i observed on the subjects at risk at $t_{(j)}$. Thus, by taking the analogue of the Kaplan–Meier estimator, we obtain

$$\hat{S}_0(t) = \prod_{t_{(j)} \leqslant t} \left[1 - \frac{d_j}{\sum_{i \in R_j} \exp(\beta' x_i)} \right] \tag{6.14}$$

This estimator in fact coincides with the K–M estimator when $\beta = 0$, i.e. in a homogeneous sample. Kalbfleisch and Prentice (1980, section 4.3) derive a step function for estimating $S_0(t)$ following the same likelihood approach as is used in Appendix A3.1 to obtain the K–M estimator. Their non-parametric maximum likelihood estimator, which requires an iterative solution in the presence of ties, gives similar results to (6.14). The estimates of $S_0(t)$ given by (6.13) and (6.14) have very similar values (for small values of u, $\exp(-u)$ is approximated by $(1-u)$), except when there are quite a few ties and in the right-hand tail of the distribution.

Under the assumption of proportional hazards, for individuals with a certain covariate vector x^*, the estimated cumulative hazard and survivor function are

$$\hat{\Lambda}(t, x^*) = \hat{\Lambda}_0(t) \exp(\beta' x^*)$$

$$S(t, x^*) = [\hat{S}_0(t)]^{\exp(\beta' x^*)} \tag{6.15}$$

The estimator $\hat{\Lambda}_0(t)$ has been proven, in the absence of ties, to have a distribution which is approximately normal in large samples. The asymptotic variance of $\hat{\Lambda}_0(t)$ is given in Appendix A6.1. In the case of one covariate x, it is consistently estimated by

$$var(\hat{\Lambda}_0(t)) = \sum_{t_{(j)} \leqslant t} \frac{1}{\left(\sum_{i \in R_j} \exp(\beta' x_i) \right)^2} + I^{-1}(\hat{\beta}) \left[\sum_{t_{(j)} \leqslant t} \frac{\sum_{i \in R_j} x_i \exp(\beta' x_i)}{\left(\sum_{i \in R_j} \exp(\beta' x_i) \right)^2} \right]^2 \tag{6.16}$$

where β is estimated by maximizing the partial likelihood and $I^{-1}(\hat{\beta})$ is the estimated variance of $\hat{\beta}$. By using the joint distribution of $\hat{\boldsymbol{\beta}}$ and $\hat{\Lambda}_0(t)$, confidence intervals can be calculated around the estimated survivor function in (6.15) for subjects characterized by having any particular vector of covariates $\mathbf{x}^*$.

The following section gives a simple example of the calculation of the survival curves predicted by the Cox model in two treatment groups. When using the estimated survival curves for description, it should be noted that $\hat{S}_0(t)$ and the curves derived from it by (6.15) are calculated using the entire sample, and not only the subgroup identified by a particular pattern of the covariate vector. These curves are constrained by the model assumptions and by the estimated parameters $\hat{\boldsymbol{\beta}}$ and are thus related to each other by well-defined quantities. If the model assumptions do not hold, they may give a very misleading picture of the effect of treatment or of the contrast between different prognostic features. An important use of these quantities is in turn to verify the model assumptions, as will be seen in Chapter 7.

It is worth noting that the baseline hazard is usually taken to indicate the hazard related to some kind of standard conditions and is not necessarily the hazard corresponding to a subject with all covariate values equal to zero. This may not be realistic in many cases, as for example when a continuous variable like age is included in the model. Often the value zero of each covariate is taken to correspond to the average value of the covariate on the entire sample. This means that all covariates x_{ki} of subject i are rescaled by subtracting the mean value $\bar{x}_k$ and $\hat{\Lambda}_0(t)$ is in fact $\Lambda_{\bar{x}}(t)$, where $\bar{\mathbf{x}}$ is the vector of the K mean values of the covariates. To show that this rescaling has been done, model (6.1) may be rewritten as

$$\lambda(t, \mathbf{x}_i) = \lambda_{\bar{x}}(t) \exp[\boldsymbol{\beta}'(\mathbf{x}_i - \bar{\mathbf{x}})]$$

However, this is only a matter of notation and we will retain the one used in (6.1) knowing that $\mathbf{0}$ may indicate either a vector of null values or the vector $\bar{\mathbf{x}}$ of averages.

6.4   AN EXAMPLE

6.4.1   Computing estimates

The data set on 42 patients with leukemia, described in section 3.1, is used here to show in detail the estimation procedures presented in previous sections. The covariate vector $\mathbf{x}$ reduces in this case to a single binary variable x with values $x = 0$ for placebo and $x = 1$ for treatment 6-MP. The distribution of time from remission to relapse is being modelled by defining the hazard function as

$$\lambda(t, x) = \lambda_0(t)e^{\beta x} \qquad (6.17)$$

Table 6.1 Calculation of the partial log-likelihood on data from Freireich and Gehan. Time from remission to relapse is in weeks and the covariate indicates treatment ($x = 0$ for placebo and $x = 1$ for 6-MP).

j	Time (weeks) to failure $t_{(j)}$ (1)	Time (weeks) censored t^* (1a)	No. of events d_j (2)	No. censored (2a)	Covariate values for subjects who fail (3)	Covariate values for subjects who are censored (3a)	Set at risk spotted by treatment Placebo n_{Bj} (4)	Set at risk spotted by treatment 6-MP n_{Aj} (5)	Contribution to the log-likelihood l_j (6)
1	1		2		0,0		21	21	$0 - 2\ln(21 + 21e^{\beta})$
2	2		2		0,0		19	21	$0 - 2\ln(19 + 21e^{\beta})$
3	3		1		0		17	21	$0 - \ln(17 + 21e^{\beta})$
4	4		2		0,0		16	21	$0 - 2\ln(16 + 21e^{\beta})$
5	5		2		0,0		14	21	$0 - 2\ln(14 + 21e^{\beta})$
6	6		3		1,1,1		12	21	$3\beta - 3\ln(12 + 21e^{\beta})$
—		6*		1		1	—	—	
7	7		1		1		12	17	$\beta - \ln(12 + 17e^{\beta})$
8	8		4		0,0,0,0		12	16	$0 - 4\ln(12 + 16e^{\beta})$
—		9*		1		1	—	—	
9	10		1		1		8	15	$\beta - \ln(8 + 15e^{\beta})$
—		10*		1		1	—	—	
10	11		2		0,0		8	13	$0 - 2\ln(8 + 13e^{\beta})$
—		11*		1		1	—	—	
11	12		2		0,0		6	12	$0 - 2\ln(6 + 12e^{\beta})$
12	13		1		1		4	12	$\beta - \ln(4 + 12e^{\beta})$
13	15		1		0		4	11	$0 - \ln(4 + 11e^{\beta})$
14	16		1		1		3	11	$\beta - \ln(3 + 11e^{\beta})$
15	17		1		0		3	10	$0 - \ln(3 + 10e^{\beta})$
—		17*,19*,20*		3		1,1,1	—	—	
16	22		2		0,1		2	7	$\beta - 2\ln(2 + 7e^{\beta})$
17	23		2		0,1		1	6	$\beta - 2\ln(1 + 6e^{\beta})$

Seventeen distinct failure times were observed on the entire data set and are shown in increasing order of magnitude in column (1) of table 6.1. For each failure time $t_{(j)}$, column (2) shows the number d_j of events (relapses) and column (3) reports the value of the covariate for each subject who fails, i.e. the treatment arm to which he was assigned. Columns (2a) and (3a) report the same quantities for subjects with censored observations at the times shown in column (1a). Censored times greater than the last failure time are not reported since they are not required for the calculation of the partial likelihood. Columns (4) and (5) show the composition of the risk set at failure time $t_{(j)}$ by subdividing the subjects who are still alive and under observation just before $t_{(j)}$, according to the treatment arm. Having these ingredients, the calculation of the contribution l_j of each observed failure time to the log-likelihood according to (6.8a) is straight-forward. Consider the simple case of one failure observed at time $t_{(3)}$ = three weeks: the contribution l_3 is the difference between the value of the covariate of the subject who fails, in this case 0, multiplied by β, and the log of the sum of the quantities $e^{\beta x_i}$ for each of the 38 subjects who are at risk of failure at three weeks. If the failure occurs on a subject with covariate $x = 1$ (6-MP), as for $t_{(9)}$ = ten weeks, the first term in the difference is β. In the presence of tied observations, the first term in the difference sums the contributions βx of each failure, and the second term considers the contribution of the set at risk as many times as there are tied failures. The information from the failures observed in the sample is represented according to (6.8a) in the log-likelihood, $LL(\beta) = \sum_{j=1}^{17} l_j$, which depends on the unknown regression coefficient β. In order to estimate β we set to zero the score $U(\beta)$, the first derivative of $LL(\beta)$ with respect to β, and apply an iterative process, to solve the resulting equation in β. In this case, by summing the contributions (6.9a), in which n_{Aj} and n_{Bj} indicate the number of subjects at risk at $t_{(j)}$ in the 6-MP and placebo group, respectively, one obtains

$$U(\beta) = \sum_{j=1}^{17} \frac{dl_j}{d\beta} = \sum_{j=1}^{17} \left[s_j - d_j \frac{n_{Aj}e^{\beta}}{n_{Aj}e^{\beta} + n_{Bj}} \right] \tag{6.18}$$

The negative of the second derivative of $LL(\beta)$ is, according to (6.11a)

$$I(\beta) = \sum_{j=1}^{17} d_j \left[\frac{n_{Aj}e^{\beta}}{n_{Aj}e^{\beta} + n_{Bj}} - \left(\frac{n_{Aj}e^{\beta}}{n_{Aj}e^{\beta} + n_{Bj}} \right)^2 \right] \tag{6.19}$$

The Newton–Raphson technique is based on the Taylor series expansion of $U(\beta)$ and the value of $U(\beta)$ at the point $\hat{\beta}$ is approximately given by

$$U(\hat{\beta}) = U(\beta^*) - I(\beta^*)(\hat{\beta} - \beta^*) \tag{6.20}$$

where β^* is the starting point chosen for the expansion. We wish to find the value $\hat{\beta}$ that satisfies the equation $U(\hat{\beta}) = 0$; thus, by setting $U(\hat{\beta}) = 0$ in (6.20), we obtain an expression for $\hat{\beta}$:

$$\hat{\beta} = \beta^* + U(\beta^*)I^{-1}(\beta^*) \tag{6.21}$$

An iterative process is applied which starts by choosing a value for β^*, usually $\beta^* = 0$ or, if there is a good guess on $\hat{\beta}$, a value in the vicinity of it. At the first step, given β^*, a value for $\hat{\beta}$ is calculated by equation (6.21). This value is then used in a second step, instead of β^*, in the right hand side of the equality (6.21), to obtain a second value for $\hat{\beta}$. The process stops at convergence to a particular value of $\hat{\beta}$, having fixed the desired degree of approximation. Another criterion is to use the value of the likelihood function at each new value $\hat{\beta}$ as a criterion for convergence. In this case, the process stops when the likelihood stabilizes at a certain value, with a given approximation. This criterion is better behaved in the presence of collinearity between regressors. The iterative approach assumes that the log-likelihood function has first and second derivative, which is true in most cases.

We start the iterative process for the estimation of β in the Cox model applied to the given data set. The initial value is chosen to be $\beta^* = 0$ and consequently

$$U(0) = \sum_{j=1}^{17} s_j - d_j \frac{n_{Aj}}{n_{Aj} + n_{Bj}} = \sum_{j=1}^{17} s_j - d_j p_{Aj} \tag{6.22}$$

and

$$I(0) = \sum_{j=1}^{17} d_j [p_{Aj} - p_{Aj}^2] \tag{6.23}$$

where $p_{Aj} = n_{Aj}/(n_{Aj} + n_{Bj})$ is the proportion of patients in the 6-MP group here indexed with A. For each j ($j = 1, \ldots, 17$), table 6.2 gives the corresponding contribution to $U(0)$ and $I(0)$ calculated on the 42 subjects. The values obtained are $U(0) = -10.2505$ and $I(0) = 6.5957$ which give a first estimate for $\hat{\beta} = -1.5541$. When this value is substituted into (6.16), a second estimate, $\hat{\beta} = -1.5088$, is obtained. By continuing the iterative process, the third, fourth and fifth step give for $\hat{\beta}$ the values $-1.509\ 191$, $-1.509\ 192$ and $-1.509\ 192$ respectively. Thus we conclude that, with an approximation of the order of 10^{-5}, the process has converged after three iterations to give the maximum likelihood estimate for $\hat{\beta} = -1.509\ 19$.

By (6.17), $e^{\hat{\beta}}$ estimates the relapse rate of the treated group ($x = 1$) with respect to the placebo group ($x = 0$):

$$\frac{\lambda(t, 1)}{\lambda(t, 0)} = e^{\hat{\beta}} = 0.2211$$

This result indicates that treatment with 6-MP is likely to reduce the failure rate to nearly a quarter of the rate experienced by untreated patients. However,

Table 6.2 Computation of $U(\beta^*)$ and $I(\beta^*)$ with $\beta^* = 0$ for the estimation of treatment effect on the data set from Freireich *et al.* (1963) according to the Cox model.

j	$p_{Aj} = \dfrac{n_{Aj} \exp \beta^*}{n_{Aj} \exp \beta^* + n_{\beta j}}$	$s_j - d_j p_{Aj}$	$d_j(p_{Aj} - p_{Aj}^2)$
1	$21/42 = 0.5$	$0 - 2 \times 0.5 \quad = -1.000$	$2(0.500 - 0.25) \quad = 0.500$
2	$21/40 = 0.525$	$0 - 2 \times 0.525 = -1.050$	$2(0.525 - 0.276) = 0.499$
3	$21/38 = 0.553$	$0 - 1 \times 0.553 = -0.553$	$1(0.553 - 0.305) = 0.247$
4	$21/37 = 0.568$	$0 - 2 \times 0.568 = -1.135$	$2(0.568 - 0.322) = 0.491$
5	$21/35 = 0.6$	$0 - 2 \times 0.6 \quad = -1.200$	$2(0.600 - 0.360) = 0.480$
6	$21/33 = 0.636$	$3 - 3 \times 0.636 = \quad 1.091$	$3(0.636 - 0.405) = 0.694$
7	$17/29 = 0.586$	$1 - 1 \times 0.586 = \quad 0.414$	$1(0.586 - 0.344) = 0.243$
8	$16/28 = 0.571$	$0 - 4 \times 0.571 = -2.286$	$4(0.571 - 0.327) = 0.980$
9	$15/23 = 0.652$	$1 - 1 \times 0.652 = \quad 0.348$	$1(0.652 - 0.425) = 0.227$
10	$13/21 = 0.619$	$0 - 2 \times 0.619 = -1.238$	$2(0.619 - 0.383) = 0.472$
11	$12/18 = 0.667$	$0 - 2 \times 0.667 = -1.333$	$2(0.667 - 0.444) = 0.446$
12	$12/16 = 0.750$	$1 - 1 \times 0.750 = \quad 0.250$	$1(0.750 - 0.563) = 0.188$
13	$11/15 = 0.733$	$0 - 1 \times 0.733 = -0.733$	$1(0.733 - 0.538) = 0.196$
14	$11/14 = 0.786$	$1 - 1 \times 0.786 = \quad 0.214$	$1(0.786 - 0.617) = 0.168$
15	$10/13 = 0.769$	$0 - 1 \times 0.769 = -0.769$	$1(0.769 - 0.592) = 0.178$
16	$7/9 = 0.778$	$1 - 2 \times 0.778 = -0.556$	$2(0.778 - 0.605) = 0.346$
17	$6/7 = 0.857$	$1 - 2 \times 0.857 = -0.714$	$2(0.857 - 0.735) = 0.245$

$$U(0) = \sum_{j=1}^{17} [s_j - d_j p_{Aj}] = \quad 10.2505, \quad I(0) = \sum_{j=1}^{17} d_j(p_{Aj} - p_{Aj}^2) - 6.5957$$

for this figure to be meaningful, a proper test on the null hypothesis, or the related confidence intervals, should be calculated, as will be shown in a later section.

Table 6.3 shows the computation of the baseline survival curve $\hat{S}_0(t)$ according to (6.14). In this case $S_0(t)$ represents the expected survival of subjects on placebo. The corresponding curve for the 6-MP group is obtained on the basis of the estimate $\hat{\beta}$ by (6.15). These two curves obtained by the Cox model under the assumptions of PH and of a multiplicative effect of treatment are plotted in figure 6.6. They are contrasted there with the Kaplan–Meier curves estimated on the two groups. The good agreement of the results indirectly supports the assumptions underlying the Cox model.

6.4.2 Relationship between the score test and the log-rank test

When the Cox model is applied to analyse the influence of a single binary covariate, as in this example, the score test of the null hypothesis $H_0: \beta = 0$ takes

Table 6.3 Survival curves for patients treated with placebo ($\hat{S}_0(t)$) and 6-MP ($\hat{S}_1(t)$) estimated according to the Cox model.

	Estimated probability of survival for groups	
j	Placebo $=\hat{S}_0(t)$	6-MP $=\hat{S}_1(t)=\hat{S}_0(t)^{0.2211}$
1	$\left(1-\dfrac{2}{21+21\times0.2211}\right)\times1 \quad = 0.922$	0.982
2	$\left(1-\dfrac{2}{19+21\times0.2211}\right)\times0.922=0.844$	0.963
3	$\left(1-\dfrac{1}{17+21\times0.2211}\right)\times0.844=0.805$	0.953
4	$\left(1-\dfrac{2}{16+21\times0.2211}\right)\times0.805=0.727$	0.932
5	$\left(1-\dfrac{2}{14+21\times0.2211}\right)\times0.727=0.649$	0.909
6	$\left(1-\dfrac{3}{12+21\times0.2211}\right)\times0.649=0.532$	0.870
7	$\left(1-\dfrac{1}{12+17\times0.2211}\right)\times0.532=0.498$	0.857
8	$\left(1-\dfrac{4}{12+16\times0.2211}\right)\times0.498=0.370$	0.803
9	$\left(1-\dfrac{1}{8+15\times0.2211}\right)\times0.370=0.337$	0.786
10	$\left(1-\dfrac{2}{8+13=0.2211}\right)\times0.337=0.275$	0.752
11	$\left(1-\dfrac{2}{6+12\times0.2211}\right)\times0.275=0.212$	0.709
12	$\left(1-\dfrac{1}{4+12\times0.2211}\right)\times0.212=0.180$	0.684
13	$\left(1-\dfrac{1}{4+11\times0.2211}\right)\times0.180=0.152$	0.659

Table 6.3   (*Continued*)

j	Estimated probability of survival for groups	
	Placebo $= \hat{S}_0(t)$	6-MP $= \hat{S}_1(t) = \hat{S}_0(t)^{0.2211}$
14	$\left(1 - \dfrac{1}{3 + 11 \times 0.2211}\right) \times 0.152 = 0.124$	0.630
15	$\left(1 - \dfrac{1}{3 + 10 \times 0.2211}\right) \times 0.124 = 0.100$	0.601
16	$\left(1 - \dfrac{2}{2 + 7 \times 0.2211}\right) \times 0.100 = 0.044$	0.501
17	$\left(1 - \dfrac{2}{1 + 6 \times 0.2211}\right) \times 0.044 = 0.006$	0.324

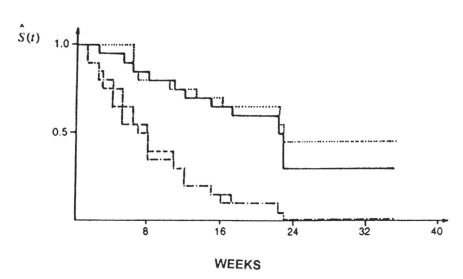

Figure 6.6   Survival curves for the placebo group according to K–M (------) and the Cox model -·-·-) and for the 6-MP group according to K–M (......) and the Cox model (———).

the form

$$Q_R = \frac{U(0)^2}{I(0)} \tag{6.24}$$

which is deduced from (5.35) for $\beta = 0$.

The expressions for $U(0)$ and $I(0)$ were reported in (6.22) and (6.23) respectively. A closer look at their form allows us to relate the score test on β to the log-rank test for treatment comparison (section 4.2). The number d_j of ties at $t_{(j)}$ is in principle the sum of the number of subjects failing at that time in both the treated $(x = 1)$ and untreated $(x = 0)$ groups, that is $d_j = d_{Aj} + d_{Bj}$. Let n_j be the total number of subjects at risk at $t_{(j)}$, that is $n_j = n_{Aj} + n_{Bj}$. With the binary covariate x defined as above, the sum s_j of the covariate values of the d_j subjects turns out to be the number of failures in the treated group: $s_j = d_{Aj}$. In addition, as in section 4.2, the expected number of failures in group A under the hypothesis of no treatment effect is $E(d_{Aj}) = d_j(n_{Aj}/n_j) = d_j p_{Aj}$ provided that we condition the expectation on the observed total number of failures and on the observed sets at risk in groups A and B at $t_{(j)}$. As a consequence, (6.22) may be written as a sum of (observed $-$ expected) terms, one for each failure time, as follows:

$$U(0) = \sum_{j=1}^{17} [d_{Aj} - E(d_{Aj})]$$

and thus the numerator of (6.24) coincides with the numerator of the log-rank statistics (4.6). The quantity $I(0)$ in (6.23) may be rewritten as

$$I(0) = \sum_{j=1}^{17} d_j \frac{n_{Aj}(n_j - n_{Aj})}{n_j^2} = \sum_{j=1}^{17} \frac{d_j n_{Aj} n_{Bj}}{n_j^2}$$

The jth term in this sum is very similar to that of the variance (4.3) of d_{Aj} calculated under the hypothesis of a hypergeometric distribution. Its value is very near to that in (4.3) if n_j is large and there are few ties. It actually coincides with (4.3) when there are no ties, i.e. $d_j = 1$ for every j. Equivalently, in this latter case, expression (6.23) takes the form of a variance, as for (6.11)

$$I(0) = \sum_{j=1}^{17} \{E(d_{Aj}^2) - [E(d_{Aj})]^2\} = \sum_{j=1}^{17} var(d_{Aj})$$

since $E(d_{Aj}) = 1 \cdot p_{Aj}$. Thus, in absence of tied failure times, the score test (6.24) on the null hypothesis coincides with the log-rank test (4.6) for the two sample comparison of failure rates. This observation shows that the score test based on partial likelihood has a nice interpretation as a rank test on a series of contingency tables, one for each failure time. On the other hand, it provides an insight into the properties of the log-rank test, which in fact performs well against alternative hypotheses within the proportional hazard class (section 4.4). The correspondence between the score and log-rank tests is not limited to the two sample case shown here. It also holds when several groups are compared (section 4.7), provided that G groups are identified by $G - 1$ indicator variables in the Cox regression model.

In the example seen here, the score test statistic takes the value $Q_R = 15.930$ which is obtained from the values of $U(0)$ and $I(0)$ given in table 6.2. The corresponding Q_{M-H} does not coincide with Q_R, due to the presence of tied observations ($Q_{M-H} = 16.793$), but the two statistics lead to similar conclusions on treatment effect.

As a final consideration on the strict relationship between the score test and the log-rank test, note that if $\beta \neq 0$, $U(\beta)$ in (6.18) may still be written as

$$U(\beta) = \sum_{j=1}^{17} [d_{Aj} - d_j p_{Aj}]$$

where now we have the exponentially weighted proportion $p_{Aj} = n_{Aj} e^{\beta}/ (n_{Aj} e^{\beta} + n_{Bj})$. The quantity $d_j p_{Aj}$ is the expected value of d_{Aj} when β is not zero and coincides with the expected value obtained in Appendix A4.3 under the alternative hypothesis $\theta \neq 1$ when the constant of proportionality of hazards θ is expressed as e^{β}.

6.5 THE "IMPLEMENTATION" OF THE COX REGRESSION MODEL

We discuss here the coding of the regressors to be included in model (6.1), omitting for now the discussion on criteria for inclusion (Chapter 9). When implementing the Cox regression model, the choice of the coding system depends on the type of the variable to be considered in the vector **x** on the model assumptions described in section 6.1. Both qualitative and quantitative variables may be analysed with (6.1).

A simple example of qualitative dichotomous covariate indicating two different treatments (or gender) was given in previous sections. The coding 0–1 was adopted for the two categories in both cases, since this coding produces a simple and easy to interpret formulation of the rate ratios. For any binary covariate, coding 0–1 is the most commonly used. However, other codes made up of consecutive integers, for example 1–2, could be adopted as well; if the lower value corresponds to the baseline category, the estimate of the regression coefficient would be the same. As far as treatment is concerned, the reference or baseline category is commonly chosen to be the placebo arm or, in its absence, the treatment which is considered the standard one. However, any choice that is

Table 6.4 Two dummy variables are used to indicate three different treatments administered to ovarian cancer patients in a randomized clinical trial. Treatment P is chosen here to be the reference one.

Treatment type	x_1	x_2
CAP	1	0
CP	0	1
P	0	0

regarded as most convenient can be made, as long as the reference category is clearly indicated in the presentation of the results. In the example of figure 6.1, the reference category corresponds to treatment A and the relative effect of B is that of nearly halving the failure rate, given that $\beta = -0.7$. If the model had been applied exchanging codes ($x = 0$ for treatment B), the estimated coefficient would have been $\beta = 0.7$, which equivalently indicates that the failure rate roughly doubles for A with respect to B.

Consider now a qualitative variable which has c categories, with c greater than 2. An example occurs in the randomized trial on ovarian cancer patients which compares three different treatment regimens (see figure 1.4). Suppose we identify the three treatments with codes 0 to 2, assigning 0 to the mono-chemotherapy (platinum = P), 1 and 2 to the polychemotherapies CP and CAP (C = cyclophosphamide, A = adriamycin), respectively. The model with this covariate definition assumes the same constant linear increase ($\beta > 0$) or decrease ($\beta < 0$) between the log hazards related to the treatment groups, as represented in figure 6.5. Unless the trial is designed to estimate and test such a linear trend, the analysis carried out with this variable would be wrong. To correctly estimate treatment effect, two variables are needed to represent two independent contrasts between one treatment chosen as reference and the remaining two. These are usually defined as separate 0–1 variables and are called *indicator* or *dummy variables*. Table 6.4 shows the code chosen to represent treatment in the analysis: the reference category is treatment P and, in the presence of *both* dummies, x_1 represents the contrast of CAP vs P while x_2 represents the contrast of CP vs P. In general, $c - 1$ indicators are necessary to describe c categories.

Consider now a quantitative variable, such as age or the value of some biological features, for instance the white blood count at diagnosis of leukemia or the serum bilirubin and albumin concentrations at diagnosis of biliary cirrhosis. When raw values of a quantitative variable are included in the model, a constant linear increase (or decrease) in the log hazard is assumed, for each unit increase in the value of the variable. Sometimes, appropriate mathematical transformations of the raw values allow for this assumption to be approximately satisfied. Examples of logarithmic transformations of the serum bilirubin and albumin concentrations are given in prognostic indices proposed to predict the outcome of patients with cirrhosis (Christensen, 1987). Often the linear dependence of the log-hazard on the quantitative variable is not believed to hold through its entire range; in this case the predictor may be extended, for instance, to include a quadratic term whose coefficient is tested to detect a possible departure from the linear relationship. Other ways of testing for the log-linear assumption against different types of alternatives can be considered (Thomsen, 1988). For example, let the predictor include the quantitative variable x of interest together with $c - 1$ dummy variables expressing its categorization of x in c classes. Then it can be tested whether the coefficients of these dummy variables are zero; if this hypothesis is rejected, the appropriateness of the log-linear relationship is to be doubted. A more sophisticated approach to this problem consists in using a spline function to model the relationship between response (log-hazard) and predictors (Harrell et al., 1988; Durrleman and Simon, 1989).

In adopting a categorization of a quantitative variable into c classes, $c - 1$ cutpoints must be defined and c groups of subjects are generated, each identified by a specific range of the covariate value. There are different ways of choosing the cutpoints. Substantially we may distinguish between approaches that rely on the data at hand or on the clinical and biological background. An example of the former approach is the definition of cutpoints on the basis of the percentiles of the variable distribution in the sample. If quartiles are calculated, three cutpoints are identified so that each group contains about a quarter of the sample. This method has the advantage of defining equally sized groups, which allows for more precise estimates of the parameters. The latter approach relies on previous knowledge of the disease course for defining cutpoints. For example, the definition of age classes in clinical trials on breast cancer would want to roughly separate women over 50 or 55 years old from younger women, because of the relationship of this particular age with change in menopausal status. In diseases which have been well studied, there might be categories of known prognostic factors which can be considered as fairly standard. An example is the categorization of white blood cell count at diagnosis of leukemia as a measure of "tumor burden", for which it is internationally agreed that cutpoints at 50 000 and 100 000 units should be considered. If such guidelines exist, they are worth adopting, at least in the main analysis, so that published results of different studies become comparable.

In practice the cutpoints of a variable are often chosen as those values which partition patients into groups with most striking differences in outcome. Arguing subsequently on the basis of these categories to demonstrate the prognostic value of the variable itself is questionable. This approach causes the usual hypothesis testing on this variable to be invalid, since the null hypothesis of equivalence is formulated for groups which were identified, by multiple testing, as those with the greatest difference in outcome. Thus suitable approaches for computing the correct p-value related to the obtained classification rule are necessary (Lausen and Schumacher, 1992). On the other hand, exploring the data may be useful to define classification rules and to identify risk groups at different prognosis, though results derived from such an approach should be critically considered and possibly validated on other data sets (Chapter 9).

Having defined c categories for a continuous variable, there remains the choice between considering them as an ordered or purely qualitative classification. Often the latter approach is adopted, which imposes less constraints, and thus $c - 1$ dummy variables are included in the model. However, we may want to maintain the ordered structure of the underlying variable in order to avoid overfitting the data. If for example, the categorization of age is done in a number c of 5- or 10-year classes, we may want to code the variable with values 1 to c and perhaps include also a quadratic term. Analogous problems arise in coding ordinal variables, i.e. discrete variables with a natural ordering of their categories. An example is the Karnofsky index which is frequently used to measure the general health status of a patient with a scale identifying 11 categories: 0 (very good conditions), $10, 20, \ldots, 100$ (very poor conditions). These values are arbitrarily defined and their aim is ranking the patients while not representing a true dimension of equally spaced categories. If the question arises whether thresholds exist in the relationship of response with the ordinal variable, this may be tested by applying the coding system suggested by Walter et al. (1987) and shown in table 6.5 for the Karnofsky index. In fact, when all 10 dummy variables are included in the regression model, the coefficient of $x_c (c = 1, 2, \ldots, 10)$ estimates the log-hazard of subjects in the category $c + 1$ minus the log-hazard of those in category c. Conversely, if one variable only, x_c, is included in the model, its coefficient contrasts the log-hazard of subjects in the set of categories $c + 1$ and above with the log-hazard of subjects in categories c and below. Thus, a forward procedure of variable inclusion may be used to identify, at subsequent steps, the possible thresholds.

Table 6.5 Coding system for an ordinal variable, Karnofsky index, useful to test for the presence of a threshold in the relationship between response and variable.

Karnofsky scale	Category	Dummy variables x_c					
		x_1	x_2	x_3	$\cdots$	x_9	x_{10}
0	1	0	0	0	$\cdots$	0	0
10	2	1	0	0	$\cdots$	0	0
20	3	1	1	0	$\cdots$	0	0
.	.	.	.	.	$\cdots$	.	.
.	.	.	.	.	$\cdots$	.	.
.	.	.	.	.	$\cdots$	.	.
.	.	.	.	.	$\cdots$	.	.
90	10	1	1	1	$\cdots$	1	0
100	11	1	1	1	$\cdots$	1	1

6.6 A FURTHER EXAMPLE

Confidence intervals and hypothesis tests for $\boldsymbol{\beta}$ can be performed in the "usual" way by resorting to the large sample properties of the partial likelihood. Here we show the application of the tests outlined in section 5.6.3 to part of the data from the clinical trial on ovarian cancer (OC) treatment (figure 1.4). A total of 197 women with advanced OC, III and IV FIGO stage tumor, were randomized in the first 20 months of recruitment (between 1/1/81 and 31/8/82), to one of the three treatment regimens. Randomization was carried out in the strata defined by residual tumor size (RTS) after first surgery and stage of tumor (S). The distribution of these characteristics by treatment group is given in table 6.6. Residual tumor size is categorized here as above or below 2 cm of maximum observed diameter. Follow-up was completed in these patients as of 1 March 1986. The trial planned, as a secondary analysis, the evaluation of treatment effect on progression-free interval. Time to failure was thus calculated from randomization to progression or death from any cause, whichever came first.

A Cox regression model is applied here; it includes treatment, coded as in table 6.4 by means of two dummy variables x_1 and x_2, and the two prognostic factors mentioned above. This model reflects the study design and adjusts the estimate of treatment effect by the two relevant prognostic factors, also accounting for the small imbalance shown in table 6.6. Residual tumor size is coded as a 0–1 variable x_3 with values $x_3 = 0$ for subjects with RTS $\leqslant 2$ cm (reference category) and $x_3 = 1$ otherwise. Another 0–1 variable x_4 identifies the stage of the tumor, $x_4 = 0$ for S = III (reference category) and $x_4 = 1$ for S = IV. The unrestricted model fitted to the data is reported in table 6.7, in the upper part (model I). Two regression coefficients, β_1 and β_2, express the effect of treatment on the distribution of time to progression. In order to test whether treatment has an overall effect, we consider the null hypothesis $H_0: \boldsymbol{\beta}_0 = (0, 0, \beta_3, \beta_4)$ in which the regression coefficients for treatment are set to zero. The corresponding restricted model

Table 6.6 Characteristics of 197 patients with advanced ovarian cancer randomized to receive three different chemo-therapeutic treatments, CAP, CP or P.

Characteristics	Treatment		
	CAP no. 68	CP no. 60	P no. 69
Residual tumor size (RTS)			
$\leqslant 2$ cm	25	22	21
> 2 cm	43	38	48
FIGO stage			
III	50	54	55
IV	18	6	14

Table 6.7 Evaluation of treatment effect on progression-free survival for 197 patients with ovarian cancer by means of the Cox model.

Model	Model $\lambda(t, \boldsymbol{x})$	Value of the log-likelihood	Variables		Estimates of regression coefficients β_k	Standard error of β_k
I	$\lambda_0(t) \cdot \exp(\beta_1 x_1 + \beta_2 x_2 + \beta_3 x_3 + \beta_4 x_4)$	−704.3069	x_1	treatment	−0.5709	0.1976
			x_2		−0.3786	0.1974
			x_3	residual tumor size	0.7628	0.1879
			x_4	Figo stage	0.3864	0.2069
II	$\lambda_0(t) \cdot \exp(\beta_3 x_3 + \beta_4 x_4)$	−708.6491	x_3	residual tumor size	0.7958	0.1866
			x_4	Figo stage	0.3149	0.1998

is that in the lower part of the table (model II). The value of the maximum likelihood is reported for both models, together with the estimated regression coefficients and their standard errors.

The likelihood of the sample will increase as additional covariates are included in the model in the attempt to explain the variability of the outcome. The joint contribution of variables x_1 and x_2 to the likelihood is evaluated by applying the likelihood ratio test (5.32) which gives the statistic

$$Q_{LR} = 2(LL_I - LL_{II}) = 2(-704.3069 + 708.6491) = 8.68$$

where subscripts I and II indicate the models in table 6.7. The statistic Q_{LR} is asymptotically distributed as a chi-square with two degrees of freedom and leads to the rejection of the null hypothesis of no overall treatment effect at $p = 0.013$ (and thus significant at a prefixed α level of 0.05). This global test provides evidence of some relationship between treatment and progression-free interval. The fact that both estimates of the regression coefficients are negative suggests that the overall effect of polychemotherapies on the progression-free interval consists in reducing the hazard. There would be no need to adopt the restricted model II if the Wald statistic (5.33) were used to test the hypothesis H_0. Based on the results of the full model I, we find that the estimated covariance matrix of the regression coefficients is

$$\mathbf{I}^{-1}(\hat{\boldsymbol{\beta}}) = \begin{bmatrix} 0.039\,049\,6 & 0.016\,768\,2 & 0.002\,532\,4 & -0.006\,613\,1 \\ 0.016\,768\,2 & 0.038\,947\,5 & 0.000\,146\,4 & 0.003\,960\,2 \\ 0.002\,532\,4 & 0.000\,146\,4 & 0.035\,315\,8 & -0.007\,499\,0 \\ -0.006\,613\,1 & 0.003\,960\,2 & -0.007\,499\,0 & 0.042\,796\,8 \end{bmatrix}$$

Thus the test statistic based on the inverse of the (2×2) submatrix corresponding to the coefficients $\hat{\beta}_1$ and $\hat{\beta}_2$ is

$$\begin{pmatrix} -0.5709 \\ -0.3786 \end{pmatrix}' \begin{bmatrix} 31.416\,61 & -13.525\,91 \\ -13.525\,91 & 31.498\,96 \end{bmatrix} \begin{pmatrix} -0.5709 \\ -0.3786 \end{pmatrix} = 8.91$$

Thus $Q_W = 8.91$, which leads to an analogous conclusion as Q_{LR}. Also the score test (5.35) points in the same direction, with $Q_R = 9.09$.

In order to analyse the single treatment effects, we need to consider the coefficients β_1 and β_2 that, given the coding in table 6.4, respectively contrast the outcome of treatments CAP and CP to that of P. Specifically, the hazard rate for subjects treated with CAP relative to those treated with P is $e^{-0.5709} = 0.57$, residual tumor size and stage being the same. This relationship is formally shown by considering the hazards of two subjects who undergo different treatments (CAP and P), but who have identical class of both residual tumor size and stage (x_3^* and x_4^*). The estimated hazard ratio (HR) from the full model I is

$$\hat{\text{HR}}(\text{CAP vs P}) = \frac{\lambda_0(t)\exp(\hat{\beta}_1\cdot 1 + \hat{\beta}_2\cdot 0 + \hat{\beta}_3\cdot x_3^* + \hat{\beta}_4\cdot x_4^*)}{\lambda_0(t)\exp(\hat{\beta}_1\cdot 0 + \hat{\beta}_2\cdot 0 + \hat{\beta}_3\cdot x_3^* + \hat{\beta}_4\cdot x_4^*)}$$

which gives $e^{\hat{\beta}_1}$ since all other terms cancel out.

Note that the meaning of β_1 depends on the remaining variables in the model. In particular, suppose that x_2 is dropped from the model. Then β_1 has a completely different interpretation, since $e^{\hat{\beta}_1}$ represents the estimated HR between the subjects treated with CAP and the subjects of the two treatment groups CP and P collapsed together. This should warn against the uncontrolled use of computer programs with automatic procedures of variable selection, especially in the presence of dummy variables.

The evaluation of treatment effect has to be done by considering the confidence intervals. Thanks to the large sample properties of the ML estimator $\hat{\beta}$ (section 6.2), the 95% confidence intervals (CI) may be calculated based on the Gaussian distribution as follows:

$$\text{CI}(\beta_1) = -0.5709 \pm 1.96 \times 0.1976 = (-0.1836, -0.9582)$$

From this interval, which is symmetric around the estimate $\hat{\beta}_1$ the confidence interval (0.38, 0.83) for the hazard rate ratio HR is obtained by exponentiation, and thus is not symmetric around $\hat{\text{HR}} = 0.57$. The HR of subjects treated with CP relative to those treated with P is estimated by $\text{HR} = e^{\hat{\beta}_2} = 0.6848$ and the related 95% CI is (0.47, 1.01). Thus the effect of CP seems compatible with the null hypothesis of no difference from P, with 95% confidence, while this is not so for the effect of CAP as compared to P.

This approach is equivalent to the application of the (two-tailed) Wald test on β_1 and β_2 separately. The calculation of the Wald statistics is straightforward, given the values of $\hat{\beta}_1$ and $\hat{\beta}_2$ and their standard errors in table 6.7. We obtain by (5.34):

$$\text{for } \mathbf{H}_0\text{:}\beta_0 = (0, \beta_2, \beta_3, \beta_4), \quad \sqrt{Q_{\text{W}}} = \frac{-0.5709}{0.1976} = -2.89 \quad (p = 0.004)$$

$$\text{for } \mathbf{H}_0\text{:}\beta_0 = (\beta_1, 0, \beta_3, \beta_4), \quad \sqrt{Q_{\text{W}}} = \frac{-0.3786}{0.1974} = -1.92 \quad (p = 0.055)$$

where $\sqrt{Q_{\text{W}}}$ asymptotically behaves like a Gaussian standard deviate.

A third comparison is possible, which is not independent from the two considered above. It is the evaluation of the hazard ratio between CAP and CP treatment groups, indicated by HR(CAP vs CP), and of the pertinent CI. This can be done by considering that:

$$\hat{HR}(\text{CAP vs CP}) = \frac{\hat{HR}(\text{CAP vs P})}{\hat{HR}((\text{CP vs P})} = e^{\hat{\beta}_1 - \hat{\beta}_2}$$

The appropriate standard error can be calculated by considering the covariance matrix of $\boldsymbol{\beta}$, and that $var(\hat{\beta}_1 - \hat{\beta}_2) = var(\hat{\beta}_1) + var(\hat{\beta}_2) - 2cov(\hat{\beta}_1, \hat{\beta}_2)$. The results are $SE(\hat{\beta}_1 - \hat{\beta}_2) = 0.2109$ and $\sqrt{Q_W} = -0.91$ ($p = 0.36$).

Both the analysis on the overall effect of treatment and that on the single effects allow for a type 1 error $\alpha = 0.05$. However, for the two types of analysis to be consequent, if a level $\alpha = 0.05$ is chosen for the overall test, a correction should be made to the α level adopted for testing each single treatment effect. For example, according to Bonferroni, the individual significance levels need not be greater than $\alpha/3$ for the overall significance level to be α. Since the Bonferroni correction is relatively conservative, other methods of adjusting the α level in multiple testing may be considered (Schweder and Spjotvoll, 1982; Prentice *et al.*, 1984; Dixon and Divine, 1987). It should be noted that the overall test on $H_0: \beta_0 = (0, 0, \beta_3, \beta_4)$ suitably accommodates the multiplicity of treatment and does not depend on the choice of the reference category. Such a test is valid whether or not the multiplicative assumption on the hazard holds, although it is more powerful if the assumption is approximately met. What is important is that the overall test has shown heterogeneity and, according to the single tests, this is accounted for mainly by the contrast of CAP vs P, the other two contrasts, CP vs P and CAP vs CP, not being statistically significant.

6.7   ADJUSTED SURVIVAL CURVES

The Cox regression model allows us to obtain an estimate of the treatment effect adjusted by prognostic covariates. This adjustment is useful, especially in non-randomized comparative studies, where imbalance of relevant prognostic factors in treatment groups is often the rule. In this case, the adjusted estimate of treatment effect may be substantially different from the unadjusted one. The problem faced here is to devise a graphical representation of the survival curves by treatment group which accounts for the adjustment, given that the usual K–M curves reflect the unadjusted analysis.

Consider for example figure 6.7 representing the K–M survival curves for a hypothetical data set containing the survival times of patients who were treated with either a standard (S) or an experimental (E) regimen: this figure suggests that treatment E is related to a longer survival. However, in the data set, a characteristic is present which has two modalities, A and B, and is markedly unbalanced in the treatment groups, as shown in table 6.8. If this characteristic is relevant to prognosis, and modality A is related to a longer

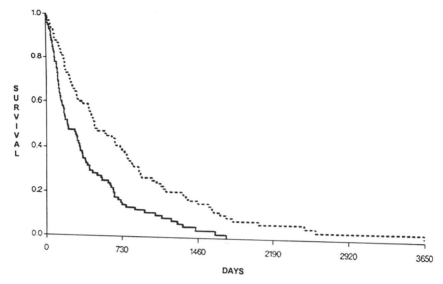

Figure 6.7   K–M survival curves for a hypothetical set of patients treated with a standard treatment (S, ——) and an experimental one (E, ------).

survival, we might ascribe to treatment E an improvement in outcome which is in fact due to the favourable patient selection.

An unadjusted estimate of the relative effect of E is obtained by fitting a Cox model with one covariate only, for treatment, as shown in table 6.9, in model I. Treatment E seems to be related to a nearly 50% reduction of the hazard rate (HR = 0.53). However, the adjusted estimate obtained by fitting model II indicates that the hypothesis of no treatment effect cannot be rejected on the basis of the observed survival data. The difference in outcome is mostly explained by the prognostic factor, with the modality B being related to a threefold increase of the hazard of death. The likelihood ratio statistic for the prognostic factor ($Q_{LR} = 57.96$, one degree of freedom) shows that it significantly contributes to explaining the survival experience, while the additional contribution of treatment is negligible ($Q_{LR} = 0.34$, one degree of freedom).

At this point, it would be useful to have a graphical representation of the survival curves by treatment according to the adjusted analysis. It is possible to do this by considering that, in the absence of censoring, the Kaplan–Meier survival curve calculated on N subjects can be obtained by averaging the N survival curves $\hat{S}_i(t)$ calculated separately on each individual i ($S_i(t)$ has value 1 until the individual i is alive and drops to zero at time of death):

$$\hat{S}_{K-M}(t) = \frac{1}{N}\sum_i \hat{S}_i(t)$$

Table 6.8 Hypothetical data set with patients treated with two different regimens S and E, and with a marked unbalance of a characteristic in treatment groups. Frequency table of treatment by characteristic with modalities A and B.

Treatment	Characteristic		Total
	A	B	
S	20	80	100
E	70	30	100
Total	90	110	200

Table 6.9 Different Cox regression models fitted to the hypothetical data set in table 6.7.

Model	Covariates	Regression coefficients	$\widehat{HR}$	Q_W	Q_{LR}
I	Treatment				
	$S(x_1 = 0)$				
	$E(x_1 = 1)$	-0.6358	0.53	17.01	16.98
II	Characteristic				
	$A(x_? = 0)$				
	$B(x_2 = 1)$	1.2081	3.35	40.32	57.96
	Treatment				
	$S(x_1 = 0)$				
	$E(x_1 = 1)$	-0.1019	0.90	0.34	0.34

Consider now that the Cox model produces an estimate of the survival curve (sections 6.3 and 6.4) for the "typical" individual in any of the four categories identified by treatment and prognostic factor. We extend the idea above to obtain the adjusted survival curves for groups S and E:

$$\bar{S}_S(t) = f_A \hat{S}_{SA}(t) + f_B \hat{S}_{SB}(t)$$

$$\bar{S}_E(t) = f_A \hat{S}_{EA}(t) + f_B \hat{S}_{EB}(t)$$

(6.25)

where f_A and f_B are the relative frequencies of subjects with modality A and B respectively in the sample ($f_A = 90/200$, $f_B = 110/200$), and $\hat{S}_{SA}(t)$ and $\hat{S}_{SB}(t)$ are the probabilities of surviving beyond t predicted on the basis of the Cox model (II) for subjects who received the standard treatment S and with the characteris-

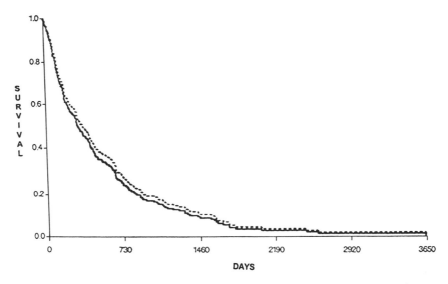

Figure 6.8   Direct adjusted survival curves obtained for the hypothetical set of patients treated with a standard treatment (S, ——) and an experimental one (E, ------).

tic A and B respectively: $\hat{S}_{SA}(t) = \hat{S}_0(t)$ and $\hat{S}_{SB}(t) = [\hat{S}_0(t)]^{\exp \hat{\beta}_2}$; $\hat{S}_{EA}(t)$ and $\hat{S}_{EB}(t)$ are the corresponding curves in the experimental group: $\hat{S}_{EA}(t) = [\hat{S}_0(t)]^{\exp \hat{\beta}_1}$ and $\hat{S}_{EB}(t) = [\hat{S}_0(t)]^{\exp (\hat{\beta}_1 + \hat{\beta}_2)}$.

The technique adopted in (6.25) is called "direct adjustment" (Makuch, 1982): in fact the "corrected" survival curve is a weighted average of the curves specific to the strata defined by the prognostic factor, with weights defined by a common reference population. The term "direct adjustment" is used in epidemiology when two sets of death rates are compared with respect to a reference population and thus its use in this context seems appropriate.

The "corrected" survival curves by treatment obtained on the basis of model II are shown in figure 6.8. This "extreme" example highlights the usefulness of the estimates (6.25): the graphical representation contributes additional evidence of the non-significant effect of treatment which may be difficult to convey on the basis of regression estimates and test statistics to a non-expert audience. The reference population is represented here by the entire sample and the adjusted curves $\bar{S}_S(t)$ and $\bar{S}_E(t)$ may be interpreted as the probability of surviving beyond t for an individual drawn at random from the reference population and treated with S and E, respectively.

A generalization of (6.25) to a number G of treatments and M of strata defined by one or more prognostic factors is straightforward. The direct adjusted survival curve for treatment $g \, (g = 1, 2, \ldots, G)$ is

$$\bar{S}_g(t) = \sum_{m=1}^{M} f_m \hat{S}_{gm}(t) \qquad (6.26)$$

Here f_m is the proportion of the entire study population in stratum m and $\hat{S}_{gm}(t)$ is the probability of surviving beyond t estimated by (6.15) for treatment g and stratum m. The variances of the estimates (6.26) are shown in Appendix A6.1. In particular, as discussed by Gail and Byar (1986), the variance of the difference between the adjusted estimates of any two treatment groups may be used to test the null hypothesis of no treatment effect.

Another adjustment procedure considers the survival function predicted by the Cox model for a patient having characteristics equal to the average of the values of the characteristics in the sample. For simplicity, let the first covariate x_1 in vector $\mathbf{x}$ be the identificator of two treatment groups and the remaining covariates $x_2, \ldots, x_K$ be those related to prognostic factors. The survival predicted for a hypothetical subject with mean vector $\bar{x}$ of covariates is

$$\hat{S}(t, \bar{\mathbf{x}}) = [\hat{S}_0(t)]^{\exp \hat{\boldsymbol{\beta}}' \bar{\mathbf{x}}}$$

where $\hat{S}_0(t)$ is the baseline survival function obtained in (6.13) or (6.14) for $\mathbf{x} = \mathbf{0}$, and $\hat{\boldsymbol{\beta}}$ are the estimated regression coefficients. The survival estimates for patients with "average characteristics" in treatment groups $S(x_1 = 0)$ and $E(x_1 = 1)$ are obtained as

$$\hat{S}_S(t) = [\hat{S}(t, \bar{\mathbf{x}})]^{\exp [\hat{\beta}_1 (0 - \bar{x}_1)]} \quad \text{and} \quad \hat{S}_E(t) = [\hat{S}(t, \bar{\mathbf{x}})]^{\exp [\hat{\beta}_1 (1 - \bar{x}_1)]} \quad (6.27)$$

Note that, in the example seen above, the two adjusted curves can be obtained from the survival curve $\hat{S}(t, 0, \bar{x}_2)$ predicted by the Cox model for the subjects with treatment S (identified by $x_1 = 0$) and mean covariate $\bar{x}_2 = 0.55$ as follows:

$$\hat{S}_S(t, \bar{x}_2) = \hat{S}(t, 0, \bar{x}_2) \quad \text{and} \quad \hat{S}_E(t, \bar{x}_2) = [\hat{S}(t, 0, \bar{x}_2)]^{\exp(-0.1019)}$$

Although it is used by some authors, $\hat{S}(t, \bar{\mathbf{x}})$ is not recommended since it has no natural interpretation. The difference between the adjustment procedures (6.26) and (6.27) is that in the former the averaging is done on the survival curves directly while in the latter it is done on the prognostic factors. The curve $\bar{S}_g(t)$ in (6.26) is to be preferred since it makes allowance for the variability of the covariate values in the sample, whereas $\hat{S}_g(t)$ in (6.27) does not (Thomsen *et al.*, 1991).

The direct adjustment technique may prove useful in any study requiring a comparative analysis adjusted by prognostic or confounding factors. For instance, the presentation of results from historical comparisons may benefit from the use of this technique, while not overcoming the limitations of these types of studies (section 2.1). If a group of patients with a given disease is being followed up, (6.26) may be used to calculate the survival we may expect in this group on the basis of the results of a historical regression analysis conducted on another group of patients with the same disease (Thomsen *et al.*, 1991). The comparison of observed and expected survival curves can be performed by applying the one-sample log-rank test (Harrington and Fleming, 1982). The software for applying the direct adjustment technique for treatment comparison in a trial or in a study with a historical control is available with a procedure developed by Silvestri *et al* (1993).

Direct adjustment of survival curves is a general procedure which can be applied to a variety of regression models, and not only to the proportional hazards Cox model.

6.8   EXTENSIONS OF THE BASIC COX MODEL

The basic Cox model (6.1) we have discussed up to now has three characteristics: (1) proportional hazards; (2) fixed regression covariates; and (3) multiplicative effect of covariates on the hazard.

In this section, extensions of model (6.1) which partly modify these characteristics are presented. Specifically, the introduction of *stratification* relaxes assumption (1); the inclusion of *time-dependent covariates* modifies feature (2) and finally the use of a link other than the exponential modifies assumption (3).

6.8.1   Stratification

Suppose that different levels of a given factor produce hazard functions which seriously depart from proportionality. An extension of model (6.1) may accommodate this by considering the stratification of the data into subgroups, each identified by a level of the factor, and applying the model

$$\lambda_m(t, \mathbf{x}) = \lambda_{0m}(t) \exp(\boldsymbol{\beta}'\mathbf{x}) \qquad (6.28)$$

where the suffix m indicates the stratum ($m = 1, \ldots, M$) and M is the number of levels of the factor (Kay, 1977; Kalbfleisch and Prentice, 1980, pp. 87–89).

Model (6.28) assumes that individuals within the mth stratum who have different covariates $\mathbf{x}_1$ and $\mathbf{x}_2$ still have proportional hazards:

$$\frac{\lambda_m(t, \mathbf{x}_1)}{\lambda_m(t, \mathbf{x}_2)} = \exp[\boldsymbol{\beta}'(\mathbf{x}_1 - \mathbf{x}_2)]$$

However, individuals in different strata are permitted to experience non-proportional hazards since the baseline hazard functions $\lambda_{01}(t), \ldots, \lambda_{0M}(t)$, one for each stratum, are arbitrary functions of time and are left unrelated. Note that the regression parameters $\boldsymbol{\beta}$ do not depend on the strata, thus the effect of the covariates is assumed to be the same in all strata. If $\boldsymbol{\beta}$ were permitted to vary, the strata would be considered as different data sets to be analysed separately.

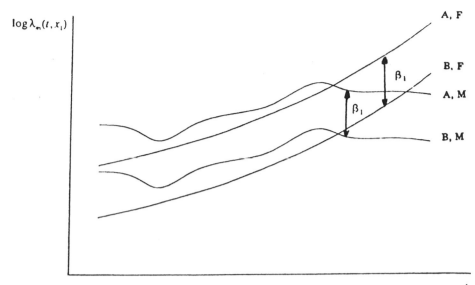

Figure 6.9   Plot of the logarithm of the hazard functions where strata ($m = 1, 2$) are defined by gender (F or M) and the regression covariate x_1 is treatment ($x_1 = 0$ for treatment A and $x_1 = 1$ for treatment B). In this context the Cox model (6.28) would be appropriate.

Consider an example with two binary covariates, say treatment (A or B) and gender. Unlike figure 6.2, where the PH assumption holds for both covariates, figure 6.9 shows a hypothetical setting where model (6.28) would be appropriate. Gender determines two strata for which the PH assumption is not met and the effect of treatment is represented by β_1 in both strata.

In (6.28) the regression coefficients $\boldsymbol{\beta}$ are estimated by maximizing a partial likelihood function which generalizes that in (6.7) and is

$$L(\boldsymbol{\beta}) = \prod_{m=1}^{M} L_m(\boldsymbol{\beta})$$

Each factor $L_m(\boldsymbol{\beta})$ is the partial likelihood built according to (6.7) within each stratum with the contributions calculated at each distinct failure time observed in the stratum. The loss of efficiency encountered in the estimation of $\boldsymbol{\beta}$ when stratification is used unnecessarily is quite limited (Kalbfleisch and Prentice, 1980, p. 112). On the other hand, from a practical viewpoint, unnecessary stratification precludes the estimation of the effect of the stratifying factor.

Hypothesis testing on regression coefficients can be carried out with the usual tests based on the properties of the likelihood function. In particular, given the relationship exploited in section 6.4.2, the score statistic calculated under the hypothesis of no treatment effect coincides with the stratified log-rank test of section 4.5 in the absence of ties. Separately in each stratum, the baseline cumulative hazard $\Lambda_{0m}(t)$ and the function $S_{0m}(t)$ are estimated by (6.12) and by (6.13) or (6.14), provided that the failure times are those observed on the subjects in the stratum m.

The Cox model with stratification can be used to test the validity of the PH assumption (section 7.1).

6.8.2   Time-dependent covariates

A time-dependent covariate is an explanatory variable whose value may change over time, thus it has value $x_i(t)$ for individual i at time t. The basic Cox model may be extended to include such a covariate in the linear predictor: the hazard function shall depend on the value $x_i(t)$ at time t, on the value $x_i(t')$ at time t' and so on. Estimation of the corresponding regression coefficient can still be made on the basis of the partial likelihood function, suitably modified to account for the changing value of $x(t)$. This will be briefly shown at the end of the present section, after discussing some aspects of the application of the Cox model with time-dependent covariates. What is crucial to the correct use of this model is the understanding of the type of dependence of the covariate value on time. We broadly distinguish two major types of time-dependent covariates which, according to Kalbfleisch and Prentice (1980, pp. 122–127), are identified as *external* or *internal* covariates. The values that an *external* covariate takes in time are not influenced by the life experience of the individual in the study, since the values are generated by a "mechanism" which is external to the individual. Conversely, the value of an *internal* covariate at time t carries information about the life experience of the individual up to that time. Thus an external covariate is not involved with the failure process, while an internal one is. Of course, both types of covariates may influence the failure process in the study, and affect prognosis.

External covariates

An external covariate whose behaviour in time is *defined* by a predetermined law or mechanism is, for example, the age of an individual. Usually, age is considered as a fixed covariate with value measured at the beginning of the observation (diagnosis, treatment start) since trials have a limited duration. However, a study of long duration could require the inclusion of age as a time-dependent covariate if it is thought its influence on the hazard changes remarkably over time. An important use of defined time-dependent covariates, which will be discussed in section 7.1, is for testing the assumption of proportional hazards in the Cox model with fixed covariates.

Another type of external covariate, termed *ancillary*, has values which result from a stochastic process external to the individual. For example, in industrial life-testing, a stress factor which amounts to a quantity $x(t)$ randomly determined by some external process is applied to a component at time t. A medical example is that of a covariate which describes the shift of a patient from control group to treatment group, if such shift is randomly determined by some external factor. In this case the covariate is an indicator defined, for each individual i, as

$$x_i(t) = \begin{bmatrix} 0 & \text{if subject } i \text{ is in the control group at } t \\ 1 & \text{if subject } i \text{ is in the treatment group at } t \end{bmatrix} \qquad (6.29)$$

Consider a programme of allogeneic bone marrow transplantation devised for patients with leukemia who experience a relapse in first remission and achieve second remission. At achievement of second remission, patients are screened for admission to the programme according to defined eligibility criteria. When a donor becomes available, the assignment is *randomly* made to an eligible patient among those who satisfy the predefined criteria on donor–recipient tissue matching. The reason for patients in the programme not to undergo transplant is failure to stay in remission until a suitable donor is available. If interest focuses on the comparison of the failure rates of transplanted and non-transplanted patients, the two treatment groups cannot be considered as if they were defined on entry. This procedure would be inappropriate since it introduces a selection bias favouring transplant: the transplanted group would include patients surviving in remission long enough for a donor to be available and meanwhile an undue proportion of bad prognosis patients, failing to survive in remission the waiting period, would be assigned to the non-transplanted group.

The use of a time-dependent indicator such as that defined in (6.29) allows one to correctly account for the mechanism of treatment allocation. Specifically, time is chosen to run from the date of admittance to the programme, as the time origin must be a point where all individuals are comparable. For a transplanted patient, $x(t)$ has value zero from the date of admission to the programme until the date of transplant, when it takes value one; for a non-transplanted patient, $x(t)$ has value zero at entry and does not change thereafter.

Disregarding fixed covariates, for the moment, the Cox model (6.1) with $x(t)$ taken to satisfy the log-linear dependence of hazard on covariates is

$$\lambda(t, x(t)) = \lambda_0(t) \exp[\beta x(t)] \qquad (6.30)$$

This expression defines the hazard function for individual i as

$$\lambda_i(t, x(t)) = \lambda_0(t)$$

if at time t after admission to the programme, he has not been transplanted, and as

$$\lambda_i(t, x(t)) = \lambda_0(t) e^{\beta}$$

if, by that time, he has been transplanted. Thus, a relatively simple relationship is taken to hold by (6.30) between the transplanted and control group, that is the transplant, when applied, multiplies the baseline hazard by a constant c^{β}. If this assumption holds, an unbiased estimate of the effect of transplantation on survival is obtained: if $\beta > 0$, transplantation is associated with a worse outcome, while if $\beta < 0$ transplantation is of benefit. Model (6.30) could be generalized to study in strata the sets of compatible patients, all waiting for a commonly acceptable donor. The stratified Cox model would be:

$$\lambda_m(t, x(t)) = \lambda_{0m}(t) \exp[\beta x(t)]$$

with the stratum m identified by a specific pattern of the matching variables.

It is worth noting that the condition of random assignment is important to ensure that the time-dependent covariate is "external". Otherwise "unspoken" internal criteria may be applied to choose among potential recipients, if more than one, who satisfy all the matching criteria. For example, in the absence of randomization, a physician may preferentially choose to transplant a patient who had a less severe illness or was more responsive during treatment. Thus the resulting estimate of transplant effect could be biased, in spite of the use of a time-dependent covariate to correct for the waiting time to transplant. In this case $x(t)$ cannot be regarded as a truly external covariate since, in some ways, its value is influenced by a latent variable affecting survival. A more extended discussion on the problems related to the use of model (6.30) for the analysis of transplant data is postponed to the example.

Internal covariates

An internal covariate takes values which result from a stochastic process generated by the individual on which the covariate itself is measured. Thus the value of the covariate $x_i(t)$ at each t is determined by the life experience of individual i up to that time point.

An example arises in clinical trials in which a patient's response to treatment is periodically monitored and classified in terms of tumor mass. The problem of comparing the survivorship of responders and non-responders was discussed in section 4.6, where a non-parametric test (Mantel and Byar, 1974) was presented. The basic idea developed there, that responders must be considered non-responders from the start of observation to the detection of response, is easily accommodated with a time-dependent variable defined as in (6.29). In this case the time for $x(t)$ to switch its value from 0 (non-responder) to 1 (responder) corresponds to the time at which response was detected.

Another example of internal covariate arises in trials where the investigator is interested in analysing the outcome accounting for the cumulative dosage of drugs actually received. This is a time-dependent variable, just because the longer individuals survive, the higher the cumulative dosage tends to be. At any specified point in time, the value $x_i(t)$ represents the dosage received by individual i from treatment start to t.

Variables which are thought to be relevant to prognosis are often measured repeatedly during a study. For example, in a trial aiming at evaluating the effect on survival of different treatments for the control of hypertension, blood pressure may be measured at the beginning of treatment and subsequently at every follow-up visit. Thus we have a variable $x_i(t)$ which is the value of the blood pressure of individual i measured at the most recent visit before t. The time-dependent variable may be included in the model when it is thought that the hazard function at a specified time point is influenced more by the blood pressure measured on the subject near this point than by that measured at the beginning of treatment.

The Cox model with internal time-dependent variables is sometimes misused, and considerable care must be taken in interpreting the results of a model including such covariates. As a simple example, suppose that two cancer treatments induce different percentages of initial response to treatment, but that the failure rate within responders and non-responders is the same for both treatments. Suppose, furthermore, that those who fail to respond initially have a much higher failure rate. The linear predictor in the Cox model is taken to include a fixed-dummy variable for treatment and a time dependent variable for response, this latter being included with the purpose of adjusting the estimate of treatment effect. This implies the comparison of the instantaneous failure rates in the two treatment groups, within the sets of subjects with the same covariate values at t (see the partial likelihood construction in the next section). Thus it is likely that the estimated regression coefficient for treatment will not be much different from zero, while the coefficient of the time-dependent variable will show that responders have a better survival. However, it would be misleading to conclude that the two treatments do not differently influence survival. The model alone is not sufficient to describe the relationship between treatment effect, response and survival, although it can be of aid in understanding the mechanism by which the treatment is acting. An integrated approach could consider the results of two other separate analyses, evaluating the effect of treatment on response, taken as the dependent variable, and on overall survival, without adjusting for response.

As a general warning, the investigator must be aware that, in therapeutical trials, an internal time-dependent covariate takes values after treatment assignment and the values may be influenced by the treatment administered. Consequently, the inclusion of $x(t)$ into the model may mask the treatment difference on survival.

Internal time-dependent covariates may be useful in the analysis of competing risks, as will be illustrated in section 10.8.

Inference with time-dependent covariates

The distinction between external and internal time-dependent covariates is conceptually important in that it may help in avoiding gross errors in the modelling process. However, the estimation procedure based on the partial likelihood treats both types of covariates in the same way. If a covariate $x(t)$ is included in the model, all information from data at time t is taken conditionally on the actual value of the variable at t. The logarithm of the partial likelihood is formulated as in (6.8); for sake of simplicity, the vector of regression variables is written as $\mathbf{x}(t)$ to indicate that it may contain one or more time-dependent covariates, the remaining ones being constant in time

$$LL(\boldsymbol{\beta}) = \sum_{j=1}^{J} \left\{ \boldsymbol{\beta}' \mathbf{x}_j(t_j) - \log \left[\sum_{i \in R_j} \exp(\boldsymbol{\beta}' \mathbf{x}_i(t_j)) \right] \right\}$$

The sum is over the distinct failure times and the first term is the contribution of the subject failing at $t_{(j)}$ with $x_j(t_j)$ being his vector of covariate values at $t_{(j)}$. In the second term, the sum runs over all subjects i whose failure times are equal to or greater than $t_{(j)}$. In the presence of ties, $LL(\boldsymbol{\beta})$ above modifies as (6.8a).

The likelihood collects the information only on the instantaneous failure rate, given the actual realization of the time-dependent variable at every $t_{(j)}$. In the presence of internal time-dependent covariates, the approach of conditioning on the realization $x(t)$ implies considering the updated value of x at t and neglecting the information on how the covariate value has been changing up to time t. However, the history of the covariate x may be a relevant aspect of the problem being studied. This happens, for instance, in studies which analyse the relationship between survival and the value of serial cancer markers (Gail, 1981) or in studies which aim at modelling the hazard of developing AIDS after HIV infection as a function of the history of markers such as the $CD4 + T$ cells/μl count (Brookmeyer and Gail, 1993, Chapter 5). In such cases the model should include a function of the time-varying variable which describes a potentially useful aspect of its history. For example, it could be useful to answer the question: "is the hazard at time t higher for those patients who have a high value of the marker six months earlier?". This can be done by including the variable $Z(t) = x(t - t^*)$ in the linear predictor (with $t^* = 6$ months) instead of $x(t)$. If we were interested in studying whether the increase (or decrease) in the value of x between time $(t - t^*)$ and t or the slope of the change in that interval influences survival at t, we would apply the Cox model with the functions $z(t) = x(t) - x(t - t^*)$ or $z(t) = (x(t) - x(t - t^*))/t^*$, respectively, in the predictor. More than one function of the history of x could be considered in the same model, to condition the hazard rate estimate at t on different aspects of the history of x before t.

Large sample tests on regression coefficients are performed with the approach shown in section (5.6.3). In particular the score test on a single time-dependent indicator of the type in (6.29) is equivalent in principle to the Mantel–Byar test of section 4.6 and, in the absence of ties, the two test statistics have the same value.

The baseline survivor function $S_0(t) = \exp[-\int_0^t \lambda_0(u)\,du]$ can still be estimated by (6.13) or (6.14) provided that, in each term of the product, the value of the covariates $\mathbf{x}(t)$ at $t_{(j)}$ is taken. However, there is no clear and useful interpretation of $S_0(t)$ in the presence of internal time-dependent covariates (see also section 4.6). With external covariates only, $\hat{S}_0(t)$ can still be used to describe the survival experience of subjects at the standard condition $\mathbf{x} = \mathbf{0}$.

Altman and De Stavola (1994) discuss the practical problems encountered when analysing the effect on survival of time-dependent covariates and give an up-to-date review of the software which allows such analysis with the Cox model. It is worth mentioning that the study of survival in the presence of time-dependent covariates has led to the application of multistate models in which transition probabilities between different states of the disease may be modelled under PH (Andersen, 1986).

An example on transplantation

The aim of this example is to illustrate the interpretation of the regression coefficient of a time-dependent covariate in different models. The data set comes from a retrospective study which is being conducted in Italy to evaluate the impact of bone-marrow transplantation in the treatment of recurrent leukemia. Here we consider the records on 155 children with acute lymphoblastic leukemia who were diagnosed as having a first relapse in the bone-marrow in eight AIEOP (Associazione Italiana Ematologia Oncologia Pediatrica) centres in Northern Italy, between 1/2/80 and 31/12/89 and achieved second remission thereafter with an induction treatment. A subsequent maintenance chemo-therapeutic regimen was the standard treatment; however, 36 patients who had a suitable donor underwent an allogeneic transplantation according to the GITMO (Gruppo Italiano Trapianto Midollo Osseo) schedule. These data do not come from a transplant programme with a precise definition of admission criteria and random allocation of donors. Thus, many potential biases exist which do not allow firm conclusions on the effect of transplant as compared to standard treatment of a leukemic relapse. This is true even if regression methods or stratified analysis are applied to adjust for treatment imbalances on known prognostic variables. Here, however, we will not stress substantive conclusions, but only describe the mechanics of the application of the Cox model with a time-dependent covariate as in (6.29) for transplantation and with two regression covariates, namely duration of first remission and time from second remission to transplantation.

For all patients, failure time T is calculated from the date of second remission to relapse or death, whichever comes first, and the cut-off date for the analysis is 31 December, 1992. Time from second remission to transplantation, when this occurred, is regarded here as the "waiting time" to transplant and is indicated with W. The values of T and W in the data set are in days.

Table 6.10 Results of different models fitted to the example on transplant: estimated regression coefficients and, in parentheses, standard errors.

Model	$-2LL^a$	Transplant status $x_1(t)$	Duration of first remission x_2	Interaction terms $x_3(t) = x_1(t) \cdot x_2$	Interaction terms Waiting time to transplant, $x_4(t)$
1	1104.79	−0.3974 (0.2501)			
2	1067.98	−0.4583 (0.2517)	−1.1492 (0.1972)		
3	1067.24	−0.6221 (0.3245)	−1.2305 (0.2208)	0.4269 (0.4923)	
4	1063.30	−1.0971 (0.4123)	−1.1689 (0.1979)		0.0042 (0.0018)
5	1101.17	−0.9836 (0.4168)			0.0037 (0.0018)

[a]The value of minus twice the log-likelihood in the absence of covariates is $-2LL = 1107.49$.

Consider the application of model (6.30) with $x_1(t)$ defined as in (6.29) to indicate the transplant status as represented in figure 6.10. The estimated regression coefficient and the standard error, in the first line of table 6.10, provide mild evidence of a better outcome for transplanted patients: as regards the null hypothesis $\beta_1 = 0$, there is a 10% level of significance on a two-tailed test according to the likelihood ratio test.

Duration of the first remission is regarded as a prognostic indicator since, clinically, a relapse occurring during treatment or soon after treatment stops is usually interpreted as the expression of a more aggressive disease then a relapse occurring later on. As the scheduled front-line therapy lasted two years overall, duration of first remission is categorized into a binary variable: $\leqslant 30$ months ($x_2 = 0$) and > 30 months ($x_2 = 1$) from the beginning of treatment at first diagnosis of leukemia. Model 2 in table 6.10 is

$$\lambda(t, \mathbf{x}(t)) = \lambda_0(t) \exp[\beta_1 x_1(t) + \beta_2 x_2)$$

It measures the effect of transplantation adjusting by duration of first remission and the result is in agreement with that of model 1. This analysis shows that late first relapses are related to a one-third decrease of the failure rate with respect to early first relapses ($\hat{\beta}_2 = -1.1492$; the likelihood ratio test for $\beta_2 = 0$ gives $Q_{LR} = 36.8$, $p < 0.001$).

Consider now model 3 which adds to model 2 a covariate $x_3(t) = x_1(t) \cdot x_2$ taking value zero until a patient is non-transplanted and value x_2 after transplant. The coefficient β_3 of this variable may be regarded as measuring the interaction between transplantation and duration of first remission. Unlike in models 1 and

$x(t)$

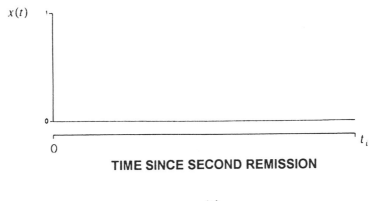

TIME SINCE SECOND REMISSION

(a)

$x(t)$

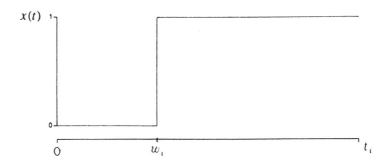

TIME SINCE SECOND REMISSION

(b)

Figure 6.10 Graphical representation of the time-dependent covariate (6.29) (a) for a non-transplanted patient with failure time t_i and (b) for a patient transplanted at time w_i and with failure time t_i.

2, in this model β_1 alone cannot be taken as the estimate of transplant effect, since this estimate depends also on the value of x_2 and is given by $\hat{\beta}_1 + \hat{\beta}_3 x_3(t)$. While the main effect of x_2 is still significantly different from zero (the Wald test for $\beta_2 = 0$ gives $Q_W = 31.06$, $p < 0.001$), the interaction term does not reach significance (for $\beta_3 = 0$, $Q_{LR} = 0.74$) and the overall estimate of transplant effect can thus be retained from model 2.

In models 1 to 3, the effect of transplantation on relative hazard is modelled assuming that it is independent of waiting time W. In other words, the change in $\lambda(t)$ given by the factor e^{β_1} is taken to be the same regardless of when transplantation occurs, that is for patients with different waiting times w_i. It is only when this assumption is true that the time-dependent covariate $x_1(t)$ adjusts for the waiting time bias.

On the other hand, it is likely that a longer duration of second remission is a marker of a less severe disease and of a better prognosis in standard treatment. Thus we are prompted to ask whether the relative effect of transplant is different for patients with different waiting times. In this data set the cumulative pre-transplant hazard, empirically estimated on all patients, with censoring of transplanted patients at w, shows a decrease in the slope after 7–8 months from second remission (figure 6.11). This estimate assumes that transplantation is an independent censoring mechanism, which is likely not to be the case given that these data are not protected by randomization. If, for instance, transplants were performed preferentially on the patients who, among the possible recipients, were healthier or had waited longest, the decline in the hazard, especially for longer durations, is likely to be more marked than that suggested by the curve in figure 6.11. To account for waiting time in the analysis, model 2 is extended to include a variable which is zero before transplant and changes to its actual value w_i at the time individual i is transplanted. This produces model 4, in which this latter variable can be viewed as an interaction term between $x_1(t)$ and waiting time W. In this case, necessarily, model 4, unlike model 3, includes an interaction term but not the corresponding main effect. In other words, as non-transplanted patients cannot be classified by waiting time, the comparison between the two groups cannot be done within "strata" defined by the same value w. This problem arises for any other variable, such as tissue matching, which applies only to transplanted patients. The coefficients of these time-dependent variables can be interpreted only if it is assumed that they affect post-transplant failure but not the overall failure, which may be reasonable for a tissue-matching variable but less so for the waiting time. In models 4 and 5, take β_4 to be the coefficient of $x_4 = x_1(t) \cdot w$: here β_4 has the meaning of a slope and β_1 corresponds to an intercept of a linear function in w. Thus any significant test on β_1 would be meaningless in these models. In model 4 it is estimated that the longer the waiting time, the less is the improvement, if any, achieved with transplant over the standard treatment ($\hat{\beta}_1 = -1.0971$; $\hat{\beta}_4 = 0.0042$ and the test for $\hat{\beta}_4 = 0$ gives $Q_{LR} = 4.68$, $p < 0.05$). For instance the hazard rate ratio for transplant versus chemotherapy is estimated to be $\hat{HR} = 0.49$ and $\hat{HR} = 0.71$ for $w = 90$ days and $w = 180$ days, respectively.

The interpretation of the coefficient of the interaction term in the absence of the corresponding main effect is unclear because no comparable control group is identified. The problem of selecting comparable controls for groups of trans-

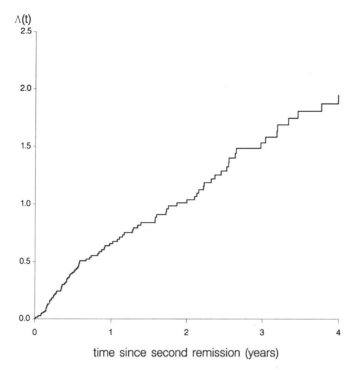

Figure 6.11 Aalen–Nelson estimate of the cumulative pre-transplant hazard for 155 children with leukemia in second remission after medullary relapse.

planted patients characterized by different waiting times could be faced as follows.Transplantees could be stratified according to waiting time intervals and controls for each stratum identified by randomly selecting among non-transplantees individuals who had survived at least as long as the interval waiting time. The stratified M–H test (4.13) or equivalently the stratified Cox model (6.28) can then be applied for comparing transplanted and non-transplanted patients.

We refer the reader who is interested in the special topic of the analysis of transplantation data to the interesting debate in the literature of the last two decades on data from the Stanford Heart Transplantation Program. In particular, the Cox model with time-dependent covariates was used in Crowley and Hu (1977) and Kalbfleisch and Prentice (1980, pp. 135–140), while parametric regression models were applied by Aitkin, *et al.* (1983), whose paper is followed by thoughtful comments by Gail (1983), Crowley and Storer, (1983) and Arjas (1986).

6.8.3   A family of proportional hazards models

Model (6.1) assumes there is a log-linear dependence of hazard on covariates. A general formulation of a proportional hazards model does not necessarily include the exponential link $\exp(\boldsymbol{\beta}'\mathbf{x})$ and a broad family of PH models is defined by

$$\lambda(t, \mathbf{x}) = \lambda_0(t)g(\mathbf{x}) \qquad (6.31)$$

Here $g(\mathbf{x})$ is any function which satisfies the requirement $g(\mathbf{0}) = 1$ and $\lambda_0(t)$ is the baseline hazard for an individual under the standard conditions $\mathbf{x} = \mathbf{0}$. If function $\lambda_0(t)$ is left unspecified, model (6.31) is semiparametric and the partial likelihood is adopted for inference. The hazards of any two individuals are proportional and their ratio is a constant depending on covariates only. In analogy with (6.4) and (6.6) this ratio is

$$\frac{\lambda(t, \mathbf{x})}{\lambda(t, \mathbf{0})} = g(\mathbf{x})$$

and

$$S(t, \mathbf{x}) = [S_0(t)]^{g(\mathbf{x})} \qquad (6.32)$$

where $S_0(t)$ is the baseline survivor function. Also, the general expression for the density function is

$$f(t, \mathbf{x}) = g(\mathbf{x})f_0(t)[S_0(t)]^{g(\mathbf{x})-1}$$

A model of the family (6.31) is, for instance:

$$\lambda(t, \mathbf{x}) = \lambda_0(t)(1 + \boldsymbol{\beta}'\mathbf{x}) \qquad (6.33)$$

Here a one-unit increase in the value of a covariate x_k is related to the *addition* of the quantity β_k in the proportionality constant. Figure 6.12 plots the logarithm of the hazard functions according to model (6.33) with one variable only, x_1, taking values from 0 to 4 and with $\beta_1 > 0$. By comparing this figure with figure 6.5 which illustrated model (6.1) with the same covariate x_1, similarities and differences in the assumptions of the two models should be clear. As with the basic Cox model, hazard functions are parallel on a logarithmic scale. However, while in figure 6.5 the exponential link produces the same difference β_1 between any two log-hazards, here the distance between successive functions declines as the covariate value increases.

Group	Values of covariate x_1	$\lambda(t, x_1)$ $= \lambda_0(t)(1 + \beta_1 x_1)$
0	0	$\lambda_0(t)$
1	1	$\lambda_0(t)(1 + \beta_1)$
2	2	$\lambda_0(t)(1 + 2\beta_1)$
3	3	$\lambda_0(t)(1 + 3\beta_1)$
4	4	$\lambda_0(t)(1 + 4\beta_1)$

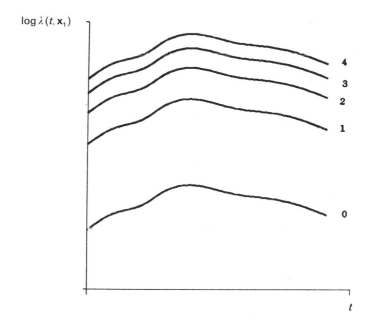

Figure 6.12 Plot of the logarithm of the hazard functions according to a PH model with additive link for one variable x_1 taking values from 0 to 4 and with $\beta_1 > 0$.

It is very difficult to gain empirical evidence to support the use of a particular function for $g(\mathbf{x})$. Thus simplicity is the criterion of choice in the modelling process and this explains why the log-linear assumption is that most widely used. This latter is easier to deal with also from a technical point of view, in that maximization of the likelihood function when $g(\mathbf{x})$ is bound to be positive for any value of $\boldsymbol{\beta}$ gives fewer problems in convergence. We applied model (6.33) to the data set in the example of section 6.6, represented in table 6.6. When fitting the model with two variables for residual tumor size and FIGO stage, there were in fact convergency problems in the estimation of the regression coefficient of the

Table 6.11   Results of models (6.33) and (6.1) applied to the data set on 197 patients with advanced ovarian cancer. One covariate x for FIGO stage ($x = 0$ for Stage III and $x = 1$ for Stage IV) is considered and the estimate of the hazard ratio (HR̂) given.

Model	Estimate of			
	Log-likelihood	Regression coefficient	Standard error	HR̂
$\lambda(t, x) = \lambda_0(t)(1 + \beta x)$	-718.59	0.6263	0.3206	1.6263
$\lambda(t, x) = \lambda_0(t)\exp(\beta x)$	-718.59	0.4864	0.1971	1.6264

first variable. The model with FIGO stage only as covariate was then fitted to the data set and its results are given in table 6.11. As the link functions are different, the estimates of β and of its standard error differ in the two models. The hazard ratio estimate in model (6.33) is

$$\text{HR(Stage IV vs Stage III)} = 1 + \beta = 1.6263$$

and coincides with the estimate $\hat{\text{HR}} = e^\beta = 1.6264$ given by the Cox model (6.1), apart from rounding errors. This is in fact to be expected if the two models include one dichotomous covariate, as the constant of proportionality between the two groups has just been parametrized differently. On the other hand, for a continuous covariate, the two parametrizations reflect different assumptions on the relationship between groups and are likely to lead to different estimates of HR.

APPENDIX A6.1   ASYMPTOTIC VARIANCE OF $\hat{\Lambda}_0(t)$ AND $\hat{S}_0(t)$

Consider the baseline cumulative hazard estimate $\hat{\Lambda}_0(t)$ given in (6.12). The stochastic process $\sqrt{N}(\hat{\Lambda}_0(t) - \Lambda_0(t))$ is shown (Tsiatis, 1981b) to converge weakly to a Gaussian process with zero mean and variance function that can be consistently estimated by

$$\hat{\sigma}_0 = N \left\{ \sum_{t_{(j)} \leq t} \frac{1}{\left(\sum_{i \in R_j} \exp(\hat{\boldsymbol{\beta}}'\mathbf{x}_i) \right)^2} + \hat{\psi}'(t)I^{-1}(\hat{\boldsymbol{\beta}})\hat{\psi}(t) \right\}$$

where

  (1) $\mathbf{x}$ is a vector of K covariates;

(2) $\hat{\Psi}'(t) = (\hat{\psi}_1(t), \dots, \hat{\psi}_K(t))$ has kth component

$$\hat{\psi}_k(t) = -\sum_{t_{(j)} \leqslant t} \left[\sum_{i \in R_j} x_{ki} \exp(\hat{\beta}'\mathbf{x}_i) \Big/ \left(\sum_{i \in R_j} \exp(\hat{\beta}'\mathbf{x}_i) \right)^2 \right]$$

(3) $\hat{\beta}$ is the vector of K estimates obtained by maximizing the partial likelihood;

(4) $\mathbf{I}^{-1}(\hat{\beta})$ is the variance–covariance matrix of $\hat{\beta}$, i.e. the inverse of the observed negative second derivative matrix of the log partial likelihood.

Hence the variance of the cumulative hazard $\hat{\Lambda}_0(t)$ at time t is estimated by $\hat{\sigma}_0/N$ which reduces to (6.16) when only one covariate x is included in the model.

Consider now the variance of the survival function estimated by (6.15) for subjects with a certain vector $\mathbf{x}^*$ of K covariates. Asymptotic results for $\sqrt{N}(S(t, \mathbf{x}^*) - \hat{S}(t, \mathbf{x}^*))$ are derived by applying the delta method (see Appendix A3.2). The variance of $S(t, \mathbf{x}^*)$, generalized to consider the presence of a few tied observations d_j at $t_{(j)}$, can be estimated by

$$var(\hat{S}(t, \mathbf{x}^*))$$

$$= [\hat{S}(t, \mathbf{x}^*)]^2 \exp(2\hat{\beta}'\mathbf{x}^*) \left[\sum_{t_{(j)} \leqslant t} \frac{d_j}{\left(\sum_{i \in R_j} \exp(\hat{\beta}'\mathbf{x}_i) \right)^2} + \hat{\rho}'(t, \mathbf{x}^*) \mathbf{I}^{-1}(\hat{\beta})\hat{\rho}(t, \mathbf{x}^*) \right]$$

where $\hat{\rho}(t, \mathbf{x}^*)$ is a vector of K components, the kth one being

$$\hat{\rho}_k(t, \mathbf{x}^*) = \sum_{t_{(j)} \leqslant t} d_j \left[\left(x_k^* - \frac{\sum_{i \in R_j} x_{ki} \exp(\hat{\beta}'\mathbf{x}_i)}{\sum_{i \in R_j} \exp(\hat{\beta}'\mathbf{x}_i)} \right) \cdot \frac{1}{\sum_{i \in R_j} \exp(\hat{\beta}'\mathbf{x}_i)} \right]$$

Gail and Byar (1986) derive the covariance of the estimated survival function at t for subjects with different covariate vectors $\mathbf{x}_1$ and $\mathbf{x}_2$, on the basis of the asymptotic properties of $S(t, \mathbf{x}^*)$. The empirical estimate is

$$cov\{\hat{S}(t, \mathbf{x}_1), \hat{S}(t, \mathbf{x}_2)\} = \hat{S}(t, \mathbf{x}_1) \cdot \hat{S}(t, \mathbf{x}_2) \exp[\hat{\beta}'(\mathbf{x}_1 + \mathbf{x}_2)]\mathbf{V}_1'\boldsymbol{\Sigma}\mathbf{V}_2$$

where

(1) for $i = 1, 2$, $\mathbf{V}_i' = (1, \hat{\Lambda}_0(t)x_{1i}, \dots, \hat{\Lambda}_0(t)x_{Ki})$;

(2) $\boldsymbol{\Sigma}$ is the estimated $(K+1) \times (K+1)$ covariance matrix of $\hat{\Lambda}_0(t)$ and $\hat{\beta}$,

which is obtained as

$$\Sigma = \begin{bmatrix} \Sigma_{11} & \Sigma_{12} \\ \Sigma_{21} & \Sigma_{22} \end{bmatrix} = \begin{bmatrix} var(\hat{\Lambda}_0(t)) & \hat{\mathbf{\Psi}}'(t)\mathbf{I}^{-1}(\hat{\boldsymbol{\beta}}) \\ [\hat{\mathbf{\Psi}}'(t)\mathbf{I}^{-1}(\hat{\boldsymbol{\beta}})] & \mathbf{I}^{-1}(\hat{\boldsymbol{\beta}}) \end{bmatrix}$$

with $\hat{\mathbf{\Psi}}'(t)$ defined as above.

When $\mathbf{x}_1 = \mathbf{x}_2 = \mathbf{x}^*$ the formula $cov\{\hat{S}(t, \mathbf{x}_1), \hat{S}(t, \mathbf{x}_2)\}$ reduces to the variance of $\hat{S}(t, \mathbf{x}^*)$ shown before. Gail and Byar use this estimator to calculate the covariance of the survival curve adjusted for prognostic factors in a Cox model (section 6.7).

7

Validation of the Proportional Hazards Models

7.1 MODEL SELECTION

The complex process of modelling cannot be fully formalized into a standard strategy and only some guidelines may be given to help the modeller. Usually, in starting the process, we select to fit models from a particular class which is considered to be relevant to the kind of data we have. Then the steps which we will usually do are the following:

(1) check if the assumptions of the selected class of models hold in the data set, preliminary to model fitting;
(2) define a basic plan for selecting covariates to be included in the linear predictor, if many explanatory variables are candidates for inclusion;
(3) assess the goodness-of-fit, that is the adequacy of the model to describe the outcome variable after fitting the model itself.

We should be satisfied with the results of the check in step (1) in order for the process to continue with the following steps; these, in turn, may contribute to reviewing some aspects of the original model and prompt us to repeat the full process with a suitably modified model, and so on, until "convergence" to a model which we find useful for the data we have. Conversely, if point (1) is not fulfilled, we should either seek some data transformation to satisfy the assumptions or select an alternative class of models which could be more appropriate; the three steps will then be repeated in the new setting.

By fitting a model we wish to substitute for a set of row data a simple theoretical pattern depending on a number of parameters which is small relative to the number of the original data. We must recognize, however, that "even if we can define exactly what we mean by a optimum model in a given context, it is most unlikely that the data will indicate a clear winner among possible models, and we must expect that clustered around the 'optimum' model will be a set of others almost as good and not statistically distinguishable" (McCullagh and Nelder, 1989, p. 23).

In this chapter we will mainly focus on point (1) and (3) above, while point (2) will be developed in the more general context of Chapter 9. The proportionality of hazards is the fundamental assumption of any model in family (6.31) and it is worth checking as thoroughly as possible in the first run. Secondly, the approximate form of the dependence of hazard on the explanatory variables, i.e. the form of $g(\mathbf{x})$, can be examined. Methods for checking these assumptions include graphical procedures as well as tests against general or specific departures from the PH and from a specific $g(\mathbf{x})$. Finally, residuals or, more generally, regression diagnostics, may be useful for assessing goodness-of-fit and are most often examined by means of graphical display.

7.2 GRAPHICAL METHODS FOR CHECKING MODEL ASSUMPTIONS

A simple graphical check of the proportional hazards assumption which was already used in Chapter 4 (figure 4.2) is discussed first. Any PH model in family (6.31) satisfies relationship (6.32) or, equivalently, the equality

$$- \log S(t, \mathbf{x}) = g(\mathbf{x})[- \log S_0(t)] \tag{7.1}$$

which expresses the proportionality of the cumulative hazard functions of two individuals with covariate vectors $\mathbf{x}$ and $\mathbf{0}$, given the relation (5.7). Consequently, the logarithm of these functions exhibits a constant distance $\log g(\mathbf{x})$:

$$\log [- \log S(t, \mathbf{x})] = \log g(\mathbf{x}) + \log [- \log S_0(t)] \tag{7.2}$$

In particular, the distance is $(\boldsymbol{\beta}'\mathbf{x})$ for the basic Cox model (6.1). Consider two different covariate patterns, i.e. two sets of values $\mathbf{x}_1$ and $\mathbf{x}_2$ for the vector $\mathbf{x}$. Under PH, the functions $\log[- \log S(t, \mathbf{x}_1)]$ and $\log[- \log S(t, \mathbf{x}_2)]$ would both exhibit a constant distance from the reference cumulative hazard $\log[- \log S_0(t)]$ and would therefore be parallel. This property can be used for a graphical check of the PH in the presence of K covariates. Consider the M subgroups defined on the basis of all the observed combinations of the values taken by the K variables, so that the mth subgroup will contain individuals with common covariate pattern $\mathbf{x}_m$. In each subgroup, the survival curve $S_m(t)$ is estimated by a non-parametric method (product limit). If the plots of the M log cumulative hazards $\log[- \log \hat{S}_m(t)]$ against t appear nearly parallel, the hazards of the M groups can be assumed to be proportional given that (7.2) holds. Thus a model of the family (6.31) with the K covariates can be fitted.

An alternative type of plot, which may be found easier to interpret, consists in plotting $\log [- \log \hat{S}_m(t)]$ for each pattern $\mathbf{x}$ of covariates different from $\mathbf{0}$ versus

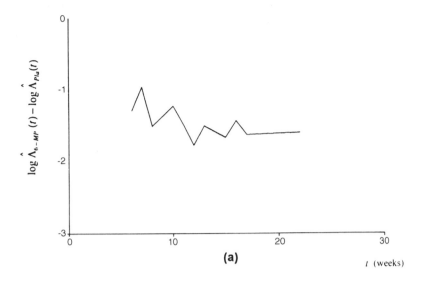

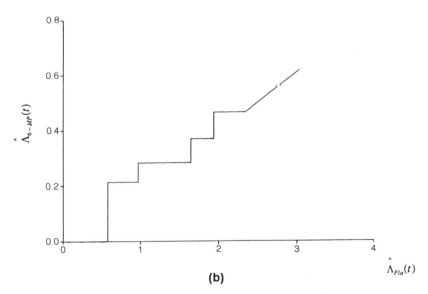

Figure 7.1 Two types of plot alternative to that in figure 4.3 for investigating the PH assumption with respect to treatment in the Freireich *et al.* (1963) data set. (a) Plot of the difference between the estimated log cumulative hazards of the 6-MP group and the placebo group versus time; (b) plot of the estimated cumulative hazard of the 6-MP group versus that in the placebo group.

$\log[-\log S_0(t)]$. On the basis of (7.2) we shall see roughly straight lines, all with slope one and shifted by a constant.

Figure 4.2 showed the first type of graphical check related to one dichotomous covariate only and thus resolved in the plot of two curves. Graphical checking for PH in the Freireich *et al.* (1963) example is also done here by other types of plots which, in the case of a two-groups comparison ($M = 2$), may be easier to interpret (figure 7.1). One consists in plotting the difference between log estimated cumulative hazards against time which, in the presence of PH, should result in a line roughly parallel to the time axis by (7.2). Alternatively, the estimated cumulative hazard, $-\log \hat{S}(t, \mathbf{x})$, of one group plotted against that in the other group should by (7.1) give approximately a straight line through the origin, with slope close to the estimated constant of proportionality. Some kind of smoothing of these curves might help in the interpretation. The conclusion that the PH assumption is reasonable for the analysis of treatment effect is supported by both plots, as it was by the plot in figure 4.2.

Most commonly, in clinical studies, there are several candidate covariates to be analysed. To keep the graphical representation simpler, one could explore the PH for each covariate separately. However, this procedure could miss the relationship between the hazards of groups identified by different patterns of a set of covariates. On the other hand, problems in considering these plots may arise due to the paucity of the data in each subgroup when several covariates are simultaneously considered and thus the number M of subgroups is large.

Table 7.1   Characteristics of 476 women from the trial on advanced ovarian cancer in figure 1.4, by treatment arm.

Characteristics	Treatment		
	CAP (157)	CP (162)	P (157)
Residual tumor size after first surgery			
$\leqslant 2$ cm	55	50	46
2–10 cm	56	58	56
> 10 cm	46	54	55
FIGO grade			
1	25	25	22
2	51	68	47
3	81	69	88
Karnofsky index			
$\leqslant 70$	27	22	23
> 70	130	140	134

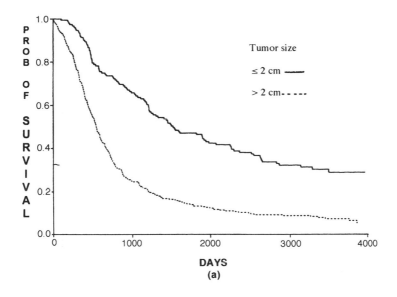

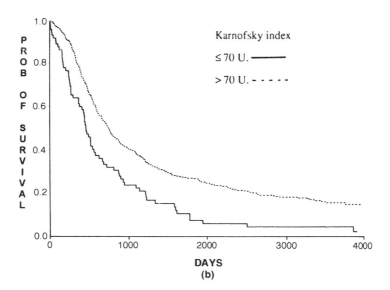

Figure 7.2   Kaplan–Meier survival curves by (a) residual tumor size (⩽ 2 cm; > 2 cm) and (b) Karnofsky index (⩽ 70, > 70 Units) on 476 women with advanced ovarian cancer.

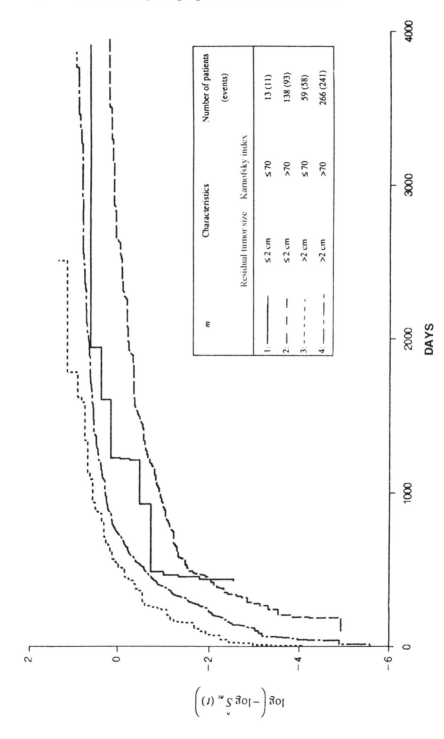

Figure 7.3   Plot of the function $\log[-\log \hat{S}_m(t)]$, where $\hat{S}_m(t)$ is the K–M survival estimate, in the four subgroups of patients identified by residual tumor size and Karnofsky index, both categorized into two classes.

We consider here an example from the advanced ovarian cancer trial in figure 1.4 consisting of the data on 476 women having complete records on various prognostic factors (table 7.1). The data set has been recently updated and the analysis on survival is performed here at the cut-off date of 31 December 1992. By then, a total of 403 deaths were observed, and four patients were lost to follow-up. Note that in this trial the follow-up is fairly long, ranging from a minimum of 7 years to a maximum of 12 years.

Let us explore the adequacy of the proportional hazard assumption for two variables, i.e. residual tumor size after first surgery, expressed as the measure of the maximum diameter observed, and Karnofsky index. When individually considered, these variables are related to prognosis, as shown by the Kaplan–Meier survival curves of figure 7.2.

Four subgroups ($M = 4$) are generated if the two covariates are taken with the simplest categorization into two classes. The product limit survival estimate $\hat{S}_m(t)$ is calculated for each subgroup and the corresponding functions $\log[-\log \hat{S}_m(t)]$, $m = 1, \ldots, 4$, are plotted in figure 7.3. The curve, based on 13 patients, has a quite irregular behaviour and we shall think of some kind of smooth version of it in order to interpret the graph. We can imagine a band around the curve in which various types of smooth curves lie that mimic the behaviour of the original one; if at least one is compatible with the PH assumption we can consider that no major departures from the PH are likely to be present. This seems to be the case for the plot in figure 7.3.

7.2.1   Use of the stratified Cox model

The graphical check may be impractical when the number M of subgroups is high. On the other hand, the exploration of one variable at a time, instead of the set of candidate variables together, may mask the relationship between hazards of different groups. A compromise between the two alternatives is reached by performing the PH check within the framework of the stratified Cox model (6.28).

Among K variables being investigated, each one, say x_k, can be checked for PH as follows. Let the vector $\mathbf{x}$ be partitioned as $\mathbf{x} = (x_k, \mathbf{x}^-)$ where $\mathbf{x}^- = (x_1, \ldots, x_{k-1}, x_{k+1}, \ldots, x_K)$ is the vector of $K - 1$ covariates, the kth one being excluded. For example, with x_k dichotomous (with values 0 or 1), the basic Cox model (6.1) would assume:

$$\lambda(t, \mathbf{x}) = \begin{cases} \lambda_0(t) \exp(\boldsymbol{\beta}'^- \mathbf{x}^-) & \text{for } x_k = 0 \\ \lambda_0(t) \exp \beta_k \exp(\boldsymbol{\beta}'^- \mathbf{x}^-) & \text{for } x_k = 1 \end{cases}$$

that is, the hazards of two individuals with the same $\mathbf{x}^-$ but with different values of x_k are proportional, with ratio $\exp \beta_k$. In order to verify the PH assumption on

x_k, consider the two strata identified by the values of x_k and apply model (6.28). This assumes instead:

$$\lambda(t, \mathbf{x}) = \begin{cases} \lambda_{01}(t) \exp(\boldsymbol{\beta}'^{-}\mathbf{x}^{-}) & \text{for } x_k = 0 \\ \lambda_{02}(t) \exp\boldsymbol{\beta}'^{-}\mathbf{x}^{-}) & \text{for } x_k = 1 \end{cases}$$

where the baseline functions in the two strata are left arbitrary and unrelated. The covariates $\mathbf{x}^{-}$ are assumed to satisfy the assumptions of the Cox model and the coefficients $\hat{\boldsymbol{\beta}}^{-}$ are obtained by maximizing the partial likelihood over the entire sample. The baseline survival functions are estimated separately in each stratum by (6.13) or (6.14) and their transforms $\log[-\log \hat{S}_{01}(t)]$ and $\log[-\log \hat{S}_{02}(t)]$ are plotted against time. If they exhibit a constant difference, the covariate x_k does not violate the PH assumption; thus x_k may be included in the linear predictor and the hazards in the two strata may be reparametrized with an arbitrary common baseline hazard as in the basic Cox model. When this check of PH is applied for a qualitative variable with c modalities, c stratum-specific baseline survival functions are estimated after taking out the $c-1$ corresponding dummy variables from the linear predictor. The PH assumption for a continuous variable may be roughly checked with this approach provided that the variable is categorized into a few classes.

Given a set of K covariates, the use of the stratified Cox model allows us to investigate the PH assumption for each of them while accounting for the

Table 7.2 Names and coding adopted for the covariates investigated for PH in the example on ovarian cancer data.

Characteristics	Covariates	
Chemotherapic treatment	C1	C2
CAP	0	0
CP	1	0
P	0	1
Residual tumor size	RT1	RT2
$\leqslant 2$ cm	0	0
2–10 cm	1	0
> 10 cm	0	1
FIGO grade	G1	G2
1	0	0
2	1	0
3	0	1
Karnofsky index	K1	
$\leqslant 70$	0	
> 70	1	

remaining ones. This provides a systematic procedure for examining the entire set of covariates which are candidates for analysis. It should be kept in mind, however, that the examination of one variable x_k is conditional on all the others satisfying the assumptions of PH and log-linear link. So if one variable is removed from **x** or used for stratification, it is advisable to recheck the remaining variables.

In the example on ovarian cancer data, this approach is adopted to investigate whether some variables, namely residual tumor size after first surgery, FIGO grade, Karnofsky index and treatment, satisfy the PH assumption. Names of these variables and their coding are given in table 7.2.

Figure 7.4 shows the plots of the estimated log cumulative baseline hazards in strata defined in turn by each of the covariates, the others being in the linear predictor of model (6.28). At first glance, no major violations of PH appear to be present since none of the covariates is related to crossing hazards which suggest that as time progresses, one modality turns from beneficial to detrimental (or

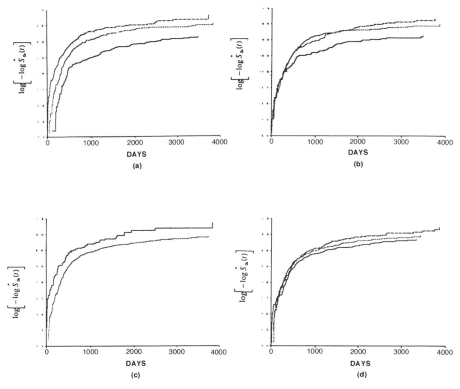

Figure 7.4 Plots of the log estimated cumulative baseline hazards against time in strata defined in turn by different variables: (a) residual tumor size, (b) FIGO grade, (c) Karnofsky index and (d) treatment. In all analyses of the individual variables, the stratified Cox model includes the remaining covariates.

vice versa) relative to others. In plots (a) and (c) the vertical distance between curves seems to show a slight decrease with time, suggesting that the effect of residual tumor size and Karnofsky index on hazard tends to diminish in the long run. On the contrary, in the strata defined by FIGO grade and treatment (plots (b) and (d)), the curves are superimposed in the first part and tend to depart from each other after a few years from surgery; a relatively large distance is observed between the log cumulative hazard of patients with FIGO grade 1 and those of patients with grades 2 and 3, who have instead a very similar prognosis.

Note that FIGO grade is expressed by means of two dummy variables in spite of the fact that it takes values from 1 to 3 and the investigator could be tempted to code it as an ordinal variable. FIGO grade is in fact a qualitative variable, and in any case plot (b) rules out the assumption of a log-linear effect on hazard since the three curves are not separated by similar amounts. In general, when an ordinal variable x is investigated, the plots of the estimated log cumulative hazards in the corresponding $M(M > 2)$ strata may be useful to explore the appropriateness of the link $\exp(\beta x)$. The multiplicative assumption in model (6.1) assumes the covariates behave as in figure 6.5. If plots are nearly parallel but not equally spaced, a transformed coding scheme for x should be attempted, such as a log-transform, to try to avoid this type of departure. The use of other types of link functions could be sometimes considered, as previously discussed (section 6.8.3).

In our example the plots suggest there might be discrepancies from the PH assumption; their relevance, however, cannot be conveyed on the basis of the plots and is worth examining with formal hypothesis tests, as it will be done in section 7.3.

The stratified Cox model is particularly useful to check the proportionality of hazards of different treatment groups while adjusting for prognostic factors which approximately satisfy PH. A method which, in our view, could be considered an extension of that just described, was presented by Dabrowska *et al.* (1992). It relies on graphical checks as well as on test statistics and will be briefly outlined here. For sake of simplicity, assume there are two treatment groups, identified with subscripts 1 and 2, and a vector $\mathbf{x}$ of fixed prognostic variables. The Cox model is fitted, with the same covariates $\mathbf{x}$, to each treatment group (strata) separately:

$$\lambda_i(t, \mathbf{x}) = \lambda_{0i}(t) \exp(\boldsymbol{\beta}_i' \mathbf{x}), \qquad i = 1, 2$$

This approach allows for the hazards ratio between treatments not to be constant while, within each treatment group, the covariate is taken to satisfy the PH assumption. Unlike model (6.28), different sets of coefficients $\boldsymbol{\beta}_1$ and $\boldsymbol{\beta}_2$ are estimated in each stratum, allowing for $\boldsymbol{\beta}_1 \neq \boldsymbol{\beta}_2$. Consider the difference between the log cumulative hazards for the two strata, given a covariate pattern $\mathbf{x}^*$:

$$\rho(t, \mathbf{x}^*) = \log \Lambda_1(t, \mathbf{x}^*) - \log \Lambda_2(t, \mathbf{x}^*)$$

where $\Lambda_i(t, \mathbf{x}^*) = \Lambda_{0i}(t) \exp(\boldsymbol{\beta}_i'\mathbf{x}^*)$ for $i = 1, 2$. This difference can be rewritten as

$$\rho(t, \mathbf{x}^*) = \rho_0(t) + (\boldsymbol{\beta}_1 - \boldsymbol{\beta}_2)'\mathbf{x}^* \qquad (7.3)$$

where $\rho_0(t) = \log \Lambda_{01}(t) - \log \Lambda_{02}(t)$ is related to $\mathbf{x}^* = \mathbf{0}$. If the hazards are proportional also with respect to treatment, the baseline log hazards differ by a constant, i.e. $\rho_0(t) = \theta$. As a consequence, $\rho(t, \mathbf{x}^*) = \theta + (\boldsymbol{\beta}_1 - \boldsymbol{\beta}_2)'\mathbf{x}^*$ is a constant in t, for any choice of the pattern $\mathbf{x}^*$. Thus the plot of $\hat{\rho}(t, \mathbf{x}^*)$ against time, obtained by substituting in (7.3) the estimates of $\boldsymbol{\beta}_1$, $\boldsymbol{\beta}_2$, $\Lambda_{01}(t)$ and $\Lambda_{02}(t)$, would be a line approximately parallel to the time axis.

Note that, if two treatments have the same effect on individuals with covariates $\mathbf{x}^*$, we expect the value of $\hat{\rho}(t, \mathbf{x}^*)$ to be near zero at any t. The authors present an estimator of the asymptotic standard error of $\hat{\rho}(t, \mathbf{x}^*)$ and draw confidence bands for $\rho(t, \mathbf{x}^*)$ to test whether one treatment is better than another, for an individual with pattern $\mathbf{x}^*$. Furthermore, plots of $\hat{\rho}(t, \mathbf{x}^*)$ which are quite apart from each other for different patterns $\mathbf{x}^*$ suggest the possible presence of an interaction between treatment and covariates. Dabrowska *et al.* (1992) develop a test statistic for the null hypothesis of no interaction $H_0: \boldsymbol{\beta}_1 = \boldsymbol{\beta}_2$ which is asymptotically equivalent to that given in Thall and Lachin (1986).

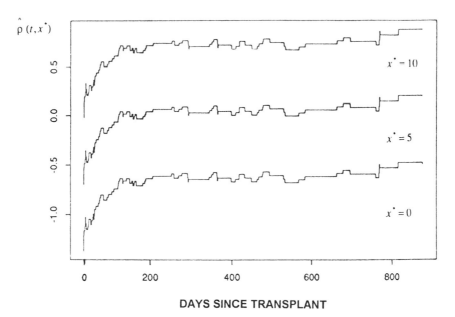

DAYS SINCE TRANSPLANT

Figure 7.5   Estimated log cumulative hazards difference comparing the cyclosporine-based treatment given with non-sequenitial (group 1) and sequential (group 2) modality, when the number of transfused blood units is set to $x^* = 0, 5$ and 10 (from Dabrowska *et al.*, 1992).

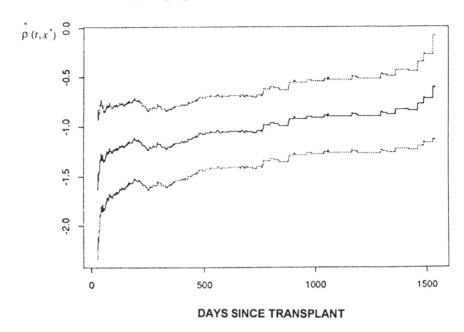

$\hat{\rho}(t,x^{*})$

Figure 7.6 Estimated log cumulative hazards difference (solid line) and 90% confidence bands (dotted lines) comparing the treatments based on prednisone (group 1) and cyclosporine (group 2) when the number of transfused blood units is set to $x^{*} = 5$ (from Dabrowska *et al.*, 1992).

Dabrowska *et al.* illustrate their approach with an example in which they study the effect of different immunosuppressive treatments following kidney transplant. The Cox model was fitted separately to the data from two groups of patients receiving a cyclosporine-based treatment according to two different modalities, indicated as non-sequential (NS) and sequential (S). The failure time was the time from transplant until failure or rejection of the graft and the only covariate in the model was the number of units of blood transfused prior to transplant. Figure 7.5 shows the estimate of (7.3) for patients with number of transfused blood units set at $x^{*} = 0$, $x^{*} = 5$ and $x^{*} = 10$. The three curves are far apart from each other and indicate the presence of a treatment–covariate interaction. The two treatment modalities are related to nearly the same outcome for patients with $x^{*} = 5$ given that $\hat{\rho}(t, 5)$ is close to the horizontal line passing through zero. Since the difference between the log cumulative hazards was estimated by taking groups 1 and 2 in (7.3) to correspond to NS and S, respectively, the plot illustrates that, for patients with $x^{*} = 0$, NS treatment is beneficial, given that $\hat{\rho}(t, 0)$ is negative, while for patients with $x^{*} = 10$, S treatment is beneficial given that $\hat{\rho}(t, 10)$ is positive.

A plot which illustrates a situation of departure from the PH assumption is shown in figure 7.6. Here the estimate of (7.3) and its 90% simultaneous confidence bands are shown for patients with $x^* = 5$ who received the treatment considered above, either non-sequential or sequential (NS + S) or an alternative schedule based on prednisone (PDN), without cyclosporine. The curve $\hat{\rho}(t, 5)$ is calculated by taking groups 1 and 2 to correspond to PDN and NS + S, respectively, after fitting the Cox model separately in the two groups. Since $\hat{\rho}(t, 5)$ is negative, PDN does better but its advantage on treatment with cyclosporine (NS + S) diminishes with time since $\hat{\rho}(t, 5)$ tends towards zero as time increases.

7.3 TESTS ON THE PROPORTIONAL HAZARDS ASSUMPTION

A lot of research work has been done to develop formal tests suitable for assessing the proportionality of hazards. Two major types of approaches to formal testing of the PH hypothesis can be distinguished: tests against a specified alternative and tests which consider a general alternative. A brief description of some tests of both types will be given here, in sections 7.3.1–7.3.3 and 7.3.4, respectively.

7.3.1 Test based on defined time-dependent covariates

A test proposed in the original paper by Cox (1972) relies on the use of *defined* time-dependent variables. Consider testing the PH assumption for the fixed covariate x_k, in the presence of the remaining $K - 1$ covariates $\mathbf{x}^-$. Define a time-dependent transform of x_k by multiplying it by a function $g(t)$ of time, $x_k(t) = x_k \cdot g(t)$, and include it in the linear predictor of model (6.1):

$$\lambda(t, \mathbf{x}) = \lambda_0(t) \exp(\beta_k x_k + \gamma x_k(t) + \boldsymbol{\beta}^{-'} \mathbf{x}^-) \tag{7.4}$$

For patients with the same pattern $\mathbf{x}^-$, the hazard ratio of two individuals with values $x_k \neq 0$ and $x_k = 0$ is $\exp(\beta_k x_k + \gamma x_k(t))$ in model (7.4) instead of the ratio $\exp(\beta_k x_k)$ assumed in model (6.1). In general, a non-zero value of γ would imply a variation in time of the hazard ratio between two individuals with a different value of x_k. The nature of this variation depends on the form chosen for $g(t)$ and, since the aim is testing for PH and not that of closely modelling the influence of x_k in time, the choice of $g(t)$ usually restricted to a few simple monotonic functions of time. Common choices for $g(t)$ are the identity function $g(t) = t$ and its logarithmic transform $g(t) = \log t$. Sometimes, either t or $\log t$ is centred around an arbitrary constant to improve interpretability and avoid instabilities in the calculation of the ML estimates. In any case, with these functions of time a value of $\gamma > 0$ ($\gamma < 0$) indicates that the hazards ratio increases (decreases) linearly in time or log time. Another relatively simple definition of $g(t)$ is a step function which is more efficient for detecting departures from the PH assumption related to hazard ratios which are fairly constant within consecutive time intervals, but change markedly between them.

The example on ovarian cancer data, with the variables of table 7.2 which underwent graphical checking in the previous section, is considered here. Model (7.4) has been applied to this data set with $g(t) = \log t - \log t^*$, with t^* defined to be the mean survival time, which is nearly three years ($\log t^* = 7$). Table 7.3 reports the estimated regression coefficients and standard errors obtained in each of the models fitted and the value of minus twice the maximized partial log-likelihood. The first column of estimated coefficients (and standard errors) reports the results of the basic Cox model while each of the following columns is obtained from a model which includes also the time-dependent transform of the regressors representing each characteristic, one at a time. The likelihood ratio test is adopted to evaluate the influence of the time-dependent covariates, and the statistics Q_{LR}, their asymptotic distribution and the related p-values are given. Only for residual tumor size is the null hypothesis of PH rejected at the conventional significance level of $\alpha = 0.05$ since the corresponding test statistic, asymptotically distributed as a chi-square with two degrees of freedom, is $Q_{LR} = 13.27$ ($p = 0.001$).

According to this analysis the hazard rate ratio related to residual tumor size varies in time and can be estimated at different points by suitably combining the coefficients of RT1, RT2, RT1(t) and RT2(t). Note that the estimated coefficients of the fixed regressors are positive, while those of their time transforms are negative and this indicates a decreasing trend in the hazard ratio as time increases, as was suspected on the basis of figure 7.4(a). For example, the hazard ratio of an individual with residual tumor of size between 2 and 10 cm with regard to an individual with residual tumor of size 2 cm or less, all the other covariates being the same, is estimated at different time points, namely at 1, 3 and 5 years since randomization, as follows:

$$\widehat{HR}\ (1\ \text{year}) = \exp[0.6432 - 0.2294(\log 364 - 7)] = 2.45$$
$$\widehat{HR}\ (3\ \text{years}) = \exp(0.6432) = 1.90$$
$$\widehat{HR}\ (5\ \text{years}) = \exp[0.6432 - 0.2294(\log 1825 - 7)] = 1.69$$

If the analysis had been performed adopting the function $g(t) = \log t$ the resulting values of the test statistics would have been the same. Changes would have occurred only in the estimate of the coefficients of the fixed covariates checked for PH as they correspond to intercepts and thus their coefficients change if $\log t$ is not centred around $\log t^*$. For example, the model with RT1(t) and RT2(t), and with $g(t) = \log t$ gives for RT1 and RT2 the estimated coefficients (and standard errors) 2.2490 (1.06) and 4.5673 (1.11), respectively. These coefficients do not have a meaningful interpretation as they correspond to the hazard ratios at time zero, while those in table 7.3 directly give the three-year estimates of HR.

Table 7.3 Check of the PH assumption in the data on ovarian cancer: defined time-dependent covariates are included, for each characteristic, in the basic Cox model, and their influence on hazard tested with the likelihood ratio statistic Q_{LR} (survival time since randomization is in days).

Characteristics	Regressors	Estimated regression coefficients and (SE)				
Treatment	C1	0.1674 (0.125)	0.1811 (0.151)	0.1697 (0.125)	0.1768 (0.125)	0.1641 (0.125)
	C2	0.2972 (0.124)	0.4529 (0.150)	0.2903 (0.124)	0.3053 (0.124)	0.3037 (0.124)
Residual tumor size	RT1	0.6943 (0.130)	0.7040 (0.130)	0.6432 (0.144)	0.7025 (0.130)	0.6992 (0.130)
	RT2	1.0740 (0.135)	1.0819 (0.135)	0.7297 (0.173)	1.0871 (0.135)	1.0773 (0.136)
FIGO grade	G1	0.3662 (0.169)	0.3804 (0.170)	0.3827 (0.169)	0.3982 (0.192)	0.3698 (0.169)
	G2	0.4609 (0.164)	0.4725 (0.164)	0.4586 (0.164)	0.5917 (0.185)	0.4704 (0.164)
Karnofsky index	K1	−0.3980 (0.135)	−0.3872 (0.135)	−0.3965 (0.135)	−0.3873 (0.135)	−0.1732 (0.186)
	C1(t)		0.0296 (0.119)			
	C2(t)		0.2223 (0.126)			
	RT1(t)			−0.2294 (0.159)		
	RT2(t)			−0.5482 (0.170)		
	G1(t)				0.0967 (0.153)	
	G2(t)				0.2328 (0.150)	
	K1(t)					0.2450 (0.129)
−2 × log-likelihood		4296.08	4292.37	4282.81	4293.13	4292.50
Q_{LR}			$3.71\,\chi^2_2$	$13.2\,\chi^2_2$	$2.95\,\chi^2_2$	$3.58\,\chi^2_1$
(p-value)			(0.157)	(0.001)	(0.229)	(0.059)

Given that the result of the test on RT1(t) and RT2(t) does not justify the PH assumption for residual tumor size, a reasonable model to adopt is the Cox model with strata defined by this variable. The results are given in the third column of table 7.4. The PH assumption related to Karnofsky index is worth testing in this new setting, given that the test in table 7.3 might generate some suspects. Thus the variable $K1(t) = K1 \times (\log t - 7)$ is included in the stratified model and the results, given in the fourth column, do not contradict the PH assumption. Thus the model of choice for the analysis of this data set can be considered the one stratified by residual tumor size. The basic and stratified Cox models, however, give very similar estimates of the regression coefficients pertinent to treatment, as can be seen by comparing the results in tables 7.3 and 7.4 and this is hardly surprising. Although the mismodelling of important prognostic factors can destroy the proportionality that might otherwise hold, the effect on the estimated treatment coefficients is unlikely to be substantial when the treatment has been allocated at random (see section 9.3.2).

Consider now the extreme situation in which PH does not hold because the hazard functions of two groups cross. This, for instance, could be expected in principle when a difficult surgical treatment is compared with a standard non-surgical treatment. Then, an early excess mortality associated with the surgical procedure and its complications is likely to occur; it is expected, however, for this procedure to be competitive, that the mortality rate falls below that of the standard treatment as time progresses. In the case of crossing hazard functions, the use of the form $g(t) = t$ can be used to find a straightforward estimate of the crossing point, under the assumption of a linear decrease or increase in time of the hazards ratio. Let x_1 be a binary covariate and $x_1 \cdot t$ be the additional time-dependent regressor with coefficient β_2. Suppose the estimates of β_1 and β_2 in model $\lambda(t, \mathbf{x}) = \lambda_0(t) \exp [\beta_1 x_1 + \beta_2 (x_1 \cdot t)]$ are significantly different from zero with opposite sign. Then the hazards in the two groups identified by x_1

Table 7.4 Results of the Cox model stratified by residual tumor size without and with a defined time-dependent covariate to verify the PH assumption for the Karnofsky index; the likelihood ratio statistic Q_{LR} is the basis of the test (survival time since randomization is in days).

Characteristics	Regressors	Estimated regression coefficients and (SE)	
Treatment	C1	0.1646 (0.125)	0.1624 (0.125)
	C2	0.2758 (0.124)	0.2802 (0.124)
FIGO grade	G1	0.3687 (0.170)	0.3744 (0.170)
	G2	0.4489 (0.164)	0.4604 (0.164)
Karnofsky index	K1	−0.3952 (0.136)	−0.2328 (0.187)
	K1(t)		0.1725 (0.129)
−2 × log-likelihood		3473.57	3471.80
Q_{LR}			$1.77\chi_1^2$
(*p*-value)			(0.18)

(for example $x_1 = 0$ for standard and $x_1 = 1$ for experimental treatment) have a crossing point. If, for example, $\hat{\beta}_1$ is positive and $\hat{\beta}_2$ negative, a higher mortality is observed in the experimental group in the early follow-up period and the reverse in the subsequent period. The hazard ratio of the experimental versus the standard treatment would be 1 at the crossing point which thus can be estimated by equating $\hat{\beta}_1 + \hat{\beta}_2 \cdot t = 0$ and obtaining the time value $t = -\hat{\beta}_1/\hat{\beta}_2$ as an estimate of the time point where the hazard functions cross.

Defined time-dependent covariates are commonly used for testing PH in model (6.1) with fixed covariates. We could, however, consider their use for checking whether PH is satisfied with regard to time-dependent ancillary indicators of the type introduced in section 6.8.2 and used to compare transplanted and non-transplanted patients. In this comparison non-proportionality could be expected since a high post-transplant mortality rate which rapidly decreases as time continues is likely to occur. Thus we are prompted to complete the analysis done in section 6.8.2 on bone-marrow transplant data by adding a term $x_1(t) \cdot t$ into the Cox model 4 of table 6.10. In the new model, minus twice the log-likelihood value is 1054.03. The estimated coefficients (and standard errors) of transplant status $x_1(t)$ and of its transform $x_1(t) \cdot t$ are -0.1100 (0.4927) and -0.0023 (0.0009), respectively. The estimated coefficients (and standard errors) of the remaining prognostic factors (duration of first remission: -1.1790 (0.1978); waiting time to transplant: 0.0054 (0.0019)) are very similar to those obtained in model 4 of table 6.10. In this analysis, the regression coefficient of $x_1(t) \cdot t$ is found to be significantly less than zero, according to the Wald test ($Q_W = 7.27$, $p = 0.007$), while it is not so for the coefficient of $x_1(t)$ ($Q_W = 0.05$), and both estimates have negative sign. Thus the hazards ratio of transplanted versus non-transplanted patients is always less than 1 and tends to decrease over time, suggesting that there is a long term advantage of bone-marrow transplant over standard chemotherapy.

7.3.2   Gill and Schumacher test

Gill and Schumacher (1987) proposed a test which has good power for testing the proportionality assumption between two groups against the alternative of a monotone increasing (or decreasing) hazards ratio. The key idea behind their test procedures is that "in non proportional hazards situations, different two-sample tests, e.g. the log-rank and a generalized Wilcoxon test, might give different answers. Our test procedures use this discrepancy to check the proportional hazards assumption and are based on the relationship between generalized linear rank tests and estimates of the proportionality constant".

Consider the following null hypothesis of PH between two groups A and B:

$$H_0 : \lambda_A(t) = \theta \lambda_B(t)$$

where θ is a positive constant. It was shown (Andersen, 1983) by means of martingale theory, that each linear two-sample rank test (4.11), in the presence of PH, is related to a consistent and asymptotically normally distributed estimator $\hat{\theta}$ of the hazard ratio. The general expression of this estimator is

$$\hat{\theta} = \frac{\sum_{j=1}^{J} w_j \hat{\lambda}_{Aj}}{\sum_{j=1}^{J} w_j \hat{\lambda}_{Bj}} \tag{7.5}$$

where the sums are over the J distinct ordered failure times observed in the two groups jointly. The hazards $\lambda_{gj} = \lambda_g(t_{(j)})$ in the two groups at each failure time $t_{(j)}$ are estimated according to (3.20), with $g = A$ or B. Denote, as usual, by d_{gj} and n_{gj} the numbers of events and of subjects at risk at $t_{(j)}$ in group g. The weights w_j depend on the type of test and are:

$$w_{1j} = \frac{n_{Aj} n_{Bj}}{n_j} \qquad \text{for the Mantel–Haenzel estimator, } \hat{\theta}^*_{\text{MH}}$$

$$w_{2j} = n_{Aj} n_{Bj} \qquad \text{for the Gehan estimator, } \hat{\theta}_{\text{G}}$$

$$w_{3j} = \frac{n_{Aj} n_{Bj}}{n_j} \cdot \prod_{l=1}^{j} \frac{n_l - d_l + 1}{n_l + 1} \qquad \text{for the Prentice estimator, } \hat{\theta}_{\text{P}}$$

The weights w_{ij} ($i = 1, 2, 3$) satisfy $w_{ij} = 0$ when the number of patients at risk at time $t_{(j)}$ in either group is equal to zero. With w_{1j}, the estimator (7.5) coincides with (4.9a).

As in the test statistics discussed in section 4.4, in estimating θ the choice of w_{2j} gives relatively more weight to the early deaths than w_{1j} while w_{3j} is an "intermediate" weighting between the two above and is less dependent on censoring than w_{2j}. Under proportional hazards these estimators of the hazards ratio will give very similar values. This is not the case under the alternative hypotheses. Gill and Schumacher base their test statistic on the difference between two different estimators and recommend using $\hat{\theta}_{\text{MH}}$ and $\hat{\theta}_{\text{P}}$ as the optimal choice or, as a second choice, requiring fewer calculations, $\hat{\theta}_{\text{MH}}$ and $\hat{\theta}_{\text{G}}$.

Let $\hat{\theta}$ in (7.5) be rewritten, with obvious notation, for the latter couple of estimators as $\hat{\theta}_{\text{MH}} = \hat{K}_{1A}/\hat{K}_{1B}$ and $\hat{\theta}_{\text{G}} = \hat{K}_{2A}/\hat{K}_{2B}$, where a suffix 1 or 2 is added to indicate the type of weights. Instead of considering the difference $\hat{K}_{2A}/\hat{K}_{2B} - \hat{K}_{1A}/\hat{K}_{1B}$, which has expected value zero under H_0, consider the quantity

$$D_{\text{G}-\text{MH}} = \hat{K}_{1B} \hat{K}_{2A} - \hat{K}_{2B} \hat{K}_{1A}$$

which should also be close to zero under H_0 and has the advantage of being symmetric under exchange of indices 1 and 2 and of groups A and B. The asymptotic variance of D_{G-MH} under H_0 is estimated by

$$var\,(D_{G-MH}) = \hat{K}_{2B}\hat{K}_{2A}\hat{V}_{11} - \hat{K}_{2B}\hat{K}_{1A}\hat{V}_{12} - \hat{K}_{1B}\hat{K}_{2A}\hat{V}_{21} + \hat{K}_{1B}\hat{K}_{1A}\hat{V}_{22}$$

where

$$\hat{V}_{ii'} = \sum_{j=1}^{J} w_{ij} w_{i'j} \frac{d_{Aj} + d_{Bj}}{n_{Aj} \cdot n_{Bj}}$$

where i, $i' = 1$, 2 indicate the type of weights. The test statistic

$$T_{G-MH} = \frac{D_{G-MH}}{[var\,(D_{G-MH})]^{1/2}} \tag{7.6}$$

is shown to have asymptotically a standard Gaussian distribution in large samples. Analogously, the test statistic T_{P-MH} is obtained by considering $\hat{\theta}_P = \hat{K}_{3A}/\hat{K}_{3B}$ instead of $\hat{\theta}_G$. These procedures for testing the PH assumption were applied to the Freireich *et al.* (1963) data set (table 3.1) and the ingredients of the calculation are reported in table 7.5. The estimators of θ obtained with different weights give very similar results, and this indirectly supports the assumption of PH for this data set. The test statistics based on the comparisons of the M–H estimator versus the Prentice estimator and versus the Gehan estimator take values $T_{P-MH} = 0.3265$ and $T_{G-MH} = 0.3138$, respectively. No monotone departures from the null hypothesis of a constant hazard ratio are thus likely to be present. Note that these test statistics do not depend on the observed value of the failure times but only on their ranks.

7.3.3   O'Quigley and Pessione test

O'Quigley and Pessione (1989) proposed a model which relaxes the assumption of proportional hazards, defining specific alternatives, but reduces to the basic Cox model as a special case. This model provides a framework in which the PH assumption may be tested against the specified alternatives, by means of a score test. Their approach requires the subdivision of the time axis into J non-overlapping arbitrarily defined intervals, $I_1, I_2, \ldots, I_J$, ordered starting from the origin. The model defines the dependence of the hazard on each covariate by means of an additional vector of parameters γ_j which, unlike β in (6.1), is allowed to vary between intervals, being constant within each interval:

$$\lambda_j(t, \mathbf{x}) = \lambda_0(t) \exp\left[(\beta + \gamma_j)'\mathbf{x}\right], \qquad j = 1, \ldots, J$$

for time t belonging to I_j. The way in which the parameters γ_j are allowed to vary determines the specific alternative to the PH assumption, which in turn

Table 7.5 Calculation of the Gill and Schumacher test on the PH assumption for the data set from Freireich *et al.* (1963), where A and B indicate the 6-MP and placebo treatment groups, respectively.

Failure time $t_{(j)}$ (1)	Number of failures d_{Aj} (2)	d_{Bj} (3)	Number at risk n_{Aj} (4)	n_{Bj} (5)	Different weights w_{1j} (6)	w_{2j} (7)	w_{3j} (8)	Different types of contributions to the quantities $\hat{K}_{1A}$ (9a)	$\hat{K}_{1B}$ (9b)	$\hat{K}_{2A}$ (10a)	$\hat{K}_{2B}$ (10b)	$\hat{K}_{3A}$ (11a)	$\hat{K}_{3B}$ (11b)
1	0	2	21	21	10.5	441	10.0116	0	1	0	42	0	0.9535
2	0	2	21	19	9.975	399	9.0471	0	1.05	0	42	0	0.9523
3	0	1	21	17	9.3947	357	8.3023	0	0.5526	0	21	0	0.4884
4	0	2	21	16	9.0811	336	7.6028	0	1.1351	0	42	0	0.9503
5	0	2	21	14	8.4	294	6.6419	0	1.2	0	42	0	0.9488
6	3	0	21	12	7.6364	252	5.5053	1.0909	0	36	0	0.7865	0
7	1	0	17	12	7.0345	204	4.9023	0.4138	0	12	0	0.2884	0
8	0	4	16	12	6.8571	192	4.1196	0	2.2857	0	64	0	1.3732
10	1	0	15	8	5.2174	120	3.0039	0.3478	0	8	0	0.2003	0
11	0	2	13	8	4.9524	104	2.5921	0	1.2381	0	26	0	0.6480
12	0	2	12	6	4	72	1.8732	0	1.3333	0	24	0	0.6244
13	1	0	12	4	3	48	1.3223	0.25	0	4	0	0.1102	0
15	0	1	11	4	2.9333	44	1.2121	0	0.7333	0	11	0	0.3030
16	1	0	11	3	2.3571	33	0.9091	0.2143	0	3	0	0.0826	0
17	0	1	10	3	2.3077	30	0.8264	0	0.7692	0	10	0	0.2755
22	1	1	7	2	1.5556	14	0.4457	0.2222	0.7778	2	7	0.0637	0.2228
23	1	1	6	1	0.8571	6	0.1842	0.1429	0.8571	1	6	0.0307	0.1842

$\hat{\theta}_{M-H} = \hat{K}_{1A}/\hat{K}_{1B} = 0.2074$; $\hat{\theta}_{G} = \hat{K}_{2A}/\hat{K}_{2B} = 0.1958$; $\hat{\theta}_{P} = \hat{K}_{3A}/\hat{K}_{3B} = 0.1971$

corresponds to $\gamma_j = 0$. Suppose we are interested in testing the PH assumption with respect to one covariate x_k among those in the $(K \times 1)$ vector **x**. According to the notation adopted in the original work, we express the $(K \times 1)$ vector γ_j as the product of a diagonal matrix ψ_j and a vector θ of constants $\gamma_j = \psi_j \theta$. The form chosen for $\psi_j = \text{diag}(\psi_{1j}, \psi_{2j}, \ldots, \psi_{Kj})$ determines the way in which the hazard ratio related to x_k may vary as time t goes along the different time intervals. The null hypothesis of PH is rephrased as $H_0 : \theta_k = 0$ while the alternative is $H_1 : \theta_k \neq 0$. As an example, consider the specification of a linear trend alternative for the covariate x_k. Define ψ_j as having all diagonal elements null except for the kth one:

$$\psi_j = \text{diag}(0, 0, \ldots, \psi_{kj}, \ldots, 0)$$

and

$$\psi_{kj} = j - 1 \quad \text{for } j = 1, 2, \ldots, J$$

As a consequence, the influence of x_k on hazard is modelled in the interval I_1 by means of β_k, in the interval I_2 by means of $\beta_k + \theta_k$, in the interval I_3 by means of $\beta_k + 2\theta_k$, and so on.

To achieve good power against an exponential decay of the effect of x_k, exponential scores are used for the J elements ψ_{kj} $(j = 1, \ldots, J)$.

With two intervals I_1 and I_2 only, and with $\psi_{k1} = -1$ and $\psi_{k2} = 1$, the alternative hypothesis considered is that hazard functions cross near the boundary between I_1 and I_2 (and produce hazards ratios of the same magnitude).

The use of the score test for H_0 is suggested since it bypasses the estimation of θ, which could create some problems. Some indications on the definition of the time intervals are given by O'Quigley and Pessione. This test of the PH assumption has the inconvenience that it relies on an arbitrary partition of the time axis. Often, the time partition is done on the basis of what the data suggest, as for example in the case of the choice of two intervals for testing the hypothesis of crossing hazards. It may become unwise, however, "...to impose too fine a structure on ψ as a result of data scrutiny rather than as a result of external considerations". The non-parametric feature of relying on the ranks of failure times appears to be an advantage of this test over a test using defined time-dependent variables which depend on the actual value of time.

7.3.4   Tests of proportional hazards against a general alternative

An omnibus test on the PH assumption was proposed by Schoenfeld (1980). It partitions both the time axis and the space of the covariate values so that mutually exclusive categories of failure times with associated covariates are formed. The statistic of the test is essentially based on the comparison of the number of events observed and of the number of those expected under the Cox model in each of the "cells" produced by this partition. This statistic is a quadratic form asymptotically distributed as chi-square. This is a general goodness-of-fit test and thus the number of observed and expected events could not agree very well either because the PH assumption does not hold or the log-linear dependence of hazard on covariates is not satisfied. The choice of different partitions leads to different tests which may help in analysing different aspects of model fit. Schoenfeld's test statistic may appear rather complicated but a more tractable test statistic based upon the same approach was derived by Moreau *et al.* (1986) and is presented here.

Consider the basic Cox model (6.1) and define a partition of the space of the covariate values and time, of dimension $K + 1$, as follows: consider a subdivision of the time axis from 0 to infinity into J consecutive non-overlapping time intervals $I_1, \ldots, I_j, \ldots, I_J$. Define then a partition of the covariate space so that (by a Cartesian product) for the jth time interval there are C_j categories, denoted by W_{jc} ($c = 1, \ldots, C_j$). The total number of cells into which the $K + 1$ dimensional space is divided is thus $C_1 + \cdots + C_j + \cdots + C_J = L$. If, following Schoenfeld, the time in the Freireich *et al.* (1963) data set is divided into two intervals, $I_1 : t \leqslant 11$ weeks and $I_2 : t > 11$ weeks, a total of $L = 4$ cells is obtained, given that the only covariate (treatment) is dichotomous and $C_1 = C_2 = 2$. Now consider the following extension of model (6.1):

$$\lambda(t, \mathbf{x}_i) = \lambda_0(t) \exp(\boldsymbol{\beta}'\mathbf{x}_i + \boldsymbol{\gamma}_j \mathbf{y}_{ij})$$

which specifies the hazard function for individual i in the time interval I_j ($t \in I_j$), where $\mathbf{y}_{ij}$ is a $C_j \times 1$ vector: its components y_{ijc} are all zero except the one, say c^*, which identifies the cell W_{jc^*} in which the individual i with covariates $\mathbf{x}_i$ lies at time t. The corresponding unknown regression coefficients are in the $C_j \times 1$ vector $\boldsymbol{\gamma}_j$, of components γ_{jc}, and the null hypothesis which reduces this model to (6.1) is

$$H_0 : \boldsymbol{\gamma}_1 = \cdots = \boldsymbol{\gamma}_j = \cdots = \boldsymbol{\gamma}_J = \mathbf{0}$$

which specifies a null value for L parameters. The log partial likelihood for the extended model can be written provided that the ordered failure times are subdivided within each time interval (assume for simplicity there are no ties). Denote by D_j the number of failures in interval I_j and with t_{rj} ($r = 1, 2, \ldots, D_j$) the ordered failure times actually observed in I_j. Then the log partial likelihood is written as:

$$LL = \sum_{j=1}^{J} \sum_{r=1}^{D_j} \left\{ (\boldsymbol{\beta}'\mathbf{x}_{rj} + \boldsymbol{\gamma}_j'\mathbf{y}_{rj}) - \log \sum_{l \in R_{rj}} [\exp(\boldsymbol{\beta}'\mathbf{x}_l + \boldsymbol{\gamma}_j'\mathbf{y}_{lj})] \right\}$$

where $\mathbf{x}_{rj}$ and $\mathbf{y}_{rj}$ are the vectors for the subject who fails at t_{rj} and R_{rj} is the set of subjects who are at risk at the failure time $t_{rj} \in I_j$. Note that LL depends on a total number of $K + L$ unknown parameters, but resorting to the score test avoids the need to estimate the L parameters γ_j (section 5.6.3). By taking the derivative of LL with regard to the L parameters γ_{jc} ($j = 1, \ldots, J$ and $c = 1, \ldots, C_j$) we obtain the relevant components of the score vector. These, evaluated under the null hypothesis, at $\boldsymbol{\beta} = \hat{\boldsymbol{\beta}}$, where $\hat{\boldsymbol{\beta}}$ are the estimates obtained for $\boldsymbol{\beta}$ under the restricted model (6.1), are

$$
U_{H0} = \left. \frac{\partial LL}{\partial \gamma_{jc}} \right|_{\gamma_j = 0, \boldsymbol{\beta} = \hat{\boldsymbol{\beta}}} = \sum_{r=1}^{D_j} y_{rjc} - \sum_{r=1}^{D_j} \left[\frac{\sum\limits_{l \in R_{rj}} y_{ljc} \exp(\hat{\boldsymbol{\beta}}' \mathbf{x}_l)}{\sum\limits_{l \in R_{rj}} \exp(\hat{\boldsymbol{\beta}}' \mathbf{x}_l)} \right]
$$

Each component of the score vector pertains to a single cell W_{jc} of the space of time and covariates and compares the total number of events $\sum_{r=1}^{D_j} y_{rjc}$ observed in the cell to the expected number of events according to model (6.1) as they were originally defined by Schoenfeld. The corresponding score statistic for testing the null hypothesis is

$$
\mathbf{U}'_{H0} \mathbf{I}^{-1}_{L \times L} \mathbf{U}_{H0}
$$

where $\mathbf{I}^{-1}_{L \times L}$ is the submatrix formed by the relevant components of the $(K + L) \times (K + L)$ inverse of the observed information matrix $\mathbf{I}$. Note that this test could be considered, as in Schoenfeld, also for testing the fit of a Cox model with time-dependent covariates. In the data set from Freireich *et al.*, with the time partition indicated above, the observed and expected number of recurrences according to Moreau *et al.* in the placebo group are as follows: in I_1, 15 and 15.275 respectively, and in I_2, 6 and 5.725 respectively. The corresponding figures for the 6-MP group are: in I_1, 5 and 4.725, and in I_2, 4 and 4.275. If Breslow's approach to ties is used in estimating $\hat{\beta}$ in (6.1), the resulting score statistic has value 0.053, which is clearly not significant.

Other tests which consider fairly general alternatives to the PH assumption were developed by Andersen (1982) and more recent developments on this issue are in Lin and Wei (1991) and Crouchley and Pickles (1993).

7.3.5   Some additional remarks

When data are investigated with models from the class (6.31), methods for checking the PH assumptions are an essential part of the analysis, since misleading conclusions may be drawn if the assumptions are incorrect. We have shown here graphical methods as well as formal tests for the PH assumption, the former being usually the most useful tools for starting the checking process. Graphical methods are more useful the more the plots they produce are expected to show a simple structure under PH, so that their interpretation is somehow less exposed to arbitrary judgement. For example, graphs that are expected to have a particularly simple form, such as that of lines with slope one or zero, ease the visual assessment of the goodness of fit. One of such graphical methods which has not been shown here but is worth mentioning is that proposed by Arjas (1988); it is based on a plot of observed versus expected failure frequencies, as suitably estimated from the model. A more sophisticated approach based on the smoothing of the cumulative hazard estimates and of hazard ratio estimates is presented by Gray (1990).

Information derived from graphical methods may help in building up tentative models for the analysis. Often, however, it is difficult to determine whether discrepancies from PH suggested by the plots are so relevant that the model must be modified or replaced by a different one. Formal tests are advocated in such situations to support the graphical investigations. These, in turn, can guide the choice of the test statistic which is most appropriate to detect certain departures from PH. Various such test statistics have been discussed here; the tests which are based on extensions of the basic Cox model are particularly useful, since they may be used to develop a suitable model.

The approach to non-proportional hazard is typically simplified to consider whether the structure in the data reflects three major types of alternatives to PH:

(1) *decaying effect*, occurring typically for the predictive effect of a prognostic factor measured at a point in time (study entry) which becomes progressively less relevant as time progresses;

(2) *diverging effect*, occurring for example when two treatments do not show a difference in the short term but only in the long term;

(3) *crossing effect*, which refers to the situation in which the hazards ratio is for instance in favour of one treatment in the short term and of the other treatment in the long run.

The examples seen in the previous sections illustrated some of these points and in the literature there are various papers which, in an estimation rather than in a testing context, have adopted different approaches to modelling situations which departed from PH. To mention some of those, stratification on follow-up time or the definition of time dependence by means of a step function is used for modelling in the presence of a decaying effect by Anderson and Senthilselvan (1982), Harrell and Lee (1986) and Cuzick *et al.* (1990). Also, the long term analysis of heart disease mortality of a cohort from the Alameda County Study illustrates the use of regression terms expressing the interactions between prognostic factors and time in a Cox model (Cohn *et al.*, 1990). A thorough discussion of different approaches to modelling the relationship of clinical covariates with outcome in a study on breast cancer, in the presence of non-monotone convergent hazards, is found in Gore *et al.* (1984). The extension of the Cox model including a frailty parameter is proposed by Andersen *et al.* (1993, pp. 672–673) and is applied in the analysis of survival of patients with melanoma. A model for testing and estimating an exponential decaying effect is proposed by Cox and Oakes (1984, pp. 76–77). A gradually diverging "treatment" effect is found in the analysis of lung cancer hazard rate for ex-smokers and current smokers (Doll, 1970; Freedman and Navidi, 1990). The crossing hazards situation is discussed in Stablein *et al.* (1981) where the Cox model with time-by-treatment interaction terms is used to analyse survival in a clinical trial on treatment of non-resectable gastric cancer.

As a final remark, it is worth noting that the use of graphical methods is also advocated to explore the type of dependence of hazard on covariates, as mentioned in section 7.2.1. Formal methods to approach the problem consider a general form for the hazards ratio which embeds both the additive and multiplicative dependence (Thomas, 1981; Guerrero and Johnson, 1982; Breslow and Storer, 1985; Moolgavkar and Venzon, 1987; Arjas and Venzon, 1988). The aim is to have the possibility, with appropriate statistics, of discriminating between the additive and multiplicative forms or, more in general, to test different specifications of the hazards ratio within a large family of functions $g(x)$ in (6.31). However, the application of these families of models, in both case-control studies and survival analysis, may not be easy because of difficulties in estimation and inference (Prentice and Mason, 1986; Valsecchi, 1992b).

7.4 RESIDUALS AND REGRESSION DIAGNOSTICS

Residuals and related diagnostics can be used to examine different aspects of model adequacy:

(1) the validity of the PH assumption;
(2) the functional form in which an experimental variable influences the outcome, given that other covariates are already accounted for in the model;
(3) the presence of single "influential" observations, i.e. observations with outlying values in the space of the covariates and with a disparate influence on the coefficient estimate;
(4) the presence of outliers, i.e. individuals whose outcomes are poorly predicted by the model.

For simplicity, figure 7.7 illustrates the idea of diagnostics (3) and (4) with reference to the classical regression context. Typically, in survival analysis using the Cox model, influential observations are often found among long survivors. Outliers are individual subjects who fail very early or very late with respect to the other subjects having similar characteristics. These aspects will be illustrated in a worked example in section 7.4.4.

In survival analysis, the definition of residuals may not be as immediate as in the context of normal-theory linear models and has stimulated a lot of recent work. Various definitions in the context of proportional hazard models have been suggested in the literature and will be sketched below, under three headings:

(1) the earlier works (Kay, 1977; Crowley and Hu, 1977; Lagakos, 1980) are related to the general definition of residuals given in Cox and Snell

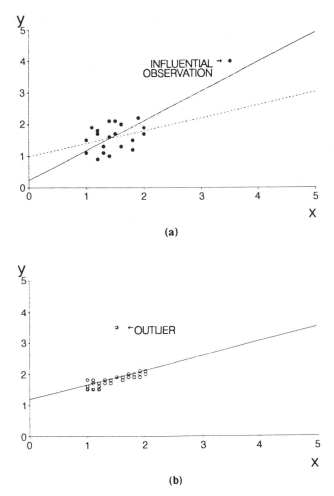

Figure 7.7 The representation of two types of aberrant observations in the simple linear model (in plot (a) the dotted line is fitted after exclusion of the influential observation).

(1968). They involved estimating, for each individual, a quantity that should behave like a unit exponential random variable had the model assumptions been adequate for the data set;

(2) a set of residuals derived by Schoenfeld (1982) to test the PH assumption;

(3) the recent progress in the definition of residuals, in the investigation of their properties and in the development of related tests of hypotheses which is based on martingale theory. Since this latter is beyond the scope of this book, a few remarks will be made on this point in relationship to the preceding ones.

7.4.1　The earlier works

The Cox model defines survival time, conditional on a covariate set $\mathbf{x}_i$, as a r.v. from a distribution with hazard $\lambda_0(t)\exp(\boldsymbol{\beta}'\mathbf{x}_i)$. Consider the general property that if the r.v. T has an arbitrary continuous distribution $F(t)$, then $F(T)$ is uniformly distributed on $(0, 1)$ and $\Lambda(T)$ has an exponential distribution with unit parameter. This property holds also for the distribution of failure time T_i conditional on a set of covariates $\mathbf{x}_i$, $F(T_i, \mathbf{x}_i)$.

The residuals are defined to be the quantities

$$\hat{r}_i = \hat{\Lambda}_0(t_i)\exp(\boldsymbol{\beta}'\mathbf{x}_i) = \hat{\Lambda}(t_i, \mathbf{x}_i), \qquad i = 1, 2, \ldots, N$$

obtained after fitting the model to a set of N observations $(t_i, \delta_i, \mathbf{x}_i)$ for both censored $(\delta_i = 0)$ and uncensored $(\delta_i = 1)$ data. Because of the general property recalled above, these residuals should be approximately distributed as possibly censored exponentials with parameter 1, if the assumed model is adequate. Thus a plot against $\hat{r}_i$ of the Aalen estimated cumulative hazard of the $\hat{r}_i$ values, $\hat{\Lambda}(\hat{r}_i)$, should exhibit a nearly constant slope.

One could ask whether the distributional property of residuals $\hat{r}_i$ still holds in the context of a semi-parametric family in which $F(t, \mathbf{x})$ is not fully specified. It is not clear what influence the partial likelihood estimates $\hat{\boldsymbol{\beta}}$ and the non-parametric estimator $\hat{\Lambda}_0(t)$ have on the unit exponential distribution presumed for residuals. In fact the suitability of this method for checking the PH assumption in semi-parametric models has been criticized. Lagakos (1980) showed, by complete enumeration for $N = 3$, that the distribution of the resulting residuals departed substantially from the exponential. Crowley and Storer (1983), throughout a simulation study, showed that important cases of deliberate misspecification of the model may be missed. As a consequence, this use of the residuals, on the grounds of their distributional properties, should be avoided.

A different use of residuals which are based on the cumulative hazard has instead proved useful. Consider the following transform of $\hat{r}_i$:

$$\hat{r}_i^* = \delta_i - \hat{\Lambda}_0(t_i)\exp(\hat{\boldsymbol{\beta}}'\mathbf{x}_i) \tag{7.7}$$

obtained by adding the mean 1 to the censored $\hat{\Lambda}(t_i, \mathbf{x}_i)$, changing sign and centring the resulting quantities about zero.

The values of $\hat{r}_i^*$ have a nice interpretation provided that the observed value of δ_i is now thought of as the number of events (0 or 1) occurring on the ith individual during follow-up. The $\hat{r}_i^*$ values appear to be the difference between the observed number of events and the number expected under the model, where this latter is expressed in terms of the risk cumulated during the follow-up period t_i. This interpretation has particular sense in the presence of a fair amount of censoring and if we think of grouping the individuals according to a covariate.

Then, in each group we have the difference between the total number of observed and expected events, these latter calculated as the sum of the individual contributions in terms of their cumulated risk. Relatively large differences within some of the groups aid in identifying patterns of covariates which are not well fitted in the model. If the model fits well, the group's residuals plotted against the group's indicators should be scattered around zero (see section 7.4.4).

The Breslow estimate of cumulative hazard up to time t_i can be rewritten as

$$\hat{\Lambda}(t_i, \mathbf{x}_i) = \sum_{j|t_{(j)} \leq t_i} \frac{d_j \exp(\boldsymbol{\beta}' \mathbf{x}_i)}{\sum_{l \in R_j} \exp(\hat{\boldsymbol{\beta}}' \mathbf{x}_l)}$$

where $t_{(j)}$ are the ordered failure times observed up to t_i. The quantity $\hat{\Lambda}(t_i, \mathbf{x}_i)$ is the sum, updated at each failure time, of the conditional probabilities that the death at time $t_{(j)}$ occurs to the individual i, with covariates $\mathbf{x}_i$, among all individuals at risk at $t_{(j)}$, where each probability is given an exponential weight depending on covariates. Suppose now that a covariate x^* strongly influences survival, for example with earlier failures occurring at higher values of x^*. It is likely that in the Cox model without x^* the residuals $\hat{r}_i^*$ plotted against the covariate x^* would not be centred around zero as expected. They could tend

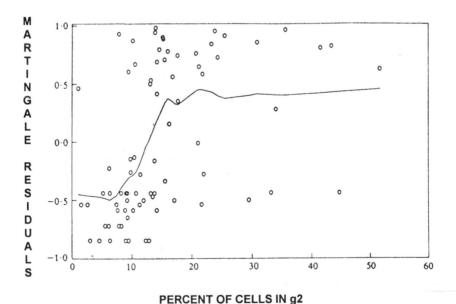

PERCENT OF CELLS IN g2

Figure 7.8 Residuals (7.7) (or martingale residuals) from a Cox model analysing data on prostate cancer plotted against the variable "percent of cells in g2 phase" when this variable is not included in the linear predictor of the model (from Therneau *et al.*, 1990, reproduced by permission of the Biometrika Trustees).

instead to be positive for larger values of x^*; in other words, subjects with high values of x^* would be seen as failing "too soon" by the model without x^*, since their cumulated hazard would be "too small". Conversely, residuals associated with lower values of x^* would tend to be negative, i.e. to identify subjects who live "too long" according to the model. The role these residuals can play is thus more on the side of determining what covariates to include in the model and in what form. A good example of this is given by Therneau *et al.* (1990) concerning a data set of patients with surgically treated stage D_1 prostate cancer. The analysis of time to recurrence of disease was done with a Cox model which excluded the variable, "percent of cells in $g2$ phase" ($g2\%$) measuring the proliferation rate of resected tissue. The individual residuals (7.7) from this fit (y-axis) were plotted against the value of the variable (x-axis) (figure 7.8) and their behaviour suggests that the variable influences the hazard. As a matter of fact, negative and positive residuals tend to cluster at lower and higher values of $g2\%$, respectively. Therneau *et al.* explored residuals with the additional aim of checking the appropriateness of a provisional cut point at 13% chosen to categorize the variable $g2\%$ into two classes. The candidate cut-point corresponded to the mean + 3 s.d. of the $g2\%$ values obtained on 60 non-cancerous tissue specimens. An aid to the interpretation of the plot comes from the smoothing function of the individual residuals and the authors comment that the result "bears out the initial guess" on the cut-point.

7.4.2 Schoenfeld's residuals

Schoenfeld (1982) exploits the score function (6.10) based on partial likelihood to define a set of partial residuals for the J deaths observed in a sample of N individuals (no ties are assumed for simplicity). Consider the contribution to the score at each time $t_{(j)}$ in terms of the entire vector $\mathbf{x}$ of K covariates by generalizing (6.9):

$$\frac{\delta l_j}{\delta \boldsymbol{\beta}} = \mathbf{x}_j - \frac{\sum\limits_{i \in R_j} \mathbf{x}_i \exp(\boldsymbol{\beta}' \mathbf{x}_i)}{\sum\limits_{i \in R_j} \exp(\boldsymbol{\beta}' \mathbf{x}_i)}$$

Note again that this difference contrasts the value $\mathbf{x}_j$ of the covariate vector for the subject who fails at $t_{(j)}$ with its expected value conditional on the set R_j of subjects at risk at $t_{(j)}$. Hence, the score statistic can be written as

$$\mathbf{U}(\boldsymbol{\beta}) = \sum_{j=1}^{J} [\mathbf{x}_j - E(\mathbf{x}_j | R_j)]$$

where the ML estimators of $\boldsymbol{\beta}$ are the solutions to the system of K equations

$U(\boldsymbol{\beta}) = 0$. Residuals are defined as

$$\hat{\mathbf{r}}_j = \mathbf{x}_j - \hat{E}(\mathbf{x}_j | R_j) \qquad (7.8)$$

and are vectors of K components, obtained by substituting in $E(\mathbf{x}_j | R_j)$ the ML estimates $\hat{\boldsymbol{\beta}}$. Unlike residuals (7.7), those in (7.8) pertain to time points and not to individuals.

The $\hat{\mathbf{r}}_j$ values are shown to be asymptotically uncorrelated with expected value zero under the correct model. Each component $\hat{r}_{kj}$ corresponds to the kth covariate, and a plot of the $\hat{r}_{kj}$ values against time can be examined graphically for departures from PH related to x_k. A large residual, either positive or negative, indicates that the event occurred at t_j is unlikely under the model, given the covariates of the individual who failed relative to those of individuals in the risk set R_j. The presence of patterns in these plots of the components $\hat{r}_{kj}$ indicates departures from the PH assumption. This property can be formally seen by considering the model with the addition of a time-dependent variable built on one of the covariates, say x_k:

$$\lambda(t, \mathbf{x}) = \lambda_0(t) \exp\left[\boldsymbol{\beta}'\mathbf{x} + \gamma g(t) x_k\right]$$

where $g(t)$ is a defined function of time. This model assumes a non-proportional effect with respect to x_k.

Schoenfeld showed that the kth element of $\hat{r}_j$ has expected value

$$E(\hat{r}_{kj}) \simeq g(t_j) var(x_{kj} | R_j)$$

with $g(t_j)$ varying about zero. The sign of $E(\hat{r}_{kj})$ depends on the sign of $g(t_j)$, given that the other factor is always positive, and thus the variation of $g(t)$ will be reflected in a plot of $\hat{r}_{kj}$ against $t_{(j)}$.

As an example, consider the calculation of the residuals (7.8) on the Freireich *et al.* (1963) data set after fitting a Cox model with one covariate x for treatment with value 0 and 1 for placebo and 6-MP, respectively. With this coding the residuals will be $-E(x | R_j)$ if the event at $t_{(j)}$ occurs in the placebo group and $1 - E(x | R_j)$ if the event occurs in the 6-MP group. Thus we expect the residuals, plotted against time, to be placed along two horizontal bands. The quantities (7.8) for this data set were calculated after breaking down tied observations by adding a small random number to each failure time and they are plotted in figure 7.9. As expected, there are more negative than positive residuals, due to the higher relapse rate in the placebo group. However, under PH, we expect both positive and negative residuals to appear distributed along the entire axis. In the plot this happens for $T > 5$ weeks, while before this time only negative residuals are present. As remarked by Schoenfeld, this indicates a possible failure of PH in this region. However, he reports that the goodness-of-fit test of section 7.3.4, calculated after dividing the time axis at $T = 5$, yields a p-value of 0.08, ignoring the *post hoc* nature of the choice of the cut-off for the time axis.

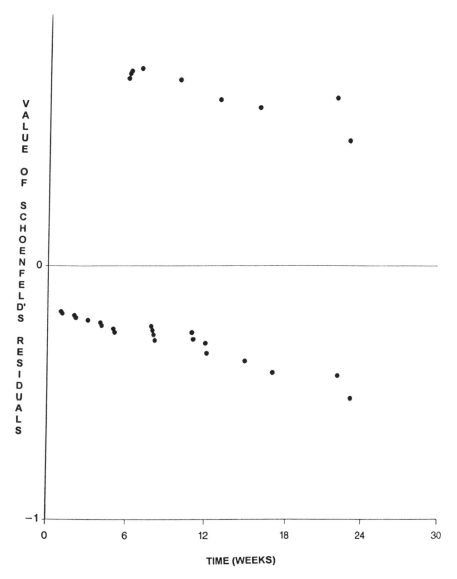

Figure 7.9 Plot of Schoenfeld's residuals for the Freireich *et al.* (1963) data set after fitting a Cox model with one covariate for treatment (0 = placebo, 1 = 6-MP).

From likelihood theory $var(x_{kj}|R_j)$ is estimated by the proper element of the observed information matrix (Kay, 1984):

$$var(x_{kj}|R_j) = \frac{\sum\limits_{i \in R_j} x_{kj}^2 \exp(\hat{\boldsymbol{\beta}}'\mathbf{x}_i)}{\sum\limits_{i \in R_j} \exp(\hat{\boldsymbol{\beta}}'\mathbf{x}_i)} - \left(\frac{\sum\limits_{i \in R_j} x_{kj} \exp(\hat{\boldsymbol{\beta}}'\mathbf{x}_i)}{\sum\limits_{i \in R_j} \exp(\hat{\boldsymbol{\beta}}'\boldsymbol{x}_i)}\right)^2$$

and the components of the adjusted partial residuals $\hat{r}_{kj}/var(x_{kj}|R_j)$ can be computed. It is perhaps easier to investigate the form of $g(t)$ by plotting these against $t_{(j)}$ or by smoothing the plot of residuals (7.8) against time, as in Pettitt and Bin Daud (1990). These authors illustrate the use of the smoothed plots for the modulation of the time-dependent influence of covariates on the hazards ratio both in a simulated and in a real example.

7.4.3   Martingale based residuals

The basis for the development of residuals into the framework of martingale theory is the difference, for any individual i, of the counting process observed on i and the integrated intensity function. In the familiar context of survival analysis where, at most, one event only is counted on the individual at time t_i and the intensity function is the hazard up to the failure time t_i (and zero thereafter), a martingale residual for model (6.1) reduces to the simple form (7.7). This type of residual is in fact a member of a broad class of martingale transform residuals defined by Barlow and Prentice (1988). Another member of this class, later called "score residuals" (Therneau *et al.*, 1990), is a quantity defined for each subject on the basis of his contribution to the score statistic. The score residuals measure the leverage exerted by each subject on parameter estimates in that they provide an estimate of the changes in the coefficients β that would occur when each of the observations were deleted. The score residuals are equivalent, apart from standardization, to the elements of the empirical influence function defined by Cain and Lange (1984) and Reid and Crépeau (1985). They are also related to the diagnostics defined by Storer and Crowley (1985) for the Cox model. Briefly, these authors consider, in a Cox model with K covariates, the change induced on the estimate β by adding an indicator of the ith subject to the linear predictor. They define a variable x_{K+1} as having value one for subject i and zero for all remaining ones, and find the one-step update in $\hat{\beta}$ occurring when this extra variable is included. This is repeated for each subject in the sample. Therneau *et al.* (1990) study the performance of the score residuals and the Storer–Crowley diagnostics as compared with the jackknife estimate of leverage and conclude that there is a fairly good agreement among these different approaches. They show an example where these influence measures are calculated with respect to the coefficient of the variable $g2\%$ estimated on the data set of prostate cancer patients. Their results are presented in figure 7.10 and seem to identify two individuals, with highest ranks in the $g2\%$ values, who have disparate influence on the estimate of the regression coefficient since each of them induces a $10-15\%$ increment in the value $\hat{\beta}$. It can be shown that the partial residuals (7.8) defined by Schoenfeld can be obtained as a special case of the score residuals (provided that these are summed over the subjects on study at each failure time).

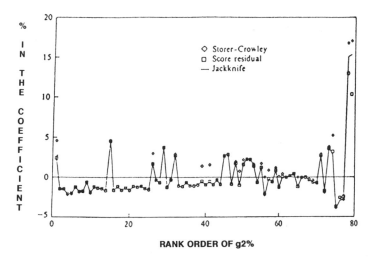

Figure 7.10 The empirical comparison of the score residuals and the Storer–Crowley diagnostics with the jackknife evaluation of influential observations in the example on prostate cancer with regard to the fitting of the variable $g2\%$ as a linear covariate (from Therneau *et al.*, 1990, reproduced by permission of the Biometrika Trustees).

Martingale theory provides a powerful formal tool for developing residuals for models generalizing the Cox model (6.1), for example with the introduction of time-dependent variables, and for parametric models as well. It also permits the construction of formal tests based on residuals for checking model assumptions.

7.4.4 Model criticism in an example

We consider the stratified Cox model fitted to the data from the randomized trial on treatment of advanced ovarian cancer, the results of which are given in the third column of table 7.4. Residuals (7.7), or martingale residuals, calculated from this model for each individual are plotted against the value of the linear predictor in figure 7.11(a). Each star represents the individual value $\hat{r}_i^*$ calculated as in (7.7), where, in this case, the estimate of the cumulative baseline hazard is different according to the stratum of residual tumor size. Each set of stars vertically aligned at a particular value of the linear predictor $\hat{\beta}'\mathbf{x}_i$ is made up of the residuals of all individuals in the sample with the same values of the dummy regressors C1, C2, G1, G2 and K1. For example, the second set of residuals concerns individuals with linear predictor -0.22, who are therefore in the CP arm, with FIGO grade 1 and Karnofsky index > 70 units (i.e., for them C1 = 1 and K1 = 1, all the other regressors being null, given codes in table 7.2). The dots plotted in each set of residuals represent the average of the $\hat{r}_i^*$ values in each group identified by a specific pattern of the covariates and thus measure the discrepancy between the proportion of events observed and that of events predicted by the model in each group. We expect these dots to be scattered around zero, without showing any particular structure, for the model to have an overall good fit. This seems to be the case in the present data set.

Another aspect of goodness of fit we shall consider is the detection of single individuals whose survival is poorly predicted by the model. These outliers may have a large positive or negative residual according to whether they die "too soon" or "too late" with respect to the model prediction. For this aim, the plot of residuals (7.7) can be misleading because the $\hat{r}_i^*$ are highly skewed since they range from minus infinity to $+1$. By looking at the plot in figure 7.11(a), it is evident how outliers among subjects who die earlier than predicted are difficult to identify since several stars are squeezed towards $+1$. On the other hand, among those who die later than predicted, the large scaling towards $-\infty$ may produce spurious outliers.

Thus it is worth searching for a transformation which generates a distribution tending to the normal one. This can be achieved by deriving the deviance residuals (Therneau *et al.*, 1990) for a semi-parametric model in analogy with the deviance residuals defined in the generalized linear model literature (McCullagh and Nelder, 1989, pp. 39–40). Without going into the details of the derivation, the deviance residuals can be written as a function of (7.7):

$$\hat{d}_i = \text{sgn}\,(\hat{r}_i^*)\{-2[\hat{r}_i^* + \delta_i \log\,(\delta_i - \hat{r}_i^*)]\}^{1/2} \qquad (7.9)$$

The argument of the logarithm is the estimated cumulative hazard $\hat{\Lambda}(t_i, \mathbf{x}_i)$, thus the residuals $\hat{r}_i^*$ which are near to 1 for individuals who die but have small predicted $\hat{\Lambda}(t_i, \mathbf{x}_i)$, are inflated in $\hat{d}_i$. On the other hand, very large values of $\hat{\Lambda}(t_i, \mathbf{x}_i)$ leading to large negative residuals $\hat{r}_i^*$ are shrunk here by the square root. Rescaling $\hat{r}_i^*$ by (7.9) does not change the values $\hat{r}_i^* = 0$, as for them $\hat{d}_i = 0$.

In particular, many censored observations with relatively short follow-up would be likely to produce a cloud of points approaching zero from negative values; according to Therneau *et al.*, a percentage of censoring greater than 40% can produce a large bolus of such points and distort the normal approximation. Figure 7.11(b) shows the deviance residuals corresponding to figure 7.11(a). In the negative part of figure 7.11(b) there are fewer individuals suspected to live much longer than predictable, and in the positive part subjects may be better distinguished.

Data records on the four individuals with larger residuals (of value in the range 2.7–3.0) and the three with smaller residuals, of value -2.4 approximately, were checked and their characteristics are shown in table 7.6. The two very early deaths (at 44 and three days from randomization) occurred in patients with a relatively large residual tumor size and were reported as due to causes other than tumor, while the two following deaths are not expected to occur "so soon" according to the model because the characteristics of the two patients are not so unfavourable (residual tumor size less than 2 cm and

Table 7.6   Characteristics of the individuals who are related to the four larger and the three smaller deviance residuals in figure 6.20.

Patient number	Deviance residual	Residual tumor size	Karnofsky index	FIGO grade	Treatment group	Survival time (days)	Status
204	2.95	2–5 cm	80	2	CAP	44	dead
306	2.86	>10 cm	80	3	CP	3	dead
435	2.80	<2 cm	100	3	CP	103	dead
68	2.77	<2 cm	100	1	CAP	189	dead
505	−2.44	>10 cm	100	2	CP	2939	alive
203	−2.40	5–10 cm	80	2	CP	4259	alive
101	−2.40	2–5 cm	100	2	CP	4045	alive

Karnofsky index equal to 100). The three outliers with negative residuals had a long survival time in spite of a relatively large residual tumor (all of them have a diameter > 2 cm) and were all in the CP treatment group. In fact, these three observations were also among the influential points (figure 7.12(a)) for the estimate of the coefficient of the dummy regressor C1 (C1 $= 1$ for treatment with CP and zero otherwise, its coefficient estimate from table 7.4 is 0.1646). According to Storer and Crowley (1985), an increase of nearly 10% in the estimated coefficient of C1, some 0.02, is expected if each individual is removed from the data set. Table 7.7 reports the results of the stratified Cox model fitted after excluding the three subjects (Model I). We observe a marked increase in the estimated coefficient of C1 (0.2317) as compared to the corresponding figure in table 7.4. In the lower part of the plot in figure 7.12(a), observations numbered

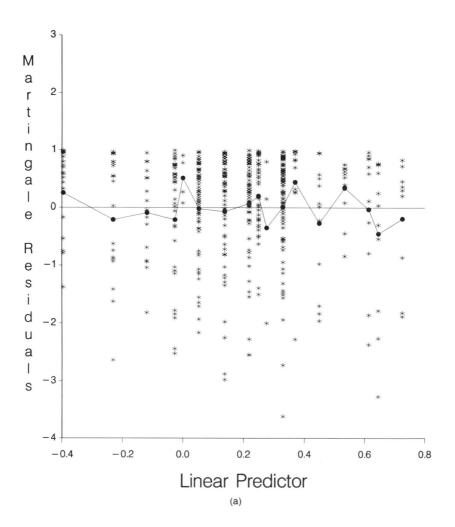

Linear Predictor

(a)

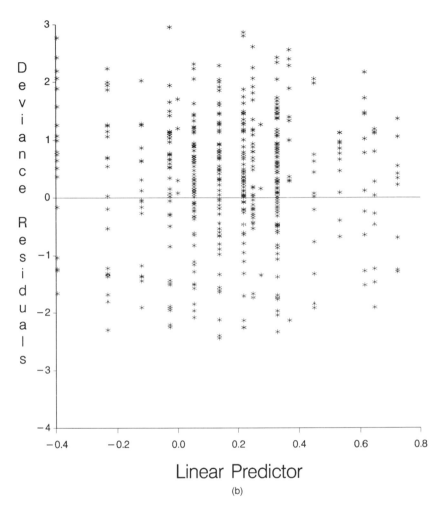

Figure 7.11   (a) Martingale residuals and (b) deviance residuals for the data set on ovarian cancer analysed with the stratified Cox model with fixed covariates C1, C2, G1, G2 and K1 in table 7.4.

75, 103 and 305 are indicated to exert a high leverage on the same coefficient, but in the opposite direction. We reviewed the data records of these individuals too, who were found to be long survivors in the CAP treatment arm, all with residual tumor size greater than 2 cm. The value of the corresponding deviance residuals is about -2.2 and thus these influential cases are similar to those discussed above in that they are also poorly fitted. Table 7.7 shows the coefficients estimated by fitting the stratified Cox model of table 7.4 to the data set excluding the three above-mentioned observations. We observe a marked

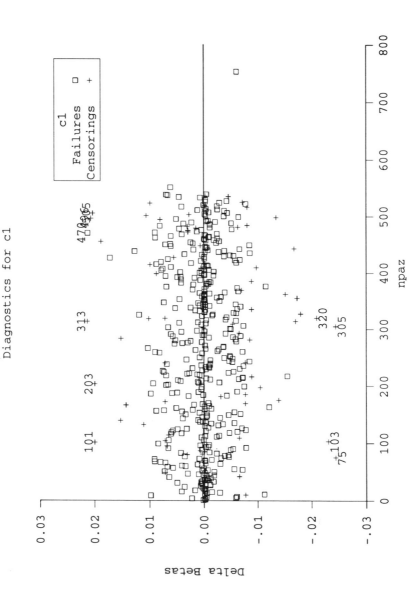

Figure 7.12   (a) Storer and Crowley diagnostics for the coefficients of regressor C1 for treatment in the advanced ovarian cancer trial.

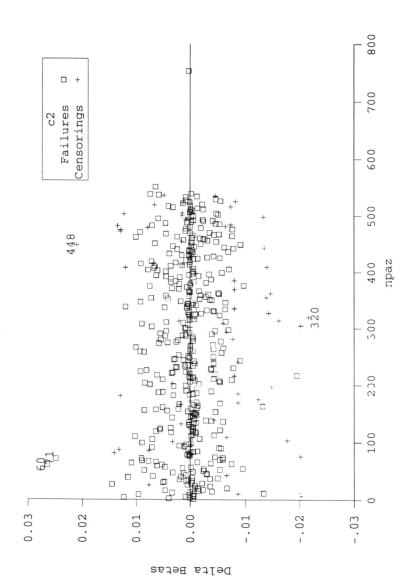

Figure 7.12 (b) Storer and Crowley diagnostics for the coefficients of regressor C2 for treatment in the advanced ovarian cancer trial.

Table 7.7 The Cox model, stratified on residual tumor size, fitted on the ovarian cancer data set which excludes the influential observations numbered 505, 203, 101 (Model I) and on the data set which excludes the influential observations numbered 75, 103, 305 (Model II).

Regressor	Estimated regression coefficient (SE) of	
	Model I	Model II
C1	0.2317 (0.125)	0.0870 (0.126)
C2	0.2871 (0.124)	0.2139 (0.124)
G1	0.4310 (0.170)	0.4176 (0.169)
G2	0.4575 (0.164)	0.4352 (0.164)
K1	−0.3678 (0.136)	−0.3615 (0.136)

decrease in the estimated coefficient of C1 ($\hat{\beta} = 0.0870$) as compared to the estimate in table 7.4.

In figure 7.12(b), the Storer and Crowley plot indicates three observations whose exclusion is approximately related to a 10% increase in the regression coefficient of C2 (this dummy has value 1 for treatment P and zero otherwise and its coefficient estimate in table 7.4 is 0.2758). Also these three individuals have an unusually long survival even in the presence of a relatively large residual tumor size.

Let us first consider the poorly fitted observations of table 7.6 which are not influential. The records on the two subjects who fail very early in spite of their low-risk characteristics were checked; the variables included in the model and other ones (age, type of surgery, tumor histotype) did not show any particular pattern which could suggest extending the model with the inclusion of other regressors. On the other hand, the exclusion from the data set of these subjects has no plausible biological justification, nor can it be advocated to find "better" estimates of treatment effect, given that their exclusion would not much change the coefficient estimates. The same consideration applies with regard to the two individuals who die early for a cause other than tumor. An analysis by cause of death could be added to the present one, but this approach should be advocated only if there is a general interest in it and data on the cause of death were carefully collected on all patients in the trial.

As far as the diagnostics on influence are concerned, a particular feature of survival data is that individuals with longest survival are most likely to be influential points. We have also found that there might be observations on long term survivors which have a high influence on the estimates of the regression coefficients but in the meanwhile are poorly fitted. The presence of long term survivors among the influential points can be understood if we think of the estimation procedure in the Cox model. At each failure time, the covariate value of the individual that fails is compared to that of all individuals at risk at that time. A long term survivor exerts influence in two ways. Firstly that individual forms part of very many risk sets (for all preceding failures). Secondly whereas early failures will be compared to a large risk set, individuals failing towards the end of the study may, depending on the censoring, be compared to a very small risk set. Two groups may initially be of similar size, but as time progresses the relative size of two groups may steadily change as individuals in the high risk group die at a faster rate than those in the other group. Eventually the risk set may be highly imbalanced with just one or two individuals from the high risk group, so that removal of one such individual will greatly affect the estimated hazard ratio. Recall that the Cox estimator does not depend on the actual failure times, but only their ranks and that the information from each failure only on the variance of the covariates in the risk set, not on the number of individuals at risk.

We are thus faced with the problem of how to deal with influential individuals of this type, whose only peculiarity is, as far as we know, that of having an unusually long survival. More in general, we should be concerned with the robustness of our estimates of treatment effect given that the partial likelihood estimator is known not to be robust (Kalbfleisch and Prentice, 1980, p. 144). It may be thought that the few influential observations which have a high influence on the estimates, and consequently may modify results on significance, should be eliminated. But excluding those who are long term survivors would result in biased estimation and jeopardize our evaluation of the results of the study. An interesting approach to the problem of robustness is to consider estimators which maximize weighted versions of the log partial likelihood:

$$LL_w(\boldsymbol{\beta}) = \sum_{j=1}^{J} w_j l_j$$

where l_j is the usual contribution corresponding to failure time $t_{(j)}$ as in (6.8) and w_j is the value taken at $t_{(j)}$ by a weight function $w(t)$ (Sasieni, 1993). In particular, two types of estimators of $\boldsymbol{\beta}$ which may be used for limiting the influence of long-term survivors, are thoroughly discussed by this author and are obtained by adopting the following weights:

(1) at each $t_{(j)}$, $w(t) = w_j$ is a quantity proportional to the number of individuals still at risk at the failure time $t_{(j)}$, for instance the Kaplan–Meier estimate of the overall survival function (ignoring the covariates) at $t_{(j)}$. As in the Prentice test (section 4.4) the resulting estimator of $\boldsymbol{\beta}$ gives more weight to early failures;

(2) $w_j = 1$ for any $t_{(j)} \leqslant t^*$ and $w_j = 0$ otherwise. This type of weight corresponds to artificially censoring all observations which go beyond time t^*. The times t^* can be chosen to correspond to given percentiles of the overall K–M survival estimate.

With the second approach we estimated the regression coefficients in the ovarian cancer data set, censoring the observations at two years (overall median

Table 7.8 Results of the Cox model stratified on residual tumor size on the data set of 476 advanced ovarian cancer patients after artificially censoring at various time points; the last column concerns the standard regression analysis (see table 7.4).

Artificial censoring time (years) Number of events	2 248	3.5 328	5 360	7 387	none 403
Regressor	$\hat{\beta}$(S.E.)	$\hat{\beta}$(S.E.)	$\hat{\beta}$(S.E.)	$\hat{\beta}$(S.E.)	$\hat{\beta}$(S.E.)
C1	0.1368 (0.160)	0.1349 (0.139)	0.1772 (0.139)	0.1635 (0.127)	0.1646 (0.125)
C2	0.1608 (0.158)	0.2190 (0.137)	0.2703 (0.132)	0.2460 (0.127)	0.2758 (0.124)
G1	0.1992 (0.231)	0.4590 (0.202)	0.3391 (0.181)	0.3495 (0.174)	0.3687 (0.170)
G2	0.2733 (0.223)	0.4705 (0.197)	0.3373 (0.176)	0.4231 (0.168)	0.4489 (0.164)
K1	−0.4184 (0.162)	−0.3587 (0.146)	−0.3948 (0.139)	−0.4034 (0.137)	−0.3952 (0.136)

survival time), 3.5 years, five years (overall 75 percentile of survival time) and seven years. This simple approach downweights the contribution of all those individuals who survive beyond the time at which data are censored and is a useful complement to the standard Cox regression analysis. The results are presented in table 7.8 and concern the fitting of the Cox model stratified on residual tumor size and with regressors C1, C2, G1, G2, and K1; the last column repeats the results already given in table 7.4 and obtained with the standard analysis. Note that the number of subjects remains 476 in every analysis, while the number of events (deaths) changes.

This type of complementary analysis is useful in assessing the soundness of the conclusions on treatment effect. In this case, where long term results are of concern, they show that the difference in survival between CAP and CP was mostly confined to those individuals who survived at least two years beyond randomization. In fact, the regression coefficient of C2, in table 7.8, tends to increase as artificial censoring moves beyond two years and the Wald statistic reaches the standard 0.05 significance level in the model based on an observation of at least five years. This shows that the relative advantage of CAP over P treatment arises with long term follow-up.

Finally, a few general considerations on the use of diagnostics are pertinent.

In survival analysis, as in ordinary regression analysis, another   potential source of high influence, which we have not met in the example above, is the presence of "extreme" covariate values. When this is the case, a check of the raw data often highlights mistakes in data reporting or input which were missed by the standard checking procedures.

As far as modelling is concerned, we strongly question the exclusion of individual observations on the basis of the analysis of diagnostics. We encourage instead their use as a means of critically evaluating the performance of a model and as an occasion for consulting with the clinicians on issues brought up by reviewing different aspects of the modelling process, for instance the inclusion/exclusion of covariates, the choice of cut-points, the length of follow-up and so on.

The number of outliers is a function of the amount of variability in the outcome not explained by the model and there might be many outliers in spite of the fact that the covariates in the model are very significant. This is a limitation which can be met also when a study was carefully enough designed so that no major known prognostic variables were ignored. Sometimes it is a matter of identifying a few more variables which are not in the model or in the data and which instead could be useful to explain at least part of the residual variability.

8

Parametric Regression Model

8.1 INTRODUCTION

In this chapter, parametric regression models for the analysis of the relationship between failure times and explanatory variables are introduced. A necessary premise to this topic is the part of Chapter 5 on the exponential model, which was developed in detail in both a homogeneous and a heterogeneous population setting. Some other parametric models will be introduced here for a homogeneous population first (section 8.2), and their properties will be discussed. Those distributions which have explicit reasonably simple forms for the hazard and survivor function will be considered in more detail and their extension to include regression variables will be given (section 8.3). In particular, the log-logistic regression model will lead to the definition of a general class of models characterized by the assumption of proportionality of the odds of failure.

In the two sections mentioned above, as well as in Chapters 5 and 6, the role of the explanatory variables **x** is investigated in terms of the change induced on the hazard and on the survivor function by different patterns of **x**. It will be shown how the parametric models considered can be rephrased as linear regression models with $\log T$ as dependent variable. In this form, the effect of covariates on the random variable $\log T$ itself is analysed. This approach leads to the definition of a general class of log-linear models called the accelerated failure time models (section 8.4).

The theoretical bases of the distributions considered in this chapter will not be discussed here and can be found in textbooks on the theory of statistics (Johnson and Kotz, 1970). Readers interested in the topic of regression models for discrete data are referred to more advanced books on survival analysis (Kalbfleisch and Prentice, 1980; Lawless, 1986).

In parametric models the approach to parameter estimation is based on the likelihood function, whose general expression, in the presence of censored data, was given in (5.19). Any pair of functions $f(t)$ and $S(t)$ derived from a specified distribution can be substituted in (5.19) to obtain the full likelihood from a sample. This general approach to parameter estimation was illustrated in detail with an exponential distribution in sections 5.5.2 and 5.6.2 and the present chapter will not consider this issue further.

8.2   SOME (CONTINUOUS) DISTRIBUTIONS OF FAILURE TIME

We outline here some distributions of the random variable T ($T \geqslant 0$) which are useful for modelling survival in a homogeneous population. Table 8.1 summarizes the distributions in terms of the functions $f(t)$, $S(t)$ and $\lambda(t)$; in all of these the parameter λ has the dimensions of the reciprocal of time and can be interpreted as a rate, while the other parameters (p, k) are dimensionless and have different interpretations in different models.

The one-parameter exponential distribution, which assumes that the instantaneous failure rate λ is independent of t, was introduced in section 5.5.1 and the corresponding functions $f(t)$, $S(t)$ and $\lambda(t)$ sketched in figure 5.1, with $\lambda = 0.65$. Two different distributions which reduce to an exponential for a particular value of a parameter are the Weibull and gamma distributions which follow.

8.2.1   Weibull distribution

The two-parameter Weibull distribution has hazard function

$$\lambda(t) = \lambda p(\lambda t)^{p-1} \tag{8.1}$$

with λ, $p > 0$. This model allows the hazard to depend on time through the exponent $p - 1$: the hazard is monotone decreasing if $p < 1$, increasing if $p > 1$, and reduces to the constant hazard λ of the exponential distribution if $p = 1$ (figure 8.1).

For the Weibull model the mean and variance have no nice closed form expression: $E(T) = \lambda^{-1}\Gamma(1 + p^{-1})$ and $var(T) = \lambda^{-2}[\Gamma(1 + 2p^{-1}) - \Gamma^2(1 + p^{-1})]$, where $\Gamma(\cdot)$ is the gamma function (Appendix A8.1).

In the applications, the Weibull distribution is convenient because of the simple form of the three functions $\lambda(t), f(t)$ and $S(t)$ where, given $\lambda(t)$ in (8.1), the others are

$$f(t) = \lambda p(\lambda t)^{p-1} \exp[-(\lambda t)^p]$$
$$S(t) = \exp[-(\lambda t)^p]$$

Table 8.1 Some parametric failure time models for a homogeneous population of individuals (number of parameters in brackets).

Distribution	Form of the hazard function	
Exponential (1) $\lambda > 0$	$f(t) = \lambda \exp(-\lambda t)$ $S(t) = \exp(-\lambda t)$ $\lambda(t) = \lambda$	constant
Weibull (2) $\lambda, p > 0$	$f(t) = \lambda p (\lambda t)^{p-1} \exp[-(\lambda t)^p]$ $S(t) = \exp[-(\lambda t)^p]$ $\lambda(t) = \lambda p (\lambda t)^{p-1}$	if $p < 1$, monotone decreasing; if $p > 1$, monotone increasing; if $p = 1$, constant ($\lambda(t) = \lambda$)
Gamma (2) $\lambda, k > 0$	$f(t) = \lambda(\lambda t)^{k-1} e^{-\lambda t}/\Gamma(k)$ $S(t) = 1 - \int_0^{\lambda t} u^{k-1} e^{-u} du/\Gamma(k)$ $\lambda(t) = f(t)/S(t)$	if $k < 1$, monotone decreasing with $\lim_{t \to \infty} \lambda(t) = \lambda$; if $k > 1$, monotone increasing with $\lim_{t \to \infty} \lambda(t) = \lambda$; if $k = 1$, constant ($\lambda(t) = \lambda$)
Log-normal (2) $\lambda, \sigma > 0$ $\Phi(z) = \int_{-\infty}^z \phi(u)du$ $\phi(z) = \dfrac{1}{\sqrt{2\pi}} \exp(-z^2/2)$	$f(t) = (1/\sqrt{2\pi}\sigma t) \exp(- [\log(\lambda t)]^2/2\sigma^2)$ $S(t) = 1 - \Phi(\log(\lambda t)/\sigma)$ $\lambda(t) = f(t)/S(t)$	non-monotonic: $\lambda(0) = 0$, then increases to maximum and decreases, with $\lim_{t \to \infty} \lambda(t) = 0$
Log-logistic (2) $\lambda, p > 0$	$f(t) = \lambda p (\lambda t)^{p-1}[1 + (\lambda t)^p]^{-2}$ $S(t) = [1 + (\lambda t)^p]^{-1}$ $\lambda(t) = \lambda p (\lambda t)^{p-1}[1 + (\lambda t)^p]^{-1}$	if $p < 1$, monotone decreasing from ∞; if $p = 1$, monotone decreasing from $\lambda(0) = \lambda$; if $p > 1$, non-monotonic: increases from 0 to a maximum at $t = \lambda^{-1}(p-1)^{1/p}$ and decreases towards 0

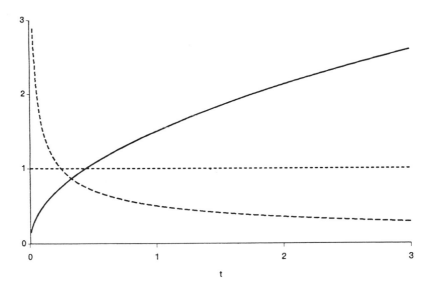

Figure 8.1 Plot of the hazard functions of Weibull distributions with the same value $\lambda = 1$ and different values of the parameter p: $p = 0.5$ (--); $p = 1$ (-----); $p = 1.5$ (———).

The applicability of the model to a data set can be empirically checked by considering the following relationship obtained from the Weibull form of $S(t)$:

$$\log[-\log S(t)] = p \log \lambda + p \log t \qquad (8.2)$$

This expresses a simple linear dependence of the logarithm of the cumulative hazard $\Lambda(t) = -\log S(t)$ on the logarithm of time. A non-parametric estimate of $S(t)$, i.e. the Kaplan–Meier $\hat{S}_{K-M}(t)$, can be used to obtain a plot of $\log[-\log \hat{S}_{K-M}(t)]$ versus $\log t$, which should give approximately a straight line for the Weibull model to be suitable. The slope of the line gives a rough estimate of p and in particular a line with slope nearly one would suggest that the simpler exponential model could be adopted. An example of this type of plot is shown in figure 8.2. The plot is obtained from a sample of 300 observations which were generated according to a Weibull distribution with parameters $\lambda = 1$ and $p = 1.5$ and, as expected, is roughly linear (the dotted line is the least squares estimated linear function of $\log[-\log \hat{S}_{K-M}(t)]$ on $\log t$ drawn to aid in interpreting the plot). The procedure adopted for generating data from a given distribution is outlined in Appendix A8.2.

We recall that, in the case of the exponential distribution being considered, an empirical check on its aptness can be done by relying on a linear relationship which is simpler than (8.2). Under the exponential model, it is

$$-\log S(t) = \lambda t \qquad (8.3)$$

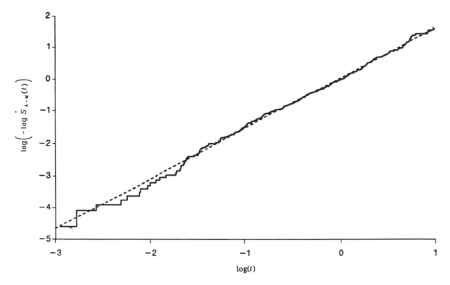

Figure 8.2 Plot of the estimated log cumulative hazard versus log time in a data set of 300 observations generated according to a Weibull distribution ($\lambda = 1$, $p = 1.5$).

thus a plot of $-\log \hat{S}_{\text{K}-\text{M}}(t)$ versus t should give approximately a straight line through the origin, the slope of which is a rough estimate of the constant failure rate (an example was given in section 5.6.4).

8.2.2 Gamma distribution

The two-parameter gamma distribution has density function

$$f(t) = \frac{\lambda(\lambda t)^{k-1} e^{-\lambda t}}{\Gamma(k)} \tag{8.4}$$

where $\Gamma(k)$ is the gamma function (Appendix A8.1) and $\lambda, k > 0$. This distribution, like the Weibull one, is a generalization of the exponential model, to which it reduces for $k = 1$. The mean is k/λ and the variance is k/λ^2. The use of this distribution as a failure time model is limited because no closed form expression is available for $S(t)$ and thus for $\lambda(t) = f(t)/S(t)$; the survivor function $S(t) = 1 - \int_0^{\lambda t} u^{k-1} e^{-u} du/\Gamma(k)$ involves an incomplete gamma integral. The hazard is monotone decreasing from infinity ($\lim_{t \to 0^+} \lambda(t) = \infty$) for $k < 1$ and monotone increasing from 0 ($\lambda(0) = 0$) for $k > 1$ and in either case tends to λ as t becomes large ($\lim_{t \to \infty} \lambda(t) = \lambda$).

8.2.3 Log-normal distribution

The two-parameter log-normal distribution has density function:

$$f(t) = \frac{1}{\sqrt{2\pi}\sigma t} \exp\left(-\frac{[\log(\lambda t)]^2}{2\sigma^2}\right) \tag{8.5}$$

This distribution is more simply expressed in terms of a linear model for the variable $\log T$, as shown in section 8.4. The survivor function

$$S(t) = 1 - \Phi\left(\frac{\log(\lambda t)}{\sigma}\right)$$

and the hazard function $\lambda(t) = f(t)/S(t)$ involve the incomplete normal integral $\Phi(z) = \int_{-\infty}^{z} \phi(u)du$ and $\phi(z) = (1/\sqrt{2\pi})\exp(-z^2/2)$. The mean and variance are $E(T) = \exp[-\log\lambda + \frac{1}{2}\sigma^2]$ and $var(T) = \exp[-2\log\lambda + \sigma^2][\exp\sigma^2 - 1]$.

The hazard is non-monotonic: it has value 0 at $t = 0$ ($\lambda(0) = 0$), increases to a maximum and then decreases, approaching zero as t becomes large ($\lim_{t \to \infty} \lambda(t) = 0$). The time point where the maximum occurs depends on the value of σ.

8.2.4 Log-logistic distribution

The two-parameter log-logistic distribution has hazard function

$$\lambda(t) = \frac{\lambda p(\lambda t)^{p-1}}{1 + (\lambda t)^p} \tag{8.6}$$

which differs from the Weibull hazard (8.1) for the denominator factor. For $p = 1$ the hazard has value λ at $t = 0$ ($\lambda(0) = \lambda$) and is monotone decreasing as t increases, and for $0 < p < 1$ the hazard is still monotone decreasing, but from infinity. For $p > 1$, it is non-monotonic: it increases from zero to a maximum, achieved at $t = (p-1)^{1/p}/\lambda$, and then decreases towards zero. In this last case the log-logistic distribution can approximate fairly well except for the tails, the log-normal distribution, with the advantage over this latter of having closed functional forms for $f(t)$ and $S(t)$:

$$f(t) = \frac{\lambda p(\lambda t)^{p-1}}{[1 + (\lambda t)^p]^2}$$

$$S(t) = \frac{1}{[1 + (\lambda t)^p]}$$

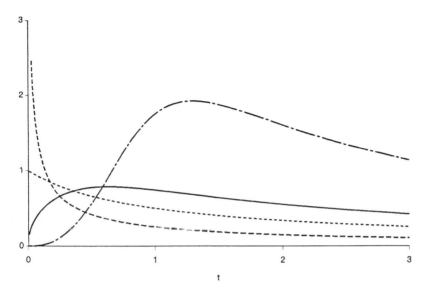

Figure 8.3 Plot of the hazard functions of log-logistic distributions with the same value of $\lambda = 1$ and different values of the parameter p: $p = 0.5$ (–––); $p = 1$ (------); $p = 1.5$ (———); $p = 3.5$ (—·—).

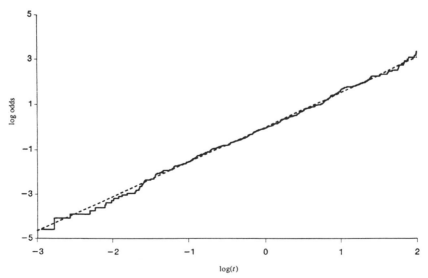

Figure 8.4 Plot of the sample estimates of the log odds of failure in a data set of 300 observations generated according to a log-logistic distribution ($\lambda = 1$, $p = 1.5$).

A plot of the hazard functions corresponding to the log-logistic distribution with different values of p is shown in figure 8.3.

The suitability of the log-logistic model for the analysis of a data set can be empirically checked using a linear relationship derived from the expression of $S(t)$ and $F(t) = 1 - S(t)$. It is

$$\log \left\{ \frac{F(t)}{1 - F(t)} \right\} = p \log \lambda + p \log t \tag{8.7}$$

that is, the log-logistic distribution corresponds to a linear model for the log odds of failure over the logarithm of time, with slope p. The Kaplan–Meier sample estimate $\hat{S}_{K-M}(t)$ can be used to calculate the log odds and a plot of them versus $\log t$ should follow approximately a straight line for the log-logistic model to be suitable. Figure 8.4 shows the plot against $\log t$ of the log odds estimated in a sample of 300 failure times generated according to a log-logistic distribution with parameters $\lambda = 1$ and $p = 1.5$ (Appendix A8.2). As expected, the relationship is approximately linear (the dotted line is the estimated least square linear function of $\log \{\hat{F}_{K-M}(t)/\hat{S}_{K-M}(t)\}$ on $\log t$ drawn to aid in interpreting the plot).

8.3   PARAMETRIC REGRESSION MODELS

We consider here the generalizations of the Weibull and log-logistic models to account for the information on various features of the individuals sampled. We wish to illustrate how to model survival in a heterogeneous population in which some explanatory variables $\mathbf{x} = (x_1, \ldots, x_k, \ldots, x_K)$ may influence the failure process.

8.3.1   Weibull regression model

The hazard function of the Weibull distribution is generally extended to include a vector $\mathbf{x}$ of covariates as follows:

$$\lambda(t, \mathbf{x}) = \lambda p (\lambda t)^{p-1} \cdot \exp(\boldsymbol{\beta}' \mathbf{x}) \tag{8.8}$$

Other functions of the linear predictor $\boldsymbol{\beta}' \mathbf{x}$ (for example $1 + \boldsymbol{\beta}' \mathbf{x}$) could be used instead of the exponential one, but this latter is considered the most natural choice, since it takes only positive values and no constraints on the values of $\boldsymbol{\beta}$ are necessary. Suppose the vector $\mathbf{x}$ is defined in such a way that $\mathbf{x} = \mathbf{0}$ indicates a set of meaningful "standard" conditions. Then, conditional on the value zero of all covariates, (8.8) reduces to the "baseline" hazard function which coincides with (8.1). For any other pattern of covariate values $\mathbf{x}^*$, the conditional hazard is multiplied by a constant $\exp(\boldsymbol{\beta}'\mathbf{x}^*)$. Three considerations follow from the Weibull regression model (8.8):

(1) the shape of the hazard depends on p, and is the same no matter what the covariate pattern is: for example, if $p > 1$, the hazard is monotone increasing in any subgroup of individuals identified by a different pattern of covariates;

(2) the effect of the covariates is to act multiplicatively on the hazard function; for any two individuals with covariate patterns $\mathbf{x}_1$ and $\mathbf{x}_2$ the hazards ratio at each time point is a constant:

$$\text{HR} = \frac{\lambda(t, \mathbf{x}_2)}{\lambda(t, \mathbf{x}_1)} = \exp[\boldsymbol{\beta}'(\mathbf{x}_2 - \mathbf{x}_1)]$$

(3) the effect of each covariate on the constant HR is multiplicative; for a one unit change in the value of x_k, HR is multiplied by a factor $\exp(\beta_k)$.

On the basis of point (2), the Weibull regression model can be considered as a parametric member of the *class of proportional hazards (PH) regression models* introduced in section 6.8.3:

$$\lambda(t, \mathbf{x}) = g(\mathbf{x})\lambda_0(t) \tag{8.9}$$

where the baseline function is specified to have the form (8.1). For $\lambda_0(t)$ unspecified, a semi-parametric model is obtained (the basic Cox model (6.1) if $g(\mathbf{x}) = \exp(\boldsymbol{\beta}'\mathbf{x})$). Properties (1)–(3) also hold for the exponential model (5.27), since this can be derived as a particular case of model (8.8), corresponding to $p = 1$.

The graphical check of the suitability of the Weibull distribution for fitting a set of data can be extended to the context of regression analysis. Given (8.8), the survivor function is

$$S(t, \mathbf{x}) = \{\exp[-(\lambda t)^p]\}^{\exp(\boldsymbol{\beta}'\mathbf{x})}$$

Thus the relationship (8.2) becomes

$$\log[-\log S(t, \mathbf{x})] = \boldsymbol{\beta}'\mathbf{x} + p \log \lambda + p \log t \tag{8.10}$$

which is still linear in $\log t$ with slope p, and with the intercept that includes the linear predictor. For any two patterns $\mathbf{x}_1$ and $\mathbf{x}_2$, the plot of the non-parametric estimates $\log[-\log \hat{S}_{K-M}(t)]$ obtained in the two subgroups of individuals whose covariates are $\mathbf{x}_1$ and $\mathbf{x}_2$ should give approximately two parallel lines for the model to be suitable. The slope of the lines provides a rough estimate of p and the difference between intercepts corresponds to the quantity $\boldsymbol{\beta}'(\mathbf{x}_2 - \mathbf{x}_1)$. For two subgroups with a unit difference in one covariate only, say x_k, all the other covariates being the same, the difference between intercepts provides a rough estimate of β_k.

The proportional hazards assumption (2) is related, as in the graphical checks for the Cox model, with the requirement of a constant distance between curves of different subgroups. The linearity, instead, is required for the Weibull form (8.1), chosen for $\lambda_0(t)$, to be appropriate. A departure from linearity suggests that another model from the class (8.9), either semi-parametric or with a different parametric definition of $\lambda_0(t)$, could be used.

8.3.2   Log-logistic regression model and the class of proportional odds models

The log-logistic distribution is commonly extended to include a vector $\mathbf{x}$ of covariates by reformulating the survivor function as follows (Bennett, 1983):

$$S(t, \mathbf{x}) = \frac{1}{1 + (\lambda t)^p \exp(\boldsymbol{\beta}'\mathbf{x})} \tag{8.11}$$

As a consequence, the distribution function is

$$F(t, \mathbf{x}) = \frac{(\lambda t)^p \exp(\boldsymbol{\beta}'\mathbf{x})}{1 + (\lambda t)^p \exp(\boldsymbol{\beta}'\mathbf{x})}$$

and the odds of failure in time is

$$\left[\frac{F(t, \mathbf{x})}{1 - F(t, \mathbf{x})}\right] = (\lambda t)^p \exp(\boldsymbol{\beta}'\mathbf{x}) \tag{8.12}$$

Conditional on the value zero of all covariates, (8.12) reduces to the "baseline" odds function, which is multiplied by a factor $\exp(\boldsymbol{\beta}'\mathbf{x}^*)$ for any individual with covariate vector $\mathbf{x}^*$.

The simplest and most intuitive way of looking at the characteristics of the log-logistic regression model (8.11) is in terms of the modelling of the odds of failure. Three considerations apply in this regard:

(1) the shape of the odds function (8.12) in monotone increasing and depends only on the parameter p;
(2) the effect of the covariates is to act multiplicatively on the odds function; for any two individuals with covariate patterns $\mathbf{x}_1$ and $\mathbf{x}_2$, the ratio of the odds of failure is a constant:

$$\varphi = \frac{F(t, \mathbf{x}_2)}{1 - F(t, \mathbf{x}_2)} \Bigg/ \frac{F(t, \mathbf{x}_1)}{1 - F(t, \mathbf{x}_1)} = \exp[\boldsymbol{\beta}'(\mathbf{x}_2 - \mathbf{x}_1)] \tag{8.13}$$

(3) the effect of each covariate on the constant φ is multiplicative; for a one unit change in the value of x_k, the odds ratio φ is multiplied by a factor $\exp \beta_k$.

Relationship (8.13) suggests a general *class of proportional odds (PO) models*. The members of this class have a distribution function $F(t)$ such that the odds function is

$$\left[\frac{F(t, \mathbf{x})}{1 - F(t, \mathbf{x})} \right] = g(\mathbf{x}) \cdot \omega_0(t) \tag{8.14}$$

where $g(\mathbf{x})$ is any function of $\mathbf{x}$, but not of t (with $g(\mathbf{0}) = 1$ and $g(\mathbf{x}) > 0$) and $\omega_0(t)$ is any function of t, not dependent on $\mathbf{x}$ ($\omega_0(t) > 0$).

In biomedical research, we are used to comparing different distributions in terms of the shape of their hazard functions. For model (8.11), the hazard function is

$$\lambda(t, \mathbf{x}) = \frac{\lambda p (\lambda t)^{p-1} \exp(\boldsymbol{\beta}'\mathbf{x})}{1 + (\lambda t)^p \exp(\boldsymbol{\beta}'\mathbf{x})}$$

The shape of the hazard depends only on the parameter p, as for (8.6). In particular, when it is non-monotonic ($p > 1$), it achieves a maximum in

$$t = \left[\frac{p-1}{\lambda^p \exp(\boldsymbol{\beta}'\mathbf{x})} \right]^{1/p}.$$

The ratio of the hazards of any two individuals with covariate patterns $\mathbf{x}_1$ and $\mathbf{x}_2$ approaches unity as t becomes large, thus the model assumes that the impact of a covariate on survival decreases with time. An example with one dichotomous covariate is illustrated in figure 8.5, where the parameters have values $\lambda = 1$, $p = 3.5$, and the regression coefficient of x is $\beta = 2$.

An empirical check of the suitability of the log-logistic model for the analysis of a data set in the presence of covariates is derived straightforwardly from (8.7). The relationship

$$\log \left\{ \frac{F(t, \mathbf{x})}{1 - F(t, \mathbf{x})} \right\} = \boldsymbol{\beta}'\mathbf{x} + p \log \lambda + p \log t \tag{8.15}$$

shows that the log odds function is linearly related to $\log t$, with slope p and intercept which depends on $\boldsymbol{\beta}'\mathbf{x}$. For the model to be suitable, we expect that a plot of the non-parametric estimate $\log\{\hat{F}_{K-M}(t)/\hat{S}_{K-M}(t)\}$ versus $\log t$ is nearly linear for any subgroup of individuals with a particular covariate pattern. Furthermore, two different patterns $\mathbf{x}_1$ and $\mathbf{x}_2$ should correspond to roughly

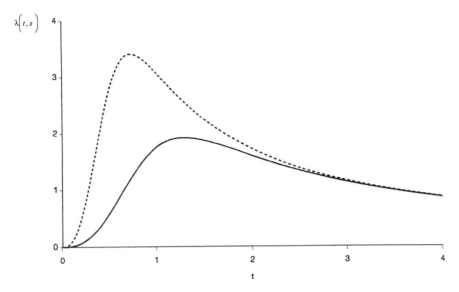

Figure 8.5 Plot of the hazard functions from a log-logistic model with parameters $\lambda = 1$, $p = 3.5$ and with one dichotomous covariate with regression coefficient $\beta = 2$ ($x = 0$: ——; $x = 1$: -------).

parallel lines, the slope of which gives an estimate of p. If the two patterns have a unit difference in one covariate only, say x_k, all the others being the same, the distance between the two lines gives a rough estimate of β_k. If a plot of (8.15) shows a departure from linearity, another model from the class (8.14), with a form of $\omega_0(t)$ different from that in the log-logistic model, could be used. If linearity is satisfied but the distance between lines is not constant, a model (8.14) assuming proportional odds is not appropriate.

8.3.3 An example

The use of the Weibull and log-logistic regression models to estimate treatment effect is illustrated on the data set from Freireich *et al.* (1963). In this example the assumption of proportional hazard was shown to be resonable (section 4.4) while the assumption of constant hazard in the two treatment groups, which was made in adopting the exponential model, was questionable (figure 5.4). A graphical check of the appropriateness of the Weibull model for fit, made on the grounds of (8.10), produces the plot in figure 8.6. Model (8.8) could be appropriate for the analysis in that linearity seems roughly satisfied in both groups, with a very similar slope. When fitted to the data set, the resulting maximum likelihood estimates of the parameters are $\hat{p} = 1.366$, $\hat{\lambda} = 0.046$ and

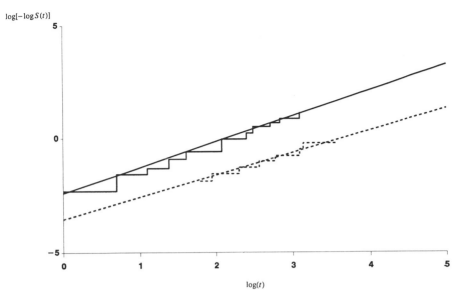

Figure 8.6 Graphical check of the Weibull model in the data set from Freireich *et al.* (1963), by treatment (------ 6-MP group and —— placebo group).

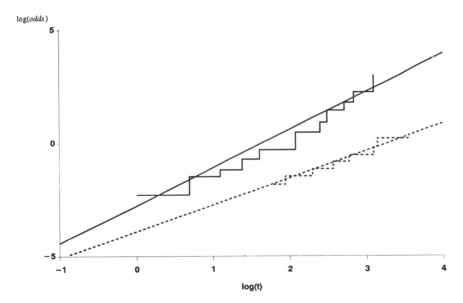

Figure 8.7 Graphical check of the log-logistic model in the data set from Freireich *et al.* (1963), by treatment (------ 6-MP group and —— placebo group).

$\hat{\beta} = -1.731$ with a likelihood value of -47.1; the hazard ratio (constant of proportionality) of 6-MP vs placebo is given by $\hat{\theta} = 0.177$.

The graphical check of the aptness of the log-logistic model, based on (8.15), is presented in figure 8.7. This does not seem to support the use of a proportional odds model since the distance between the curves in the two treatment groups tends not to be constant. Moreover, the functional form of the hazard is questionable, given that a departure from linearity is suggested by the plot, in particular in the placebo group.

8.4 LOG-LINEAR REGRESSION MODELS FOR *T*

All the survival distributions presented in section 8.2 lead to linear models for the variable $Y = \log T$ or equivalently, to log-linear models for T. These generalize easily to include the regression variables $\mathbf{x}$, as shown in table 8.2 for the parametric models most commonly applied. In particular, both the Weibull and log-logistic regression models given by (8.8) and (8.11) can be reformulated in terms of a linear model for Y; for an individual $i\,(i = 1, \dots, N)$ with survival time t_i and covariates $\mathbf{x}_i$ it is

$$y_i = \log t_i = \alpha - \sigma \beta' \mathbf{x}_i + \sigma w_i \qquad (8.16)$$

The parameters α and σ are transforms of λ and p ($\alpha = -\log \lambda$ and $\sigma = 1/p$) and the error terms w_i are i.i.d. random variables. According to whether the common distribution of these latter is the extreme value or the logistic distribution, the linear model for Y corresponds to the Weibull or log-logistic model. From table 8.2 it is easy to see that the exponential model is obtained as

Table 8.2 Some parametric regression models expressed as log-linear models for T.

	Log-linear model for $T\,(Y = \log T)$		Distribution of W
Exponential	$Y = \alpha - \beta'\mathbf{x} + W$	$\alpha = -\log \lambda$	extreme value distribution: $f(w) = \exp(w - \exp w)$
Weibull	$Y = \alpha - \sigma\beta'\mathbf{x} + \sigma W$	$\alpha = -\log \lambda$ $\sigma = 1/p$	extreme value distribution: $f(w) = \exp(w - \exp w)$
Log-normal	$Y = \alpha - \beta'\mathbf{x} + \sigma W$	$\alpha = -\log \lambda$	standard normal distribution: $\phi(w) = \dfrac{1}{\sqrt{2\pi}} \exp\left(-\dfrac{w^2}{2}\right)$
Log-logistic	$Y = \alpha - \sigma\beta'\mathbf{x} + \sigma W$	$\alpha = -\log \lambda$ $\sigma = 1/p$	logistic distribution: $f(w) = \exp w (1 + \exp w)^{-2}$

a particular case of the Weibull model when $\sigma = 1$ (or equivalently $p = 1$). Table 8.2 shows that the log-normal model is a standard linear regression model for Y in that it defines the distribution of W to be the standardized Gaussian one.

The derivation of the linear form in Y, from the density functions of T, is shown in detail in Appendix A8.3 for the Weibull model (8.8) and an analogous approach can be used for the remaining models.

Note also that the parametric regression model which assumes a baseline gamma distribution (8.4) can be formulated in terms of a log-linear model for T but is not included in table 8.2 because of its complex form.

8.4.1 The class of accelerated failure time (AFT) models

The log-linear models in table 8.2 highlight an aspect which was not previously considered, that is the dependence of the failure time on covariates. In all these models, the Weibull (and exponential), the log-normal and the log-logistic ones, the covariates act additively on the variable Y, thus multiplicatively on T. Consider the exponential model, in which one covariate only, x, with values 0 and 1, is included to represent two treatment groups:

$$\log T = \alpha - \beta x + W \quad \text{and} \quad f(w) = \exp(w - \exp w)$$

A change in the value of x does not affect the distribution of W but the model assumes that any individual with covariate $x = 0$ having survival time t_0 would have survival time $t_1 = t_0/e^\beta$ had the covariate been $x = 1$, given that $\log t_1 = \log t_0 - \beta$. A positive (negative) regression coefficient would imply a shorter (longer) survival time for individuals with $x = 1$ as compared to those with $x = 0$ and we say that the covariate accelerates (decelerates) the time to failure. This can be also seen by looking at the exponential survival functions of the two treatment groups in the upper part of table 8.3 which are related as follows:

$$S(t, 1) = S_0(te^\beta)$$

At any time t, survival $S(t, 1)$, conditional on $x = 1$, equals the baseline survival (conditional on $x = 0$) at a later (if $\beta > 0$) or earlier (if $\beta < 0$) time point te^β. Again, if $\beta > 0$ ($\beta < 0$) the group with $x = 1$ is at worst (better) prognosis than the group with $x = 0$. The relationship between the density and hazard functions of the two groups can be written as

$$f(t, 1) = e^\beta f_0(te^\beta) \quad \text{and} \quad \lambda(t, 1) = e^\beta \lambda_0(te^\beta)$$

where $f_0(\bullet)$ and $\lambda_0(\bullet)$ are the baseline density and hazard functions corresponding to $x = 0$.

Table 8.3 Survivor, density and hazard functions for an exponential regression model with one dichotomous covariate (0–1) and for a Weibull regression model (8.8) with vector **x** of covariates.

Exponential regression model

$x = 0$	$x = 1$
$S_0(t) = \exp(-\lambda t)$	$S(t, 1) = \exp[-\lambda(t \cdot e^\beta)]$
$f_0(t) = \lambda \exp(-\lambda t)$	$f(t, 1) = e^\beta \lambda \exp[-\lambda(t \cdot e^\beta)]$
$\lambda_0(t) = \lambda$	$\lambda(t, 1) = \lambda e^\beta$

Weibull regression model

$\mathbf{x} = \mathbf{0}$	$\mathbf{x} \neq \mathbf{0}$
$S_0(t) = \exp[-\lambda^{1/\sigma} t^{1/\sigma}]$	$S(t, \mathbf{x}) = \exp[-\lambda^{1/\sigma}(t \cdot \exp(\sigma\boldsymbol{\beta}'\mathbf{x}))^{1/\sigma}]$
$f_0(t) = \lambda_0(t) S_0(t)$	$f(t, \mathbf{x}) = \lambda(t, \mathbf{x}) \cdot S(t, \mathbf{x})$
$\lambda_0(t) = \dfrac{\lambda^{1/\sigma}}{\sigma} t^{1/\sigma - 1}$	$\lambda(t, \mathbf{x}) = \exp(\sigma\boldsymbol{\beta}'\mathbf{x}) \dfrac{\lambda^{1/\sigma}}{\sigma} (t \cdot \exp(\sigma\boldsymbol{\beta}'\mathbf{x}))^{1/\sigma - 1}$

A representation of the multiplicative assumption on t in terms of the corresponding random variables is $T_1 = T_0/e^\beta$ and, generalizing to a vector **x** of covariates, this model assumes that

$$T_\mathbf{x} = T_0/\exp(\boldsymbol{\beta}'\mathbf{x})$$

This relationship between failure time and covariates leads to the definition of a general class of log-linear models for T, which is called *the class of accelerated failure time (AFT) models*. All members of this class assume that the survival time of an individual with covariates **x** has the same distribution as

$$T_\mathbf{x} = T_0/g(\mathbf{x}) \tag{8.17}$$

given that T_0 is the survival time under $\mathbf{x} = \mathbf{0}$ and has a specified parametric distribution. Whatever this is, it turns out that the accelerated failure time model coincides with a log-linear model:

$$\log T_\mathbf{x} = \mu_0 - \log g(\mathbf{x}) + \varepsilon \tag{8.18}$$

where ε is a random variable with zero mean and with a distribution which derives from the form of $\lambda_0(t)$ and does not depend on the covariates **x**. The function $g(\mathbf{x})$ can have any form, provided that $g(\mathbf{x}) \geqslant 0$ and $g(\mathbf{0}) = 1$. Note that (8.17) and (8.18) are equivalent to the following relationship between the quantiles of the survivor function:

$$\log Q_\mathbf{x}(p) = \log Q_0(p) - \log g(\mathbf{x})$$

where $Q_\mathbf{x}(p) = S^{-1}(t, \mathbf{x})$, with $t = 1 - p$ and $Q_0(p) = S_0^{-1}(1 - p)$, $0 < p < 1$.

In fact all models in table 8.2 reflect the general form (8.18). Models other than the exponential one have an additional parameter σ which has the effect of rescaling the contribution of each of the covariates, but does not depend on them. Thus relationship (8.17) still holds with $g(\mathbf{x}) = \exp(\sigma\boldsymbol{\beta}'\mathbf{x})$.

The functions describing the distribution of failure time T in any model (8.18), given that the covariates act according to (8.17), satisfy the following relationships:

$$S(t, \mathbf{x}) = S_0(t \cdot g(\mathbf{x}))$$

$$f(t, \mathbf{x}) = g(\mathbf{x})f_0(t \cdot g(\mathbf{x})) \tag{8.19}$$

$$\lambda(t, \mathbf{x}) = g(\mathbf{x})\lambda_0(t \cdot g(\mathbf{x}))$$

Table 8.3 illustrates that, starting from the definition (8.8), the exponential model with a 0–1 covariate and the Weibull model with a vector $\mathbf{x}$ of explanatory variables can be rewritten in a form that satisfies relationships (8.19).

It is worth mentioning that a recently addressed issue is that of a semi-parametric approach to the analysis within the AFT framework (Wei, 1992).

8.5 RELATIONSHIP OF AFT MODELS WITH PROPORTIONAL HAZARDS AND PROPORTIONAL ODDS MODELS

Let us first focus on the AFT and PH classes of models in a two-sample problem, where $x = 0, 1$. In order to illustrate how the influence of the covariate x is modelled differently in the two classes, consider a hypothetical case in which the baseline hazard is a step function represented by the continuous line in figure 8.8. If x is related to an accelerating factor of value 2, that is $g_{\mathrm{AFT}}(x) = 2$, the group with $x = 1$ experiences a higher failure rate, which is represented by the asterisks in the upper part of figure 8.8. For example, in the time interval $[0.5, 1)$, when $\lambda_0(t)$ has constant value 0.5, $\lambda(t, 1) = 2\lambda_0(2t)$. Thus, given that $\lambda_0(t) = 1$ in the interval $[1, 2)$, $\lambda(t, 1) = 2$ in the interval $[0.5, 1)$ and this value does not change thereafter as far as $\lambda_0(t)$ remains constant for $t > 2$ (at least up to $t = 4$ for the line of asterisks in the figure to be correct). If instead x is related to a PH factor of 2, i.e. $g_{\mathrm{PH}}(x) = 2$, the hazard in the group with $x = 1$ is twice the baseline $\lambda_0(t)$, as represented in the lower part of figure 8.8. With the usual choice of the exponential function $e^{\beta x}$, the same value of the coefficient $\beta_{\mathrm{AFT}} = \beta_{\mathrm{PH}} = 0.69$ gives two different pictures of the way the hazard increases for a change in the value of x. On the other hand, it is clear from figure 8.8 that, if the baseline hazard does not change at $t = 1$, the plots under AFT and PH would coincide. Thus, depending on the shape of the baseline function, there might be models for which it is equivalent to assume PH or AFT. In fact the Weibull model has this feature, as it can be seen as an AFT (section 8.4) and a PH

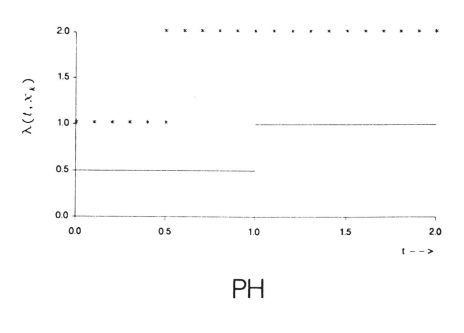

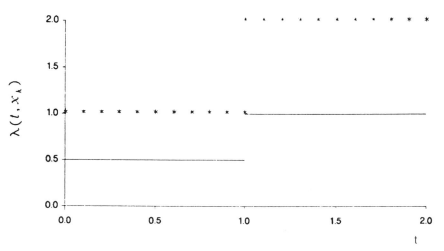

Figure 8.8   Baseline hazard (———) accelerated by a factor of 2 (∗∗∗∗∗) in AFT and multiplied by a constant 2 (∗∗∗∗∗) in PH.

(section 8.3.1) model. From table 8.3, for a dichotomous covariate x:

$$\lambda(t, 1) = e^{\sigma\beta}\lambda_0(t \cdot e^{\sigma\beta})$$

and thus the Weibull model satisfies the AFT relationships (8.19) with $g_{\text{AFT}}(x) = e^{\sigma\beta x}$. The PH assumption is satisfied too, given that the Weibull model expressed in the form (8.8) is

$$\lambda(t, 1) = e^{\beta} \cdot \lambda_0(t)$$

and thus belongs to the PH class (8.9) with $g_{\text{PH}}(x) = e^{\beta x}$. Having chosen the Weibull distribution (8.1) as the baseline, we have that one unit difference in the same covariate x has the effect of multiplying the hazard by e^{β} and this turns out to be equivalent to accelerate the failure time by a factor $e^{\sigma\beta}$.

In general, if $\lambda_0(t)$ does not have a Weibull form, this equivalence is not true and the acceleration imposed by a covariate on the failure time does not translate into a multiplication of the hazard by a constant. Intuitively, it is the simple monotonic nature of the hazard function, for any value of p (or σ), which is at the basis of this result. The coexistence of the AFT and PH property can be formally shown to hold only in the Weibull case with fixed explanatory variables (Cox and Oakes, 1984, p. 71).

As far as the relationship between AFT and PO models is concerned, a hypothetical two sample problem is again used to illustrate the assumptions on the influence of a dichotomous covariate x on the odds function. This is an increasing function of time; here, the baseline odds are assumed to have the form of a step function, as in figure 8.9 (continuous line). The effect on the odds function of an accelerating factor 2, i.e. $g_{\text{AFT}}(x) = 2$, is shown in the upper part of figure 8.9. In the group with $x = 1$ the odds of failure at time t are

$$\frac{1 - S_0(t \cdot 2)}{S_0(t \cdot 2)}$$

and considering suitable time intervals as $[0, 0.5)$ and $[0.5, 1)$, the odds denoted by the line of asterisks are obtained. If instead x is related to a PO factor of 2, with $g_{\text{PO}}(x) = 2$, the resulting odds function is represented by the asterisks in the lower part of figure 8.9. Again, we are prompted to ask whether there are parametric models from the AFT class which are also PO models, because of a specific form of their baseline survivor function. A model of this type is the log-logistic one which we have represented as both an AFT (section 8.4) and a PO (section 8.3.2) model. It can be formally shown, indeed, that this model is the only one to satisfy both the AFT and PO assumptions (Lawless, 1986).

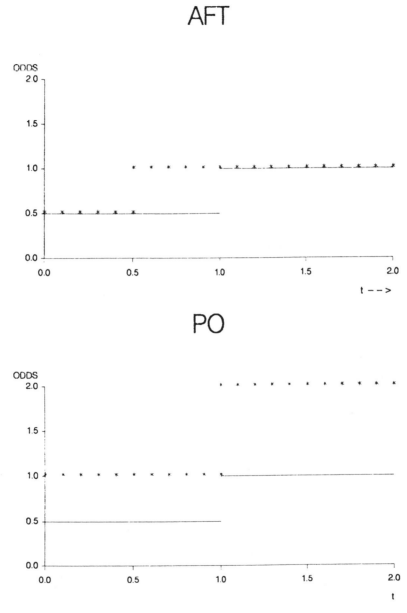

Figure 8.9   Baseline odds function (———) accelerated by a factor of 2 (∗∗∗∗∗) in AFT and multiplied by a constant 2 (∗∗∗∗∗) in PO.

8.6   AN EXAMPLE

In this section the use of a log-logistic model is illustrated by means of the example originally adopted by Bennett (1983). It concerns data from the Veterans Administration lung cancer trial, in which patients with advanced inoperable lung cancer were treated with chemotherapy. The treatment difference (test vs standard) was shown to be statistically not significant and the attention was then focused on the evaluation of the prognostic role of two covariates: Performance Status (PS) (Karnofsky index), a measure of general fitness on a scale from 0 to 100, and tumor histotype: large cell, adeno, small cell and squamous. Only the subset of 97 patients with no prior therapy was

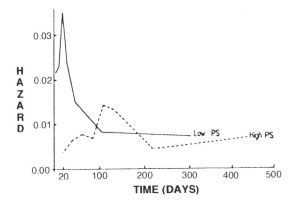

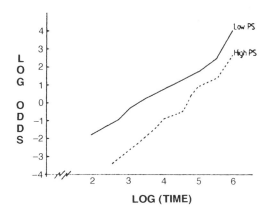

Figure 8.10   Smoothed empirical hazard function estimates (upper part) and log odds estimates (lower part) for data from the Veterans Administration lung cancer trial (from Bennett, 1983, reproduced by permission of The Royal Statistical Society).

Table 8.4 Results of fitting a log-logistic model to lung cancer data (from Bennett, 1983, reproduced by permission of The Royal Statistical Society).

Parameter	Estimate	Standard Error	Q_W	Odds ratio
p	1.841	0.163		
β_0	−5.293	0.860		
β_1	−0.055	0.010	30.25	
β_2	1.295	0.552	5.50	3.651
β_3	1.372	0.521	6.93	3.943
β_4	−0.138	0.576	0.06	0.871

Log-likelihood of the fitted model is −131.202.

considered by Bennett. After classifying patients by performance status as High (score over 50) or Low (score 50 or below), Bennett obtained the smoothed empirical hazard estimates shown in the upper part of figure 8.10; it appears that the two plots are not monotonic, suggesting the unsuitability of a Weibull model. The empirical check according to (8.15) is given in the lower part of figure 8.10; since the lines drawn for the two groups appear to be reasonably straight and parallel they provide some suggestion for the use of the log-logistic model.

The log-logistic model (8.15) was fitted by inserting into the linear predictor the performance status as a continuous variable (x_1) and the histotype by means of three dummies with "large cell" as the reference group, namely x_2 adeno (1) vs large cell (0); x_3, small cell (1) vs large cell (0); and x_4, squamous (1) vs large cell (0). The results are reported in table 8.4; note that $p \log \lambda$ of (8.15) corresponds to the intercept β_0.

The results show that, for any time t, the odds of dying by t are some four times greater for adeno or small cell histotype compared with large cell or squamous histotypes. Since performance status was inserted as a continuous variable, its effect can be assessed by choosing proper cut-off points, say 50 (partial confinement of the patient) and 80 (the patient is able to care for himself or herself) and computing the pertinent odds ratio for a given t. Considering, for example, $t = 180$ days and the small cell histotype, we have:

PS = 50: log odds = $1.841 \log 180 - 5.293 - 0.055 \cdot 50 + 1.372 = 2.88923$
PS = 80: log odds = $1.841 \log 180 - 5.293 - 0.055 \cdot 80 + 1.372 = 1.23924$

The odds ratio estimate of a patient with PS = 50 compared to that of a patient with PS = 80 is simply $\exp[-0.055(50 - 80)] = 5.21$, and this value is constant for any time t.

8.7 CONCLUDING REMARKS

There are two aspects in any of the models considered in section 8.3 which deserve consideration: the form of $g(\mathbf{x})$ and the distributional assumptions for the r.v. T, conditional on $\mathbf{x} = \mathbf{0}$.

Table 8.5 Regression coefficient estimates obtained by means of four different models fitted to a data set on breast cancer patients.

Variable and code	Cox		Weibull		Log-logistic		Log-normal	
	$\hat{\beta}$	SE($\hat{\beta}$)	$\hat{\beta}$	SE($\hat{\beta}$)	$\hat{\beta}$	SE($\hat{\beta}$)	$\hat{\beta}$	SE($\hat{\beta}$)
Metastatic nodes								
N^- 0								
N^+ 1	1.1477	0.1051	1.2117	0.1073	1.2736	0.1028	1.1978	0.1045
Tumor size								
T_1 0 1	−0.8310	0.2080	−0.8380	0.2059	−0.8850	0.2076	−0.9412	0.2126
T_2 1 0	−0.3728	0.1407	−0.3863	0.1388	−0.3151	0.1508	−0.3547	0.1540
T_3 0 0								
Age								
50–54 0								
Others 1	0.3515	0.13344	0.3908	0.1327	0.3532	0.1371	0.2873	0.1340
Intercept			−6.0034	0.1920	−5.5854	0.2022	−5.4644	0.2017
Scale parameter			0.9877	0.0420	0.7225	0.0294	1.2443	0.0459
Log-likelihood	−2589.35 (partial)		−943.76		−919.64		−915.10	

As in ordinary multiple regression analysis, the specification of $g(\mathbf{x})$ requires consideration of (i) which covariates to include and (ii) the functional form of $g(\mathbf{x})$. In applications, particularly in the medical field, background knowledge very rarely permits one to prefer one distributional form over another.

The body of evidence emerging from medical literature suggests that when the primary concern is in evaluating the effect of covariates little is to be gained by moving from semi-parametric models, like the Cox model, to the parametric ones. Consider, for example, breast cancer patient survivorship: Pocock *et al.* (1982) and Gore *et al.* (1984) found that Cox, Weibull and log-logistic models give the same view of how the prognostic factors influence survival. Similar considerations emerge from table 8.5 reporting the results of an analysis carried out on a series of 716 breast cancer patients followed up for more than 20 years (Mezzanotte *et al.*, 1987). The table reports the estimates of the regression coefficients and their standard errors, together with codes adopted to generate the regressors, obtained by fitting Cox, Weibull, log-logistic and log-normal models. It appears that the values of the regression coefficients estimated by the four models are similar. If one considers, however, the three parametric models, the values of the log-likelihood function calculated with the log-logistic and log-normal models are practically equal whilst they compare favourably to that of the Weibull model, since the larger the log-likelihood, the better is the fit of the model. Moreover, on the basis of the pattern of the estimated hazard functions and the graphical analysis of residuals, the authors concluded that the log-logistic and log-normal models were to be preferred to the Weibull one.

On the other hand in many branches of medical research, failure models are frequently used to predict the median survival time or the probability of a given individual surviving for at least 5 years, say. For the purposes of prediction, it is worth considering parametric models. For example, Mezzanotte *et al.* (1987) used the log-normal model to analyze a series of breast cancer patients and were able to partition the whole case-series in three subsets with different survival probability at 10 years from surgery: more than 0.70, between 0.70 and 0.40 and less than 0.40, respectively. This classification relies upon patient characteristics which do not reflect some peculiarity of the sample, but conform to the clinical knowledge of the history of the disease. Nonetheless it seems sensible not to ascribe too rigid an interpretation to the estimated survival functions in terms of long run frequencies since it would have been constrained either by the choice of the form of hazard function or of $g(\mathbf{x})$ or of both.

Formal methods are sometimes advocated to choose among different candidate models. One which considers both the value of the maximized likelihood L and the number of parameters p in the fitted model was introduced by Akaike (1972). The Akaike information criterion (AIC) is defined as $-2 \log L + 2p$: according to this criterion the model of choice, among various non-nested models, is the one which has the lowest value of AIC.

A general model which includes all the distributions in section 8.2 as special cases can be used to discriminate which models are more suitable to fit the data.

Such a model is the generalized F distribution discussed in Kalbfleisch and Prentice (1980, pp. 28, 64–66) and Ciampi *et al.* (1986). The method is computationally complicated and in practical applications might have little ability to discriminate among alternative models.

APPENDIX A8.1 THE GAMMA FUNCTION

The function $\Gamma(k)$ is defined by

$$\Gamma(k) = \int_0^\infty u^{k-1} e^{-u} \, du, \quad k > 0$$

and thus $\Gamma(\cdot)$ is just a notation for the definite integral above. Integration by parts yields

$$\Gamma(k+1) = k\Gamma(k)$$

and, hence, if k is an integer n, it is

$$\Gamma(n+1) = n!$$

APPENDIX A8.2 GENERATION OF RANDOM FAILURE TIMES FROM A GIVEN DISTRIBUTION

The generation of a random set of N failure times from a given distribution of T is simple if an explicit form of the inverse of the distribution $F(t)$ is available. Consider the variable Y which takes values $F(t)$ for any $t, y = F(t)$, and the inverse function $t = F^{-1}(y)$, with $0 \leqslant y \leqslant 1$. Let a random set of N observations be generated from a uniform distribution over $[0, 1]$. These are taken as the values of Y and the inverse function $F^{-1}(y)$ is applied to this set in order to obtain a set of values for T which are distributed according to $F(t)$.

The parametric models presented in section 8.1 admit the following inverse functions, expressed in terms of t and $y = F(t)$ for a given t:

(1) exponential distribution, $y = 1 - \exp(-\lambda t)$

$$t = -\frac{1}{\lambda} \log(1 - y)$$

(2) Weibull distribution, $y = 1 - \exp[-(\lambda t)^p]$

$$t = \frac{1}{\lambda} [-\log(1 - y)]^{1/p}$$

(3) log-normal distribution, $y = \Phi\left(\frac{\log(\lambda t)}{\sigma}\right)$

$$t = \exp[\sigma\Phi^{-1}(y) - \log\lambda]$$

(4) log-logistic distribution, $y = (\lambda t)^P / [1 + (\lambda t)^P]$

$$t = \frac{1}{\lambda}\left(\frac{y}{1-y}\right)^{1/p}$$

APPENDIX A8.3   DERIVATION OF THE LINEAR FORM IN $Y = \log T$ FOR THE WEIBULL DISTRIBUTION

Given the continuous random variable T, we seek the distribution of $Y = \log T$ knowing that the Weibull hazard for T, conditional on $\mathbf{x}$, is given by (8.8). The logarithmic function has the properties which allow us to apply the transformation technique, as it can be verified in any text on the theory of statistics. In particular, and omitting the index i for the individual, $y = \log t\,(t > 0)$ is a one-to-one transformation with an inverse, $t = \exp y$, which has a continuous non-zero derivative and thus

$$f_Y(y, \mathbf{x}) = \frac{\mathrm{d}}{\mathrm{d}y}(\exp y)\cdot f_T(\exp y, \mathbf{x})$$

The Weibull model has density function

$$f_T(t, \mathbf{x}) = \lambda p(\lambda t)^{p-1}e^{\boldsymbol{\beta}'\mathbf{x}}\cdot\exp[-(\lambda t)^p e^{\boldsymbol{\beta}'\mathbf{x}}]$$

and, by applying the transformation technique, it follows that

$$f_Y(y, \mathbf{x}) = p\lambda^p(e^y)^p e^{\boldsymbol{\beta}'\mathbf{x}}\cdot\exp[-\lambda^p(e^y)^p e^{\boldsymbol{\beta}'\mathbf{x}}]$$

By setting $\alpha = -\log\lambda$ we obtain

$$f_Y(y, \mathbf{x}) = pe^{-\alpha p + \boldsymbol{\beta}'\mathbf{x}}e^{yp}\cdot\exp[-e^{-\alpha p + \boldsymbol{\beta}'\mathbf{x}}e^{yp}]$$

and thus

$$f_Y(y,\mathbf{x}) = p\exp\left[p\left(y - \alpha + \frac{\boldsymbol{\beta}'\mathbf{x}}{p}\right) - \exp\left\{p\left(y - \alpha + \frac{\boldsymbol{\beta}'\mathbf{x}}{p}\right)\right\}\right]$$

If we define $\sigma = 1/p$ and

$$w = \frac{y - \alpha + \sigma\boldsymbol{\beta}'\mathbf{x}}{\sigma}$$

then $f_Y(y, \mathbf{x}) = (1/\sigma)\exp[w - e^w]$ and the variable W is found to have the extreme value distribution

$$f_W(w, \mathbf{x}) = \exp[w - e^w]$$

9

The Study of Prognostic Factors and the Assessment of Treatment Effect

9.1 INTRODUCTION

The study of prognostic factors is an important part of the analysis of clinical data and parallels the study of risk factors in epidemiology. Knowledge of clinical and biological characteristics of patients and disease which are related to differences in prognosis is important for various reasons, which are outlined in section 9.2. In particular, the issue of when and how to adjust the estimation of treatment effect in clinical studies is dealt with in section 9.3. The issue of parsimony, that is the need of identifying a minimal "core' set of relevant factors, the approach to the study of new prognostic factors and the use of prognostic factors for prediction will be dealt with in section 9.4. Finally, the evaluation of treatment effect and its interpretation in terms of clinical relevance is discussed in section 9.5.

9.2 WHY STUDY PROGNOSTIC FACTORS?

The main motivation for the development and use of regression models in survival analysis and in epidemiological studies has been the need of estimating the effect of treatment or exposure allowing for prognostic or confounding factors. In this respect regression models are a formal and elegant tool for estimating treatment effect by comparing "like with like", that is for applying the basic idea of stratification (section 4.5). The advantage is that regression models directly provide estimates of the impact of explanatory variables as well and thus can be quite useful if the quantitative evaluation of the relationship of these variables with outcome is of interest.

It is certainly true that in clinical studies the evaluation of prognostic factors is, *per se*, a matter of interest. The clinical investigator is aware, especially in oncology, that the variability in the outcome can be much more related to some of these factors than to different treatments. Thus the study of prognostic factors may provide some insight into the *natural history* of the disease, that is in the understanding of how it evolves, given a set of patient/disease characteristics. The term "natural history" does not indicate a lack of therapeutic intervention; usually we are rather explaining the course of the disease in the presence of whatever types of intervention have been done in an attempt at improving the course itself. Nonetheless this type of study may stimulate research on the *underlying biological mechanisms* and in this case the primary purpose is to understand why certain factors seem to be related to outcome.

What instead is more relevant to the process of clinical decision-making is the use of prognostic factors for the *prediction* of disease outcome. The clinical investigator is in particular willing to discriminate between patients who are likely to have very different prognoses, given some characteristics measured at disease onset or during the course of the disease. The idea usually is that once subgroups of patients at different risk of failure are identified, one can tailor the treatment strategy to the *risk group*. However, an investigator cannot claim to have identified treatment specificity just because he finds, in a clinical trial comparing placebo or a standard regimen to an experimental treatment, some subset in which the treatment difference is not significant and another in which it is significant (section 2.7). A proper analysis should consider an overall test for interaction between treatment and prognostic factors and a previous plan which limits the analysis to a number of biologically meaningful subset hypotheses (section 4.5). On the other hand, the identification of prognostic features which, regardless of the actual treatments received by patients, identify different risk groups, may guide in the decision on future treatment studies. For instance, in spite of the improvements reached in the treatment of childhood leukemia in countries which adopt similar intensive chemotherapeutic regimens, it is possible, on the basis of various factors, to identify a relatively small group of patients who still experience a very high risk of failure. This group could be considered for a study on therapies that are more demanding, in terms both of toxicity and side effects and of economical cost, as for example bone marrow transplant or other types of intensive therapies. At the other extreme, the datum that Stage I, FIGO grade 1 ovarian cancer patients have a 90% survival at five years from first surgery, without any chemotherapy, could induce not to treat future patients with these characteristics. This is because of the lack of any proven active and completely side-effect-free treatment or because of the difficulty of thinking that, with the drugs actually available, a randomized study could show a worthwhile effect of treatment over no treatment.

The knowledge of prognostic factors contributes thus to the *planning of future studies*, either randomized or not. In defining eligibility criteria, prognostic factors are used to identify patients who are likely not to be harmed by the study treatments and in the meanwhile future patients for whom the results of the study are thought to be relevant. In randomized trials with relatively broad eligibility criteria, a few factors which are known from experience to be related to large differences in outcome are worth using to stratify the randomization process, especially in studies of relatively small size (section 2.3).

Finally, an aspect which pertains more to the epidemiological side, is that the knowledge of risk factors may stimulate the conduct of intervention or prevention trials.

9.3   ADJUSTING THE ESTIMATE OF TREATMENT EFFECT

When comparing treatments, the main reason for adjusting on explanatory variables is to avoid any bias that might result because of imbalances of these variables across the treatment groups. Such imbalances need only concern us when they occur for variables which are of prognostic relevance, and are thus related to outcome.

The same reason for adjusting is present in the familiar context of linear regression where, in addition, a substantial decrease in the variance of the estimate of treatment effect can be achieved after adjustment by a covariate related to outcome. Unlike in linear regression, in non-linear models used in survival analysis there is in practice no meaningful reduction in the variance of the estimate as a result of adjustment. Moreover, in non-linear models even the omission of a balanced covariate of prognostic relevance may bias the estimate of treatment effect. These issues will be discussed in the following subsections, with reference to adjustment for unbalanced and balanced covariates in survival models, as applied to data arising from randomized clinical trials. Reports appearing in the literature on clinical trials show there are still very different attitudes toward adjustment in the estimation of the difference in the outcome of patients randomly assigned to different treatments. On the contrary, in observational studies, adjustment, or stratification, is the standard procedure since it is well known that non-randomized treatment groups are very likely not to be comparable with regard to major prognostic (or confounding) factors.

9.3.1   In the presence of unbalanced covariates

Simple randomization ensures that treatment groups could differ in their characteristics only by chance; thus the larger the trial, the more likely it is that subjects allocated to different treatments have very similar characteristics (section 2.3). There still remains the possibility that in a given trial a few specific characteristics are not split equally between groups, even if randomization was correctly performed. The distribution of characteristics by treatment arm is commonly given in any report on a clinical study, in the form of a table. It is absurd, however, to think of detecting important imbalances on the basis of significant tests performed to compare the treatment groups with respect to each explanatory variable. These tests do not provide any information on the possible influence of the imbalance on treatment outcome, since the attached p-value only assesses the probability of such imbalance having occurred by chance, when we know it did, in a proper randomization. Thus we cannot infer, from the lack of "statistical significance", at whatever level we choose to fix it (the magic 0.05), a lack of influence on outcome of an imbalance of a prognostic factor across treatment groups. This topic is thoroughly discussed by Altman (1985); his example is taken here to illustrate the problem and it serves to illustrate the scope even if it simplifies the outcome to be a binary variable, with no time and censoring involved.

Suppose that two treatments A and B are assigned by simple randomization to 400 subjects and that the trial aims at comparing the outcome in terms of the probability of dying within a specified period after entry. In addition, suppose that according to a given characteristic, two strata, 1 and 2, accounting for 75% and 25% of the subjects, are defined, and in stratum 2 the probability of dying is twice as high as that in stratum 1, regardless of treatment (no interaction is present). Table 9.1 shows the expected results of such a trial when perfect balance is achieved, that is in each treatment group exactly 25% of the subjects are in the worst prognosis stratum. Overall 50% and 40% of the patients die with treatment A and B, respectively. Treatment absolute difference is 10% in favour of B or, in other terms, the relative difference is 20% ((0.5–0.4)/0.5%) which expresses the relative benefit of B over A. Clearly, the stratified analysis of data in table 9.1 gives the same estimate of treatment difference, since this latter is $80\% - 64\% = 16\%$ in stratum 2 and $40\% - 32\% = 8\%$ in stratum

Table 9.1 Mortality in two treatment groups with a balanced prognostic factor related to a hazards ratio of 2 and with a 25% prevalence of worse prognosis patients (stratum 2) (from Altman, 1985, reproduced by permission of The Royal Statistical Society).

Prognostic factor	Treatment A	B	Total
Stratum 1	60/150 (40%)	48/150 (32%)	108/300 (36%)
Stratum 2	40/50 (80%)	32/50 (64%)	72/100 (72%)
Total	100/200 (50%)	80/200 (40%)	180/400 (45%)

$\chi^2_{\text{TREAT}} = 3.65$, $p = 0.06$
χ^2_{TREAT} (with Yates's correction) compares the overall mortality in the two treatment groups $\left(\frac{100}{200} \text{ vs } \frac{80}{200}\right)$

Table 9.2 Mortality in two treatment groups with imbalances in the prognostic factor related to a hazards ratio of 2 (from Altman, 1985, reproduced by permission of The Royal Statistical Society).

	Treatment A	Treatment B	Total
(a) Excess of worst prognosis patients in treatment group A			
Stratum 1	56/141 (40%)	51/159 (32%)	107/300 (36%)
Stratum 2	47/59 (80%)	26/41 (64%)	73/100 (73%)
Total	103/200 (51.5%)	77/200 (38.5%)	180/400 (45%)
$\chi^2_{BAL} = 3.85$, $p = 0.05$		$\chi^2_{TREAT} = 6.31$, $p = 0.015$	
(b) Excess of worst prognosis patients in treatment group B			
Stratum 1	64/159 (40%)	45/141 (32%)	109/300 (36%)
Stratum 2	33/41 (80%)	38/59 (64%)	71/100 (71%)
Total	97/200 (48.5%)	83/200 (41.5%)	180/400 (45%)
$\chi^2_{BAL} = 3.85$, $p = 0.05$		$\chi^2_{TREAT} = 1.71$, $p = 0.9$	

χ^2_{TREAT} computed as in table 9.1

χ^2_{BAL} (with Yates's correction) compares the proportions with and without the risk factor in the two treatment groups: $\frac{59}{200}$ vs $\frac{41}{200}$ in both the above tables

1 ($\frac{1}{4} \cdot 16\% + \frac{3}{4} \cdot 8\% = 10\%$). The difference is not quite significant at the $\alpha = 0.05$ level ($\chi^2_{TREAT} = 3.65$, $p = 0.06$).

Data are now perturbated to introduce two opposite cases of imbalance, in which excess worst prognosis patients are first put in treatment group A (table 9.2(a)) and then in group B (table 9.2(b)). The imbalance is taken in both cases to be just significantly different from zero, since $\chi^2_{BAL} = 3.85$ and $p = 0.05$. In this way the two cases in table 9.2 bracket all the cases in which the degree of imbalance would turn out not to be statistically significant. The overall estimates of treatment difference are markedly affected and the two cases define a range of values between 13%, obtained in case (a), and 7% in case (b) (or, in terms of the relative benefit of B over A, the range of estimates is 25% to 14%). The corresponding p-values, as given by χ^2_{TREAT}, range from $p = 0.015$ to $p = 0.19$. Thus, the omission to consider for adjustment a prognostic variable in the presence of an even modest imbalance might result in biased estimates of treatment effect. As suggested by Altman's results on similar examples on a binary outcome, the relative chance increases the more the variable is prognostic; on the other hand, ranges of the type calculated above reduce as sample size increases and are not greatly affected by the prevalence of the bad prognosis characteristic.

9.3.2 In the presence of balanced covariates

Various theoretical papers have studied the implications of omitting a covariate balanced across treatment groups in the analysis of survival data. The problem is that if for instance the *true* hazard of an individual depends on treatment (x_1) and on a covariate (x_2) according to the PH multiplicative model,

$$\lambda(t, \mathbf{x}) = \lambda_0(t) \exp(\beta_1 x_1 + \beta_2 x_2) \qquad (9.1)$$

then the PH model which omits x_2 is misspecified. Even if x_2 is a balanced covariate, its omission mixes different hazards by ignoring strata defined by x_2 and this may cause the hazards in the treatment groups not to be proportional any more. Let the incorrect model assumed to perform the analysis be:

$$\lambda^*(t, x_1) = \lambda_0^*(t) \exp(\beta_1^* x_1) \qquad (9.2)$$

The consequences of this type of misspecification on estimation of treatment effect are summarized here without going into the details of their formal derivation (under the usual assumption of non-informative censoring):

(1) the estimate $\hat{\beta}_1^*$ is biased towards zero, that is $|\beta_1^*| < |\beta_1|$ both in parametric PH models, such as the exponential model (Gail *et al.*, 1984; Gail, 1986; Chastang *et al.*, 1988) and in the Cox model (Struthers and Kalbfleisch, 1986);

(2) in both parametric and semi-parametric models, β_1^* is asymptotically unbiased for $\beta_1 = 0$;

(3) for moderately censored exponential survival data, the analysis with an exponential survival model (9.2) gives less biased estimates of β_1 than does the Cox model (Gail *et al.*, 1984; Gail, 1986).

According to these findings, the omission of a *balanced* covariate from the linear predictor can lead to a biased estimate of treatment effect. Since this may seem quite non-intuitive, it is shown to happen in a simple example, as suggested by Chastang *et al.* (1988). A set of death times is generated from a multiplicative exponential model in which two different treatments A and B and a dichotomous covariate defining two strata (1 and 2) are related to a hazard ratio of 2 (B vs A) and 6 (stratum 2 vs 1), respectively. Twelve observations are generated and each treatment arm contains six observations equally distributed in the two strata. The order in which the deaths occur is the following:

$$B_2 \quad A_2 \quad B_2 \quad A_2 \quad B_2 \quad B_1 \quad A_2 \quad A_1 \quad B_1 \quad A_1 \quad B_1 \quad A_1$$

where the subscripts indicate the membership in one of the two strata. Suppose that these data are analysed by using the Cox model, in which the estimates are based on the ranks given above. With one covariate only, x_1 for treatment, as in (9.2), the unadjusted estimate is $\hat{\beta}_1^* = 1.507$ while the adjusted estimate in a Cox model including also the indicator x_2 for stratum, as in (9.1), is $\hat{\beta}_1 = 2.125$. Thus, when strata are taken into account, the estimate of β_1 is close to the true value of 2 while the omission of a balanced covariate for stratum, which has a strong prognostic effect, results in underestimation. Given the mechanism of generation of these data, the PH assumption holds for treatment within each stratum and evidently fails to do so when the strata are ignored.

After this "extreme" example a question naturally arises whether we can discriminate among different situations in which the bias might be expected to be substantial or negligible. This is important since in practice we need some guidelines to define a strategy for the analysis. Chastang *et al.* (1988) derived a simple approximate formula for calculating the bias in the multiplicative exponential model and explored with it the dependence of the extent of bias on five quantities: the true values of β_1 and β_2, the percentage of censoring and the distributions of x_1 ($x_1 = 0, 1$ for treatments A and B, respectively) and x_2 ($x_2 = 0, 1$ for strata 1 and 2, respectively). In summary they found that, with two dichotomous covariates x_1 for treatment and x_2 for a prognostic factor, the bias $\beta_1 - \hat{\beta}_1^*$ is most affected by the value of β_2, the distribution of the prognostic factor x_2 and the percentage of censoring, while the two other quantities have a limited influence.

Table 9.3   Values of $\hat{\beta}_1^*$ obtained from a multiplicative exponential model omitting the covariate x_2, when the true model is (9.1), with different values of β_1 and β_2, $\lambda_0(t) = 0.01$, prob ($x_2 = 0$) = 0.5, 50% censoring, and 25 patients in each of the four categories defined by x_1 and x_2 (from Chastang *et al.*, 1988).

True parameters and hazard ratios				Estimated parameters and hazard ratios	
β_1	HR	β_2	HR	$\hat{\beta}_1^*$	HR
0.5	1.65	0	1.00	0.50	1.7
		1	2.72	0.47	1.6
		2	7.39	0.38	1.5
		3	20.09	0.27	1.3
		4	54.60	0.19	1.2
1.0	2.72	0	1.00	1.00	2.7
		1	2.72	0.93	2.5
		2	7.39	0.76	2.1
		3	20.09	0.56	1.7
		4	54.60	0.38	1.5
2.0	7.39	0	1.00	2.00	7.4
		1	2.72	1.87	6.5
		2	7.39	1.55	4.7
		3	20.09	1.17	3.2
		4	54.60	0.81	2.3
4.0	54.60	0	1.00	4.00	54.6
		1	2.72	3.80	44.7
		2	7.39	3.29	26.8
		3	20.09	2.63	13.9
		4	54.60	1.98	7.2

Table 9.3 shows the values of $\hat{\beta}_1^*$ for different values of β_1 and β_2 in the true model (9.1) with $\lambda_0(t) = 0.01$, same size of the two treatment groups and a 0.5 prevalence of the good prognosis patients. For a prognostic factor which is related to approximately a threefold increase of the hazard (i.e. $\beta_2 = 1$) the bias in the estimate of treatment effect from model (9.2) is modest, for all values of β_1. We see instead that the bias becomes appreciable as β_2 is equal to 2, which gives a hazard ratio of about 7, or $\beta_2 > 2$. These results were obtained in the presence of a 50% overall censoring, and thus in a setting penalized by a percentage of censoring which is related to a larger bias for a given treatment effect β_1 and for different values of β_2, as shown in figure 9.1, where $\beta_1 = 0.5$. This figure deserves an additional comment: in the exponential model, the bias tends to zero not only when censoring becomes very small, as shown also by Gail (1986), but also when censoring is very heavy. This can be explained by the fact that most of the observed deaths, when there is a lot of non-informative censoring, will occur in the worst prognosis stratum while the better prognosis patients will not be represented much among the deaths. Thus the estimation of β_1 will not suffer much from lack of information on membership in the two strata.

Finally, the dependence of the amount of bias on the prevalence of the good prognosis feature $(x_2 = 0)$ is shown in figure 9.2, for various values of β_2. The

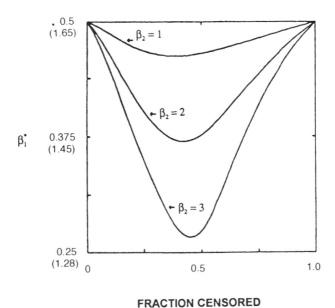

Figure 9.1   The effect of the percentage of censoring on $\hat{\beta}_1^*$ for $\beta_1 = 0.5$ and various values of β_2, with $\lambda_0(t) = 0.01$ and prob $(x = 0) = 0.5$ in model (9.1) (from Chastang *et al.*, 1988).

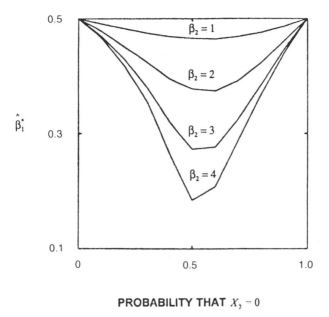

Figure 9.2   The effect of the probability that $x_2 = 0$ on $\hat{\beta}_1^*$ for $\beta_1 = 0.5$ and various values of β_2, with $\lambda_0(t) = 0.01$ and in the presence of 50% of censored data (from Chastang *et al.*, 1988).

amount of bias for all values of β_2 tends to reduce from a maximum, achieved when prob $(x_2 = 0)$ is about 0.5 or 0.6, to zero when one of the two strata is preponderant. It tends to decrease more slowly when prob $(x_2 = 0)$ is larger than when it is smaller. The reason for this asymmetry is worth commenting on: given a fixed amount of non-informative censoring (50%), if most subjects have $x_2 = 0$, censoring will affect this group heavily and there will remain similar amounts of deaths in the two groups; more information will be lost if we omit x_2. Conversely, when most subjects have $x_2 = 1$, the bias is smaller because most deaths will be within the same group with respect to x_2, namely in the one with $x_2 = 1$, and omitting x_2 will have less influence. This is also consistent with the finding in figure 9.1 that the bias decreases as the percentage of censoring increases.

The increase in the magnitude of the bias with increased prognostic value of x_2 is evident also in the two figures 9.1 and 9.2. Results on the dependence of bias on β_2 were obtained for the Cox model by Struthers and Kalbfleisch (1986) and are similar to those obtained for the exponential model except when censoring is absent. In this case $\hat{\beta}_1^*$ is biased toward zero in the Cox model, while it is not biased in the exponential model.

We now consider hypothesis testing, and in particular the performance of the score test for the null hypothesis $H_0 : \beta_1^* = 0$ when the true model is (9.1) and we omit the balanced covariate x_2 in the analysis by applying model (9.2). On the grounds of the results in Lagakos and Schoenfeld (1984), Morgan (1986) and Gail (1986), the variance of the score based on partial likelihood is not influenced by the omission of a balanced prognostic factor and neither is the nominal size of the test. However, a loss of power can occur due to the underestimation of β_1. Theoretical results on the exponential model for uncensored data show that a deflation of the score variance can occur, leading to an anticonservative test. However, this behaviour is substantially attenuated in the presence of censoring (20% or more), as usually met in practical situations.

9.3.3   Concluding remarks

We have been discussing how randomization does not always lead to asymptotically unbiased estimates of treatment effect when needed covariates are omitted. However, as pointed out by Gail *et al.* (1984), randomization assures asymptotically unbiased estimates in a surprisingly large number of practical cases. This is very important since in applied work we do not know which is the true model and we may even not have detected yet, due to our scarce knowledge of the disease mechanisms, a factor which strongly influences prognosis. In this absence of knowledge, randomization remains a valuable tool for obtaining unbiased estimates of treatment effect. Moreover, the bias is negligible, under the false model, when treatment effect β_1 is near zero. This is in accord with the fact that, based on the randomization, the significant tests of the null hypothesis $H_0: \beta_1 = 0$ have valid size, regardless of the form of the true population model and regardless of what covariates may or may not have been omitted in the analysis of a sample.

In practice, as a general approach to data analysis of randomized clinical trials, we should be protected from giving substantially biased estimates if we adjust on very strong prognostic variables, even if they are balanced across treatment groups. Results in the exponential and Cox model with one binary covariate suggest that in the presence of a fair amount of censoring the bias may become appreciable if the prognostic variable is related to a hazard rate ratio of about 7 ($\beta_2 = 2$) or greater, particularly when the two strata are roughly equally represented in the data set.

We have shown here the results concerning simple cases of one covariate only in the linear predictor, while in practice we usually deal with a large number of covariates and not all binary ones. However, the principles stated above maintain their value in these more complicated cases, where attention should be given also to the relationship among different covariates. For instance, a prognostic factor which is balanced on its own may become unbalanced across treatment groups when these are stratified according to another factor. This would occur, for example, if age is balanced by design (with stratified randomization) and the modalities of another relevant prognostic factor, such as an indicator of the severity of the disease, are differently distributed in different age classes, the older subjects having more severe disease. Provided that we are careful in specifying in advance at least a "core" of known prognostic variables which we think are to be considered for adjustment, so to avoid the dangers of exploratory analysis, we think an adjusted analysis of treatment effect should always be performed, in addition to the unadjusted one. In most of the situations encountered in practice there will be no substantial difference between the coefficient estimates from the two approaches in a randomized clinical trial. For example table 7.4, column 3, reports the adjusted estimates of the regression coefficients of the dummies C1 and C2 representing treatment contrasts in the ovarian cancer trial. If we perform an unadjusted analysis of treatment effect we obtain that the estimated coefficients and standard errors for the two dummies C1 and C2 are 0.1756 (0.124) and 0.3155 (0.123), respectively. They are similar to those obtained with the adjusted analysis; this finding, in accordance with to point (2) in section 9.3.2, may be expected in many oncological trials where treatment effect is likely to be small. However, if differences are noted between results of unadjusted and adjusted analyses, it is probably preferable to base inference on the latter one. We agree with the two reasons for this given by Chastang *et al.* (1988): "First, if the 'true model' is $\beta_1 x_1 + \beta_2 x_2$ then we avoid bias in estimating β_1, the treatment effect. But there is a philosophical problem: how do we ever know the true model? However, a sensible attitude might be that if we find a large difference between $\hat{\beta}_1$ and $\hat{\beta}_1^*$ then we may be able to show that the model with the covariate fits the data better than the model with treatment alone. Thus following usual statistical practice, we prefer the model which fits the data best because it is more reasonable to treat it as the 'true model'. The second justification for preferring the adjusted model is that with it we are answering a different and more important question specific to individual patients as characterized by their covariate values x_2 rather than averaging the treatment effect over all values of x_2 in the study population. This justification avoids the philosophical problem of referring to the difference $\beta_1 - \hat{\beta}_1^*$ as a bias. A proper view is that $\hat{\beta}_1$ and $\hat{\beta}_1^*$ are providing answers to different questions, the effect of treatment given x_2 versus the effect of treatment averaging over x_2."

9.4   THE STUDY OF PROGNOSTIC FACTORS
BY MEANS OF REGRESSION MODELS

Regression models presented in Chapters 6 and 8 are being used more and more in clinical and epidemiological studies. They have many advantages and perhaps the greatest one is that these models can relate a continuous variable to an outcome of interest. On the contrary, stratification methods have to categorize continuous variables to be able to partition individuals into strata and consequently cause some loss of information in the presence of a kind of dose–

response relationship between the explanatory variable and the outcome.

Another advantage is that regression models can more easily allow for the simultaneous assessment of the effects of several factors and each of these effects is adjusted for all the other factors included in the model. However a "price" for these advantages over stratification must be paid: regression methods usually make more assumptions than stratification methods and their mathematical ability to deal with many covariates is to be sustained with an adequate sample size. Regression models used unwisely, as it is often the case since the advent of powerful software for regression analysis, can do more damage than good. In particular, the study of prognostic factors is made much easier by the use of regression methods but valid prognostic estimates can be obtained only if the model assumptions hold at least approximately. Which regressors to include in a model and in which way is part of the general strategy for building a good model for the analysis. This strategy was discussed in detail for the semi-parametric PH regression model (section 6.5 and Chapter 7); an analogous approach is suggested for the parametric models, where in addition the assumptions on the distributional form of the failure time are to be verified (Chapter 8). When many variables are candidate prognostic factors, the process of model building faces the additional problem of selecting which variables to include in the linear predictor. This because of the universally accepted *principle of parsimony* in modelling. The effort of the investigator is to limit the complexity of the adopted model so as to be able to summarize and study with a few parameters the major systematic effects present in the data. Thus we do not wish to include in the model regression parameters we do not need for describing the major features of the relationship of outcome with treatment and patient/disease characteristics. Not only does a parsimonious model permit one to think more clearly about the data, but one that is substantially correct produces better predictions than one that includes spurious extra parameters. The inclusion of more and more covariates in a model can produce an increasingly better fit of the data set, to the extreme that a model with one parameter for each individual gives the perfect fit of the data. However, the ability of the model to accurately predict outcome in future patients deteriorates as the model becomes more and more complex. The principle of parsimony also has practical relevance: the identification of a limited set of strong prognostic predictors should serve also to avoid the collection of unnecessary data on patient/disease characteristics and to optimize the use of expensive medical tests.

The relationship of the prognostic factors with outcome can be modelled by choosing different functions that link the linear predictor to the failure rate. The exponential function, which corresponds to a multiplicative assumption, is the one most commonly adopted in regression analysis, for its tractability. Thus the relationship with outcome of a prognostic factor which identifies different groups of patients is usually described in terms of the failure rate in any group, relative to an arbitrary reference group, with the related confidence limits. When more than two groups are defined, the presentation of the relative failure rates estimated with the Cox model can benefit from the construction of the confidence limits for all groups, the baseline one included, as described by Easton *et al* (1991).

9.4.1 Problems with usual strategies for variable selection

When many explanatory variables are considered to be potential prognostic factors it is customary to use systematic selection procedures which will choose a relatively small subset of these variables by excluding those which have a limited predictive value. The most commonly used procedures are known as *stepwise selection* and are described in detail in Appendix A9.1. They rely on the fact that, in the absence of prior beliefs on the value of the coefficient of a regressor x_k, we are willing to accept that its value is zero if the inclusion of x_k into the linear predictor leads to a non-significant likelihood ratio statistic (or other test statistics), at a prefixed level α. The value of the maximized likelihood function of a model containing $k + 1$ regressors is always greater than that of the same model containing k of the regressors. In fact, we expect that more of the variability in the outcome is explained with the addition of one variable but, for this addition to be worthwhile, we want its predictive contribution to be "significant" in terms of a p-value being less than some prefixed level α, typically the same for all variables and for all sequential comparisons. As sound as it can seem, the uncritical application of the stepwise procedure for the selection of prognostic factors is open to criticism.

First of all, if the number of variables being investigated is large with respect to the sample size, spurious prognostic associations could be found because of "noise" in the data or multiple comparisons. Freedman (1983) illustrated this problem in the context of ordinary linear regression by simulation and by asymptotic calculation. He showed that, in the absence of any dependence between the outcome and the explanatory variables, a model with a small number of explanatory variables with significant coefficients and large R^2 can be easily obtained through a selection procedure if the number of investigated variables is comparable to the number of individual observations. In survival analysis, in order to investigate the relationship of outcome with many explanatory variables we need a sample conveying much statistical information, that is a sample in which many events of interest are observed. Based on empirical evidence, Harrel *et al.* (1984) proposed as a rule of thumb not to attempt a stepwise regression analysis of survival data when there are less than 10 times as many events in the sample as there are candidate regressors. This rough calculation should also account for interaction terms, if any of these is to be considered.

A second problem is that different models can be obtained on the same data set according to whether stepwise regression was based on a "forward" or "backward" selection. This can be easily understood if we consider that with forward selection the variable selected first may greatly influence the choice of the variables at the subsequent steps. A variable which is highly correlated with that selected first is likely not to be selected subsequently. This problem brings up the issue of variable importance. Results of stepwise regression are often misinterpreted, as the order of entry of the variables is taken to be their order of "importance". A biologically important prognostic factor should presumably have the same effect regardless of its prevalence in the analysed data set, but the effect might not be detected as different levels of prevalence can lead to different decisions in the statistical analysis. If, for instance, male gender is an important risk factor but in a sample there is a low prevalence of males, it is likely that gender will be related to a higher p-value than another variable which is less prognostic but whose categories are more balanced. Also, given the same set of explanatory variables, the number of variables selected on the basis of the sole significance is a function of the sample size and tends to increase as sample size increases (Harrel *et al.*, 1985).

A third problem with stepwise procedures is that the set of selected prognostic variables may change even for minor changes in the data. This lack of "stability" is well shown by Harrel *et al.* (1984) who selected markedly different sets of prognostic factors when the analysis was replicated by taking repeated samples from the same patient population. In fact there might be models with different sets of explanatory variables that are similar in their ability to predict the outcome variable. One determining reason for this is that, in most practical applications, explanatory variables are not under the control of the investigator and are often highly interrelated (multicollinearity). Thus the selection of what appears to be the unique prognostically meaningful set, without considering possible alternatives to the "automatic" process, is not a satisfactory approach. With regard to this problem, the presentation of results of a stepwise regression could benefit from the approach suggested by Hauck and Miike (1991).

Finally, the predictive ability of the selected model is overoptimistic. The estimated regression coefficients are biased away from zero since the same data set is used both to select the variables to be included and to estimate their regression coefficients in the final model (Copas, 1983). The significance levels associated with these coefficients are not strictly valid and nothing is known about the true selection level because of the complex pattern of choice of the minimum or maximum p-value.

9.4.2   Recommendations and cautions in prognostic factor analysis

There are various approaches that can be taken to avoid the problems of stepwise regression. First of all, the clever definition given by Robins and Greenland (1986) should be considered: "a statistical model is a mathematical

expression for a set of assumed restrictions on the possible states of nature".
Thus, when for instance we apply a basic Cox model, with no interactions, we
impose two restrictions on nature, that is proportional hazards and log-linear
effect of each covariate on hazard. Moreover, if for each individual we are able to
know the value of 100 covariates but we study 10 of them, we are implicitly
making restrictions on the possible states of nature we will consider. In the best
case this happens because there is a substantive theory or rationale which has
determined the study design; thus the previous knowledge of the disease and the
definition of meaningful clinical hypothesis restrict the field of interest. Any data
set, even if very large, may become insufficient if too many questions are asked
for an answer and thus the effort should be made to identify relevant questions
prior to data analysis. Any investigator has some prior beliefs on the possible
impact of a prognostic variable. Perhaps the optimal estimate of this impact
would be obtained in a Bayesian framework by updating with the information
from the observed data the prior distribution for the unknown parameters
(Robins and Greenland, 1986), but this approach would be rather complicated. In
practice, prior knowledge or beliefs are used to clarify at least which are the
relevant aspects to be considered and this usually produces a fairly limited
number of candidate variables. In this case the application of the stepwise
regression technique may not suffer so many problems. It is, however, advisable
to "force" into the model the "core" of covariates which are consolidated
prognostic factors, if any, even if they do not reach statistical significance in the
sample being analysed. In particular this is very important if the analysis aims at
evaluating the relationship to prognosis of new factors, as discussed below.
When, in spite of any effort, the number of candidate covariates remains large,
Harrell *et al.* (1984, 1985) suggest that data reduction might be achieved by
principal components analysis. This reduction is obtained with no reference to
the response variable and principal components of the available predictors are
then used as candidate prognostic factors in a stepwise analysis. They also
considered the use of clusters analysis for the same purpose, and found a better
quality of the predictions obtained with these two techniques as compared to
ordinary stepwise regression in two clinical examples. Another approach, often
used, is to perform a "univariable screening" of the candidate explanatory
variables and to subsequently consider in the regression analysis only the
variables which contributed to discriminate prognosis when singularly considered. Commonly the Kaplan–Meier survival curves for the groups of patients
classified according to each putative variable and the corresponding log-rank
test are calculated. This approach gives some ideas on the relation to outcome of
the single variables but its automatic use as a screening procedure has many
pitfalls, due to the fact that it does not account for the multivariable nature of the
problem (correlation between variables, joint effect of different variables otherwise of negligible effect, presence of interactions between variables).

If the principle of good research would require that also in the study of prognostic factors specific hypotheses be stated in advance, in practice this can be

difficult to achieve, especially when data on new potential prognostic factors are being investigated. Most of the time, the analysis on these is typically exploratory, and considers different subsets of factors, with different categorizations and their influence on different end-points and within different subsets of patients. This approach would not be objectionable if it were recognized that an exploratory analysis cannot go beyond the generation of interesting hypotheses to be tested on other data sets. Thus convincing evidence on the effect of new predictors should derive from a subsequent confirmatory study, with specific pre-stated hypotheses and adequate sample size for these to be tested, as thoroughly discussed by Simon and Altman (1994). For instance, this problem has been frequently met in recent years with the development of new markers which assess or reflect the biological behaviour of malignant neoplasms. Fielding *et al.* (1992), in discussing this aspect, summarize the main properties that must be demonstrated for a new marker to be used widely in clinical research and patient management, in three points:

(1) the factor must bear a strong relationship to patient prognosis;
(2) the factor must provide additional prognostic information beyond that provided by conventional factors;
(3) the method of determining the factor must give good reproducibility both within and between laboratories.

It is clear from requirements (1) and (2) that stepwise analysis should not be used in this case. Point (1) suggests that a marginal analysis should be performed, for example by comparing the survival experience of patients classified according to different values or categories of the new factor alone. For this point to be satisfied such a classification should produce groups that tend to have different survival experiences. The second point requires that this new factor is an independent predictor of survival, that is it maintains its ability to discriminate prognosis in the presence of other established prognostic factors. To prove this, two models should be fitted, one containing the standard prognostic factors and the other still containing these, but with the addition of the regressor(s) which express the new factor. The comparison of the fit of the two models will enable to establish if the new factor significantly adds prognostic information. For instance, the global likelihood ratio test, with as many degrees of freedom as the number of the added regressors, will allow one to detect the overall significance of the new factor.

The stability of a model obtained by a stepwise selection can be investigated by applying the bootstrap method (Efron and Tibshirani, 1986), as shown by Altman and Andersen (1989). In their paper the final model obtained by Christensen *et al.* (1985) in a data set on primary biliary cirrhosis patients was evaluated for stability. Christensen *et al.* selected six variables among 24 candidate ones, partly by using a forward stepwise procedure. The choice of the six variables included in their model was judged in the light of the frequency of inclusion of each candidate explanatory variable in 100 bootstrap samples on which stepwise selection was applied. In the 100 models obtained, the six variables included in the original model were selected most frequently (from 100% to 62%) and two more variables were entered in over a quarter of the models. Thus the original model by Christensen *et al.* was judged to be satisfactory. Furthermore, Altman and Andersen compared the bootstrap confidence intervals for the estimated survival probabilities to those obtained in the original model with the usual approach. The bootstrap confidence intervals were larger and this indicates, as expected, a risk of underestimation of the uncertainty of predictions from a single regression model which is both selected and fitted on the same data set.

The results of the bootstrap resampling procedure are directly combined with the usual stepwise techniques to produce an innovative selection strategy in Sauerbrei and Schumacher (1992), who generalize the approach presented by Chen and George (1985). These methods are appealing but expensive in computing time.

9.4.3   The predictive value of a prognostic model

As outlined in section 9.4.1, the *p*-value as a criterion to select statistically significant covariates allows for the hypothesis of no effect in the given sample to be ruled out but is not a measure of clinical importance. Moreover, finding in a data set many statistically significant variables does not necessarily mean that a lot is understood about prognosis. In linear regression, the proportion of variability in the outcome which is explained by a model is measured by the multiple correlation coefficient R^2 and therefore a greater predictive value can usually be assigned to a model with greater R^2. An analogous measure is not easily defined in survival regression models and the recent literature proposes various methods to measure the predictive strength of a given model in terms of the amount of explained variability.

As the likelihood is a measure of the explained variation, the difference between the log-likelihood $LL(\hat{\beta})$ of the selected model and the log-likelihood of the model with no covariates $LL(\mathbf{0})$ can be transformed into a measure of the proportion of explained variation, as follows:

$$1 - \exp\left[-\frac{2}{N}\left(LL(\hat{\beta}) - LL(\mathbf{0}) \right) \right]$$

A discussion on the properties of this type of measure is found in Nagelkerke (1991). Alternative measures of explained variation based on the error in the prediction of the individual survival functions are given by Korn and Simon (1990) and Schemper (1990).

The predictive value of a model should also be seen in the perspective of using this model for future patients, for example, for discriminating those at worst prognosis. This is a different issue from explained variation and has also been addressed in the recent literature. It is sensible to think, as suggested by Harrell *et al.* (1985), that the strength of a prognostic model relies in its ability to *discriminate* patients with different prognosis and also to give *reliable predictions* on the outcome of future patients. The two aspects are obviously connected but different in that the first requires that a prognostic model be able to predict, given patient/disease characteristics, who are, say, the worst, average, and lower risk patients. The second aspect has to do with the possibility of predicting that the five-year survival rate of a future worst prognosis patient will be 40% if the selected model gives that figure on the fitted data. Thus the performance of a model on an independent but similar data set is another key feature on which it would be possible to judge the predictive ability of the model itself and, in practice, its usefulness. Roughly speaking, there can be two main reasons for the performance of a model to deteriorate when applied to a new data set: either the model is being fitted to a different patient population or the model overfitted the data set on which it was generated (Charlson *et al.*, 1987). The first problem may occur, for example, when a model developed in an institution is applied to patients from another institution which differs in the referral patterns and thus selects a different patient population in which some of the indicated prognostic factors may be less relevant. The second problem, already discussed, has to do with the fact that statistical models are overoptimistic when judged on the same set used for generating them.

The assessment of the predictive value of a model as a part of the model building process can be approached with a "training-and-testing" procedure. If an adequate data set is available, this is split into a "training" sample, on which the model is developed with standard methods, and a "test" sample (or validation sample) which is used to validate the model, but not to build it. The test sample is randomly selected from the entire data set, and is usually of smaller size than the "training" sample, if not many data are available. The problem in survival analysis is to quantify the prediction error since simple measures such as the squared error in linear regression cannot be applied. Harrell *et al.* (1984) considered in particular the discriminating ability of the model on the validation sample and quantified it with an index of concordance which reduces to the area under the "receiver operating characteristic" (ROC) curve when the outcome reduces to a dichotomous variable (dead/alive). This index is the probability that, for a randomly chosen pair of subjects in the validation set, the predicted and observed outcomes are concordant. It is calculated by taking all possible pairs of patients and for a given pair there is concordance if the patient who survives longer is the one having the higher predicted survival. A pair is not counted if both patients are still alive or if only one has died and the follow-up of the other is less than the failure time of the first. When the value of this index is 0.5 the model has no discrimination ability, while for a value equal to 1.0 the model can perfectly rank patients according to the severity of their outcomes. No information is gained on the reliability of the values of the survival curves. Later works have addressed this problem more specifically. A simple graphical aid for exploring predictive ability is to plot the survival curves observed in the test set for some groups of patients with specific patterns of the covariates selected in the model. These should be very similar to those predicted by the model, had the model predictive accuracy. What we expect from theory is that, since the influence of the covariates is overestimated in the training sample, the observed curves will tend to be less separated from each other than the predicted ones. This is known as *shrinkage* and is a general phenomenon which easily occurs in regression analysis when the number of parameters in the model is high. The more marked is the shrinkage, the more is the prediction error in the model. The problem is thus not only to quantify the shrinkage but, possibly, to adjust the prediction rule based on the estimated coefficients. This can be done as follows (Van Houwelingen and Le Cessie, 1990): consider the model, say the Cox model, which has been selected as the "best" one for explaining the outcome in the training set. The estimates obtained for the regression coefficients β are used to calculate, for any individual with covariates x_i in the test sample, the value of the linear predictor $\hat{\beta}'\mathbf{x}_i$. This value is used as the only regressor in a new Cox model fitted on the test sample; the corresponding estimated regression coefficient should be about 1 for the original model to be good for prediction. The actual value of this coefficient, when lower than one, can be used to obtain an improved prediction with the model selected on the training set, provided that the regression coefficients are multiplied, i.e. shrunk, by it. In some sense, the model is "calibrated" according to its performance on a validation sample and the predicted survival curves will be shrunk accordingly. An example on modelling the survival of renal-transplanted patients is discussed in Van Houwelingen and Thorogood (1992).

Other validation strategies which are different from the training-and-testing one can be applied and are particularly useful when the sample size is small. We refer to the cross-validation technique, in which a large number of subsets of the original set of N data are considered. For instance the entire set is partitioned into K subsets each with an equal or nearly equal number of subjects $r_1, r_2, \ldots, r_K$. Then, for each $k = 1, \ldots, K$ the set which contains $N - r_k$ subjects can be seen as the training set with its corresponding test set of r_k subjects. The jack-knife or leave-one-out approach, which is related to bootstrap (Efron and Gong, 1983), is another technique used for validation. The use of cross-validation for the assessment of the predictive value of a selected model is illustrated in Van Houwelingen and Le Cessie (1990) and in Verweij and Van Houwelingen (1993).

9.4.4   The definition of risk groups

It is often desirable to summarize the information on prognosis conveyed by several variables with the definition of risk groups. These groups should be identified in such a way that subjects be homogeneous with respect to prognosis within each risk group and heterogeneous between different risk groups. There are two possible approaches to the definition of risk groups which are based on the construction of prognostic indices or on recursive partitioning. The former approach directly descends from the results of any regression model which has been selected for prognostic prediction. The linear predictor, with the estimated regression coefficients, gives a prognostic index (PI) $\hat{\boldsymbol{\beta}}'\mathbf{x}_i$ for any subject with covariate pattern $\mathbf{x}_i$. The distribution of PI in the data set is then considered and different risk groups may be defined according to the centiles of the distribution. This approach has been adopted for discriminating patients with different diseases, for example cyrrhosis (Christensen *et al.*, 1985) or ovarian cancer (Lund and Williamson, 1991; Marsoni and Valsecchi, 1991).

Recursive partitioning differs from the PI approach in that it identifies different risk groups through Boolean combinations of variables. In synthesis, the entire data set is subdivided into two groups according to the one split on a single explanatory variable which maximizes the difference in outcome among all explanatory variables that are considered and all their possible splits. Subsequently, the same process is repeated by considering all possible variables and splits in each of the two groups and so on recursively. This type of algorithm is also called a binary tree. For its application, the definitions of criteria for splitting and for stopping the recursive process are needed. Clearly these criteria determine the complexity of the tree and consequently the number of risk groups, since these are the terminal nodes of the tree. The monograph on Classification and Regression Trees (CART) by Brieman *et al.* (1984) stimulated much interest on recursive partitioning. CART was developed to deal with quantitative and binary response variables and subsequently the approach has been generalized to survival data (among the most recent papers, see Ciampi (1991) and LeBlanc and Crowley (1993)). The results of the application of a CART-like recursive partitioning to a data set on advanced ovarian cancer patients treated in several Italian centres are shown in figure 9.3. This tree was obtained by processing

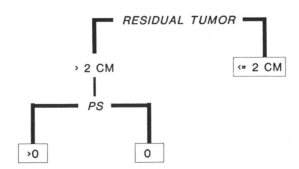

Figure 9.3   Regression tree obtained on a set of advanced ovarian cancer patients.

a training set of 478 patients (the splitting criterion was based on the partial likelihood ratio statistic) and by validating the resulting models on a test set of 239 patients (Valsecchi, 1992a). Three groups of patients with decreasing severity of outcomes from left to right were defined. The first split was determined by the size of residual tumor after first surgery at the optimal cut-point of 2 cm. The second split refines the classification of patients with residual tumor size greater than 2 cm by means of the performance score, the patients with performance score zero (corresponding to a Karnofsky index of 100) being at relatively better prognosis than the remaining patients in this subgroup.

The tree approach to the definition of risk groups is appealing to the clinician as its application not only enables the selection of a subset of prognostic factors and cut-points but also helps to untangle the essential features of a complex network of relationships. Very complicated structures can emerge from this type of exploratory analysis, when no validation is performed, but these are likely to lack predictive ability for future patients.

9.5   EVALUATION OF TREATMENT EFFECTIVENESS

In section 2.6 the difference between statistical significance and clinical relevance was discussed and it was suggested that the results of a clinical trial should be reported in terms of the actual p value and of the confidence interval of the "true" value of effectiveness, whatever the adopted measure. In this section we wish to go further both by commenting on the results of a "negative" trial and by elaborating upon some measures of benefit and harm, both absolute and relative, in relation to the statistical methods presented in previous sections.

After 16 years from the beginning, Veronesi *et al.* (1990) presented a final assessment of the results of the trial described in figure 1.3 aiming at evaluating the conservative treatment (Quart) in comparison to the standard treatment (Halsted mastectomy) in breast cancer. As regards death from all causes, the two K–M curves are reported in figure 9.4. The estimated cumulative survival in the Quart and mastectomy groups are respectively 0.92 and 0.91 at five years, and 0.80 and 0.77 at 10 years. The value of the log-rank test is $Q_{M-H} = 0.814$, $p = 0.367$. Since *a priori* the conservative treatment cannot be expected to improve survival, the fact that the survival curve for the conservative treatment is over the corresponding one for mastectomy, though not statistically significant, might be considered reassuring for the surgeon. However, before concluding that radical mastectomy appears to involve unnecessary mutilation, and, thus, change over to the conservative treatment, the surgeon may well ask: "If I adopt the conservative treatment, what is the risk that my results become worse?".

An approximate answer to this question may be given by halving the p-value reported above, namely some 18%. Furthermore, consider the M–H point

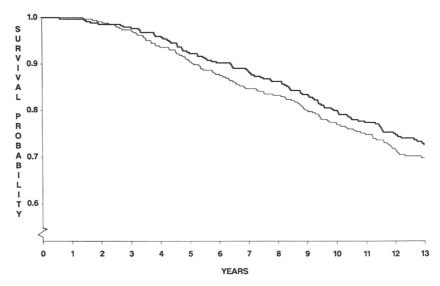

Figure 9.4   K–M survival curves of the trial described in figure 1.3. Surgical treatment:
——Halsted mastectomy; ——Quart.

estimate of the hazards ratio: $\theta = \lambda_{\text{Halsted}}/\lambda_{\text{Quart}}$; this turns out to be $\hat{\theta}_{\text{M-H}} = 1.134$ and the 95% confidence interval is: $0.860 \leqslant \theta \leqslant 1.504$. The left-hand of the interval, $(0.86, 1.0)$, is of concern for the surgeon aiming at deciding whether to adopt the conservative treatment since in this subinterval the standard treatment compares favourably to the new one.

However, it may happen that surgeons are not familiar with hazard ratios so that the above results may not be straightforwardly informative to them. As a consequence it appears useful to rephrase the previous results in terms of survival probability, say, at five and 10 years from surgery. A decreased hazard rate of some 12% (i.e. a hazard ratio of 0.882) corresponds to an increase of some 3% in the 10-year survival probability after Quart; this probability is estimated to be 0.80. Furthermore the confidence limits of θ allow us to compute the confidence limits for the 10-year survival probability in the mastectomy group by assuming the probability in the Quart group to be the one actually observed. Owing to the relationship between $S(t)$ and $\lambda(t)$ given by (5.7) and to the underlying PH assumption, the predicted 95% confidence limits for the 10-year survival probability after mastectomy are:

$$(0.8005)^{0.860} = 0.826 \quad \text{and} \quad (0.8005)^{1.504} = 0.716$$

The same computations have been repeated for the five year survival (0.92 in the Quart group) and the results are reported in figure 9.5 together with the

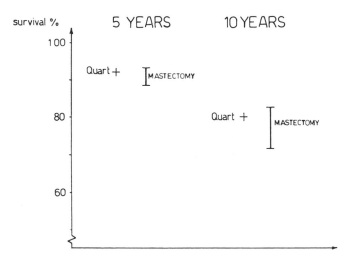

Figure 9.5   Predicted 95% confidence interval of the survival probability for Halsted mastectomy at 5 and 10 years from surgery (see text).

previous ones. For each of the two time-points, the figure makes available to the surgeon a probable range of survival probability from mastectomy compared to that from Quart: there is only a 2.5% chance of mastectomy improving the 10-year probability by more than 2.6% (82.6, 80.00) and the five-year probability by more than 1% (93.35, 92.31). These findings may be helpful in deciding to adopt the conservative surgical approach for future patients.

The effectiveness measures we will consider are based on the distribution functions for failure time (section 5.2), i.e. the *risk* of failure $F(t)$ and the failure *rate* (see section 3.6.1) or *intensity rate* (see 5.4); we shall also deal with the *odds* of failure at time t, as defined in section 8.2.5, i.e.:

$$\omega(t) = F(t)/S(t) \tag{9.3}$$

Like risk, odds is dimensionless and its lower boundary is zero, but unlike risk, it has no finite upper boundary. This basic measure is widely adopted in epidemiology and less often in clinical medicine. The relationship between odds and risk of death is made clear by the following set of values:

$F(t)$	0.0	0.167	0.33	0.50	0.67	0.90	0.95
$\omega(t)$	0.0	0.200	0.50	1.00	2.00	9.00	19.00

Measures suitable for assessing the effectiveness of a new treatment, with regard to a standard one or a placebo, may be easily obtained by computing differences and/or ratios of the corresponding basic failure measures computed on each of the two samples of patient undergoing different treatments.

9.5.1 Measures suitable for describing clinical trial results

Absolute risk reduction, risk ratio and relative risk reduction

Let $F_1(t)$ and $F_2(t)$ be the death risks in the first (standard treatment or placebo) and in the second (new treatment) sample respectively; $F_1(t) \geqslant F_2(t)$ is assumed here.

The absolute risk reduction (ARR) at time t is given by the difference:

$$\text{ARR} = F_1(t) - F_2(t) \qquad (9.4)$$

Although it provides a measure of clinical effect, ARR may hardly be incorporated into clinical practice, because it may not seem sensible to physicians to express the absolute risk reduction in terms of decimal fraction. Two companions of this measure appear to be more appealing. The first is the estimate of the number of deaths avoided (NDA):

$$\text{NDA} = \text{ARR} \times n \qquad (9.5)$$

(if $n = n_2$, the number of patients accrued in the second sample, NDA is the estimated number of deaths avoided in the second arm of the trial. Otherwise, if the researcher knows the number of patients n in the whole population who could benefit from the treatment, NDA is a measure of the therapeutic impact). The second is the number of patients needed to be treated (NNT) in order to avoid one death:

$$\text{NNT} = 1/\text{ARR} \qquad (9.6)$$

Laupacis *et al.* (1988) argue that this is a highly useful measure for clinicians, because it can be used to describe the benefit as well as the harm of a therapy and of other clinical manoeuvres. If the relative risk reduction (see later) is constant over time, the NNT decreases as the duration of follow-up increases. Thus, for a proper use of this statistic, the pertinent length of follow-up should be specified.

The risk ratio or relative risk after t time units (weeks, months, years) of follow-up, given by the ratio

$$\text{RR} = F_2(t)/F_1(t) \qquad (9.7)$$

is widely used in epidemiology to compare the risk of death in two groups of people. It has an intuitive appeal to many researchers in this field, since it seems sensible to talk about "doubling" or "halving" the risk of death.

A quantity more appealing to the physician interested in assessing the effectiveness of a given treatment appears to be the relative risk reduction, also defined as the relative difference (D_R):

$$D_R = [F_1(t) - F_2(t)]/F_1(t) = 1 - RR(t) \tag{9.8}$$

An appealing feature of D_R emerges if placebo is used in the control arm. Suppose that in a clinical trial n_1 patients are randomized to placebo and n_2 patients to the treatment group. After a given period of follow-up, let π_1 be the risk of death in the subset treated with placebo and $(1 - \pi_1)$ the probability of surviving. We expect that in the treated group $(1 - \pi_1) \times n_2$ patients would anyhow survive as a result of the "effect" of placebo. Therefore the medical treatment can only affect the status of the remaining $\pi_1 \times n_2$ patients who would die by protecting a fraction $(1 - \pi_2)$ of them. In other words the number of deaths, $\pi_1 \times n_2$, is accordingly decreased by a factor π_2 ($\pi_2 \leqslant 1$). The risk $F_1(t)$ corresponds to π_1, whereas the risk $F_2(t)$ is the product of probabilities $\pi_1 \times \pi_2$. A synopsis is given in table 9.4. From the last row of this table it appears that π_2 is given by $RR = F_2(t)/F_1(t)$, and the measure of effectiveness of the treatment under study is given by $(1 - \pi_2) = 1 - F_2(t)/F_1(t) = D_R$. The same statistic could perhaps be used for evaluating the effectiveness of an adjuvant treatment in a trial comparing surgery plus adjuvant treatment versus surgery alone.

Death rate ratio and relative rate reduction

Let λ_1 and λ_2 be the two death rates in the first and second group of patients, respectively. The rate ratio or relative rate of the second sample to the first is given by $\theta = \lambda_2/\lambda_1$ and the relative rate reduction by

$$RRR = \frac{\lambda_1 - \lambda_2}{\lambda_1} = 1 - \theta \tag{9.9}$$

Three non-parametric estimates of θ were introduced in section 4.3, proposed by Peto *et al.* (1976), Mantel and Haenzel (1959) and Anderson and Bernstein (1985) respectively.

Table 9.4 Expected deaths by treatment.

Event	Placebo	Pharmacological treatment
Deaths	$\pi_1 n_1$	$\pi_1 \pi_2 n_2$
Survivors	$(1 - \pi_1)n_1$	$[(1 - \pi_1) + \pi_1(1 - \pi_2)]n_2 = (1 - \pi_1\pi_2)n_2$
Total	n_1	n_2
Risk of death	$F_1(t) = \pi_1$	$F_2(t) = \pi_1\pi_2$

In previous sections we saw that with a dichotomous variable x adopted for identifying the treatment groups, the estimate of θ is easily obtained by

$$\hat{\theta} = \exp(\hat{\beta}) \qquad (9.10)$$

in a PH multiplicative model where $\hat{\beta}$ is the estimated regression coefficient of the variable x. These models enable the investigator to express, at every instant t of the follow-up time, the hazard of patients treated with the new treatment as $100 \times \theta\%$ that of patients treated with the standard treatment; as a consequence, the relative reduction of the hazard rate in the new treatment group is $100(1 - \theta)\%$.

The relationship between RR and θ is presented in Appendix A9.2; it suffices here to say that when the cumulative rate of death is sufficiently small, θ and RR are numerically similar.

Odds ratio and relative reduction of odds

The ratio of odds of failure in the new treatment group to that of the standard treatment group is given by

$$\varphi = \omega_2(t)/\omega_1(t) = \frac{F_2(t)}{F_1(t)} \cdot \frac{S_1(t)}{S_2(t)}$$

and the relative reduction of odds by

$$\text{RR}_\omega = 1 - \varphi \qquad (9.11)$$

The odds ratio at a given time point can be estimated non-parametrically on the basis of the K–M survival curves. On the other hand, from section 8.3.2, we recall the class of models whose main feature is that the odds functions $[\omega(t, \mathbf{x})]$ of two samples having different vectors of covariates are assumed to be proportional in time. Here, the odds ratio of failure of two treatment groups, given the regression coefficient $\hat{\beta}$ (simple or partial) of the dummy variable specifying the type of treatment in the regression model, is estimated by

$$\hat{\varphi} = \exp(\hat{\beta}) \qquad (9.12)$$

9.5.2   Confidence interval of measures of effectiveness

Absolute measures

These measures are based on the difference of two independent statistics; thus, by resorting to the asymptotic properties of these estimates, confidence intervals can be easily computed. Two formulas, (3.9a) and (3.9), were given in section 3.4.1 to compute the variance of $\hat{S}(t)$ which are appropriate for the K–M and life-table estimates of $S(t)$. Hence if we use (3.9a) and the K–M estimate the variance of the absolute risk reduction is

$$var(\hat{ARR}) = \frac{\hat{F}_1(t)[\hat{S}_1(t)]^2}{n'_1(t)} + \frac{\hat{F}_2(t)[\hat{S}_2(t)]^2}{n'_2(t)}$$

where $n'_i(t)$ $(i = 1, 2)$ is the number of patients alive just before time t minus $d(t)$, the number of failures at t. The approximate $100(1 - \alpha/2)\%$ confidence limits are

$$\hat{ARR} \pm z_{1-\alpha/2} SE(\hat{ARR}) \tag{9.13}$$

These limits multiplied by n_2 give the corresponding confidence limits of NDA (9.13a).

Relative measures

The computation of confidence intervals of hazard rates ratio and odds ratio and, thus, of the corresponding relative rate reduction and relative reduction in odds is straightforward provided that the validity assumptions of the models used are fulfilled. In fact, since all the packages commonly used to perform regression analyses give $\hat{\beta}$ and $SE(\hat{\beta})$, the $100(1 - \alpha/2)\%$ confidence interval of θ or φ is

$$\exp[\hat{\beta} \pm SE(\hat{\beta})] \tag{9.14}$$

Complementing to 1 the limits computed by means of (9.14) one attains the confidence interval of RRR or of RR_o, according to the type of model that was fitted to the data (9.14a).

The relative risk (9.7) and the relative difference (9.8) are estimated by ratios of independent statistics. Their confidence intervals cannot be computed in the straightforward way previously adopted but by resorting to Fieller's theorem, which applies to normally distributed statistics. It can be used here because of the asymptotic properties of the K–M estimator. The confidence interval of RR is now estimated by introducing the statistic

$$u = [\hat{F}_2(t)] - [RR \times \hat{F}_1(t)]$$

The variance of u is estimated by

$$var(u) = var[\hat{F}_2(t)] + RR^2 \, var[\hat{F}_1(t)]$$

By leaving RR unspecified and by setting

$$u^2 - z_{1-\alpha/2}^2 \, var(u) = 0$$

a quadratic equation in RR is obtained. Its solutions are the approximate confidence limits of RR, namely

$$\{\hat{F}_1(t)\hat{F}_2(t) \pm z_{(1-\alpha/2)} \times \sqrt{a+b-c}\}/d \qquad (9.15)$$

where

$$a = \hat{F}_1^2(t)\, var(\hat{F}_2(t))$$
$$b = \hat{F}_2^2(t)\, var(\hat{F}_1(t))$$
$$c = z_{(1-\alpha/2)}^2 \, var(\hat{F}_1(t))\, var(\hat{F}_2(t))$$
$$d = F_1^2(t) - z_{(1-\alpha/2)}^2 \, var(\hat{F}_1(t))$$

$Var[\hat{F}_i(t)]$ is computed by means of (3.9) or (3.9a). By complementing to 1 the two values given by expression (9.15), one can compute the confidence limits of D_R (9.15a).

In the case of the ratio of two binomial variables, Bailey (1987) suggested a modification of Fieller's approach in order to minimize the skewness of the random variable u. Recently, Gart and Nam (1988) reviewed and evaluated numerically the various methods proposed in the statistical literature to estimate the confidence limits of the ratio of two binomial parameters.

Table 9.5   Measures suitable for describing clinical trial results.

Measure	Formula	Expression no. for Estimator	Variance or confidence interval
Absolute			
Absolute risk reduction	$A\hat{R}R = \hat{F}_1(t) - \hat{F}_2(t)$	(9.4)	(9.13)
Number of deaths avoided	$N\hat{D}A = A\hat{R}R \times n_2$	(9.5)	(9.13a)
Number needed to be treated	$N\hat{N}T = 1/A\hat{R}R$	(9.6)	
Relative			
Relative risk	$\hat{R}R = \hat{F}_2(t)/\hat{F}_1(t)$	(9.7)	(9.15)
Relative difference	$\hat{D}_R = 1 - R\hat{R}(t)$	(9.8)	(9.15a)
Hazard rate ratio	$\hat{\theta} = \hat{\lambda}_2/\hat{\lambda}_1$	(4.9), (9.10)	(4.10), (9.14)
Relative rate reduction	$R\hat{R}R = 1 - \hat{\theta}$	(9.9)	(9.14a)
Odds ratio	$\hat{\phi} = \exp(\hat{\beta})$	(9.12)	(9.14)
Relative reduction of odds	$\hat{R}R_\omega = 1 - \hat{\phi}$	(9.11)	(9.14a)

Examples

Three examples are included; the calculations have been carried out to full arithmetical precision, as is recommended practice, although intermediate steps are shown as rounded results. Formulas and identification numbers relative to the measures of basic mortality and of effectiveness elaborated upon in this section are reported in table 9.5, together with the number specifying the formulas used to compute the pertinent confidence limits. It is hoped that this synopsis is helpful for routine applications.

The first example concerns breast cancer in women aged 50 years or more at surgery. We shall assume that the risk of death after five years in the group with surgery alone is be $\hat{F}_1(5) = 0.33$, and that the hazard rate ratio in the group treated with surgery plus tamoxifen for two years or more be 0.75. This value approximates the estimate given for this subset by the Early Breast Cancer Trialists' Collaborative Group (1992). From expression (5.14), assuming a constant hazard rate, $\hat{\lambda}_1 = -\ln(1 - 0.33)/5 = 0.080$; $\hat{\lambda}_2 = 0.080 \times 0.75 = 0.60$ and the estimate $\hat{F}_2(5) = 1 - \exp(-0.060 \times 5) = 0.259$ is obtained. All results concerning the measures of effectiveness are reported in table 9.6. The similarity of the three figures in the last column of this table suggests that, in this case, the three measures of effectiveness are not markedly different from one another for describing trial results. Moreover, the last row informs the physician that some 15 patients should be treated with the adjuvant treatment to prevent one death.

The second example relies on Freireich *et al.* (1963) data. The K–M estimates of the duration of remission reported in table 3.1 were utilized to estimate the relative risk (R̂R) and, by means of (5.14), the hazard rate ratio ($\hat{\theta}$) for two different times of follow-up, $t = 11$ and $t = 22$ weeks.

An overall estimate of θ, assuming PH, is obtained according to (4.7): from table 4.2, the numerator of the log-rank test is $U = -10.2505$; being negative, it informs the researchers that the 6-MP treated group fared better that the placebo group. The denominator is $var(U) = 6.2570$. Thus the estimate is $\hat{\theta} = \exp(-10.25/6.26) = 0.1943$.

Table 9.6 Estimates of measures of effectiveness for the example of breast cancer patients after five years of follow-up (see text).

Basic measure	Treatments		Ratio	% Reduction
	Surgery	Surgery + tamoxifen		
Risk of death	$\hat{F}_1 = 0.33$	$\hat{F}_2 = 0.26$	$\hat{R}R = 0.78$	22.0
Rate of death	$\hat{\lambda}_1 = 0.08$	$\hat{\lambda}_2 = 0.06$	$\hat{\lambda}_2/\hat{\lambda}_1 = 0.75$	25.0
Odds of death	$\hat{\omega}_1 = 0.49$	$\hat{\omega}_2 = 0.35$	$\hat{\omega}_2/\hat{\omega}_1 = 0.71$	29.0
Absolute risk reduction = 0.07				NNT = 14.3

Table 9.7 Estimates of measures of effectiveness for the data by Freireich *et al.* (1963).

Measures	Treatment		Ratio	% Reduction
	Placebo	6-MP		
Risk of death at 11 weeks (K–M estimate)	$\hat{F}_1(11) = 0.7143$	$\hat{F}_2(11) = 0.2471$	$\hat{R}R(11) = 0.3459$	65.4
Risk of death at 22 weeks (K–M estimate)	$\hat{F}_1(22) = 0.9524$	$\hat{F}_2(22) = 0.4622$	$\hat{R}R(22) = 0.4853$	51.5
Rate of death at 11 weeks (calculated by (5.14))	$\hat{\lambda}_1(11) = 0.1139$	$\hat{\lambda}_2(11) = 0.0258$	$\hat{\lambda}_2(11)/\hat{\lambda}_1(11) = 0.2265$	77.3
Rate of death at 22 weeks (calculated by (5.14))	$\hat{\lambda}_1(22) = 0.1384$	$\hat{\lambda}_2(22) = 0.0282$	$\hat{\lambda}_2(22)/\hat{\lambda}_1(22) = 0.2037$	79.6
Instantaneous rate of death (calculated by (4.7))	—	—	$\hat{\theta} = 0.1943$	80.6
Instantaneous rate of death (Cox estimate)	—	—	$\hat{\theta} = 0.2211$	78.9
Odds of death (log-logistic estimate)	—	—	$\hat{\phi} = 0.2821$	71.8
$\hat{ND}A \approx 10$				$\hat{NN}T \simeq 2$

By resorting to the Cox model, the partial likelihood estimate of the regression coefficient pertinent to the indicator of treatment is $\hat{\beta} = -1.509\,19$ (section 6.4.1), giving an estimate of the hazard rate ratio of 6-MP versus placebo equal to $\exp(-1.51) = 0.2211$. This estimate, obtained at the end of the relative process of likelihood maximization, is well approximated by the one-step estimate given above in this example where the PH assumption is tenable.

Contrary to the results reported in table 9.6, in table 9.7 the relative reduction of risk and of odds of remission are different from the reduction of rate. In this case, the measure of effectiveness one chooses is important, and since the proportional hazard assumption seems to be fulfilled, the relative rate reduction appears to be the most appropriate measure of effectiveness. On this ground, the relative risk reduction appears to highly underestimate the effectiveness of 6-MP, whereas odds reduction, as calculated after fitting the log-logistic model, tends to slightly underestimate it.

The estimate of the standard error of $\hat{\beta}$ in the Cox model is $\text{SE}(\hat{\beta}) = 0.409\,56$; by expression (9.14) the 90% confidence interval of θ is

$$\exp(-1.509\,19 \pm 1.645 \times 0.409\,56) = (0.11, 0.43)$$

It is straightforward to compute the 90% limits of the relative rate reduction as the complement to 1 of the above values $(0.57, 0.89)$. At $t = 22$ week we have $\widehat{\text{ARR}} = 0.49$; the approximate 90% confidence limits of ARR by expression (9.13) are

$$0.49 \pm 1.645 \times \sqrt{0.002\,16 + 0.0223} = (0.23, 0.75)$$

The product $21 \times 0.49 = 10.29$ enables us to estimate NDA in 22 weeks, the confidence interval of which is $21 \times (0.23, 0.75) = (5, 16)$.

The number of patients needed to be treated is $\text{NNT} = 0.49^{-1} = 2.04$.

It has been previously assessed that the relative risk reduction underestimates the effectiveness of 6-MP. Nevertheless, for the sake of completeness, by means of (9.15) the 90% confidence interval of RR at 22 weeks is computed as

$$\{0.952 \times 0.462 \pm 1.645 \times \sqrt{a+b-c}\}/d$$

where

$$a = 0.952^2 \times 0.022\,23$$
$$b = 0.462^2 \times 0.002\,19$$
$$c = 1.645^2 \times 0.022\,23 \times 0.002\,19$$
$$d = 0.952^2 - 1.645^2 \times 0.002\,19$$
$$(\text{RR}_L, \text{RR}_U) = (0.44 \pm 1.645\sqrt{0.020})/0.90 = (0.23, 0.75)$$

Thus the confidence interval of D_R is $(0.25, 0.77)$.

Consider now the example of section 8.6 concerning the survival experience of lung cancer patients, in which a log-logistic model was fitted. If, according to Bennett (1983), the variable performance status (PS) is categorized into two classes, low PS (score $\leqslant 50$) and high PS (score > 50), the estimate of the odds ratio of low PS vs high PS is: $\hat{\psi} = \exp(1.2314) = 3.43$. When the same data set is fitted with models belonging to the proportional hazards class, the following estimates of hazards ratio are attained: $\hat{\theta} = 2.19$ by the Weibull model and $\hat{\theta} = 2.14$ by Cox model. They are close to each other but, since the proportional odds model is suitable for fitting this data set, the prognostic effect of performance status is underestimated when measured in terms of hazard rate ratio.

In comparing survival patterns of treated and control patients, "risk", "rate" and "odds" of death give similar results when the underlying force of mortality is relatively low and there is a small effect of treatment (see the first example). Otherwise the choice of one of the three statistics mentioned above matters, and this choice may be suggested by a careful study regarding the model most suitable for fitting the data.

APPENDIX A9.1   PROCEDURES FOR VARIABLE SELECTION

Stepwise selection is a term defining three different types of procedures which are described here. They all involve the sequential comparison of statistical test criteria to critical values based on the assumed probability distribution for the test statistic. The test statistic can be, for instance, the likelihood ratio between the current model and the candidate model at each step. The likelihood ratio statistic is generally considered the best to use, but it requires considerably more computation than the score and Wald statistics. As sample size increases, however, the three statistics tend to agree closely. The critical values correspond to a predetermined α-level, the same for all subsequent steps and for all regressors.

Forward (or step-up) procedure: the procedure starts from the null model, i.e. the model without covariates, and subsequently adds variables to build up a progressively more complex model, until further addition does not significantly improve the model as a whole. The criterion sequentially adopted is the following: identify the single regressor which is not in the current model and for which the value of the test statistic is highest; add this regressor to the current model if the corresponding *p*-value is less than the prefixed α, and refit; otherwise, stop the procedure. The current model at the stop is the final one.

Backward (or step-down) procedure: the procedure begins with the full model, i.e. with all candidate covariates, and eliminates variables in turn, until further deletion would significantly reduce the ability of the model, as a whole, to explain outcome variability. The criterion sequentially adopted is the following: identify the single regressor in the current model for which the value of the test statistic is lowest; drop this regressor from the model if the corresponding p-value is greater than α, and refit; otherwise stop the procedure. The current model at the stop is the final one.

Stepwise procedure: the procedure is an extension of the forward procedure where at each step provision is made for inclusion and deletion of variables. The procedure starts with the null model and proceeds as the forward approach with the addition that, at each step, after the inclusion of a new variable, the test statistics are calculated for all the "old" variables already included in the regressor as a result of the preceding steps. Any old variable which leads to a non-significant statistic, according to a level α^* (usually $\alpha^* > \alpha$), is removed from the model. Thus all old variables in the current model must correspond to p-values not greater than α^*. At this point, the process reconsiders the possible inclusion of another new variable, as in the forward procedure, and stops if none of the variables which are candidates for inclusion reaches significance. The current model at the stop is the final one.

Another selection procedure which is less often applied in survival analysis is the *all subset selection.* In synthesis, this considers, with K candidate regressors, the 2^K possible models, from the null model to the full model, through all the submodels with any subset of the K regressors. The criterion for selecting the model is based on the value of Akaike's function which depends on the log-likelihood LL_p and the number p of regression parameters in the model (see Lawless and Singhal (1987) for details).

APPENDIX A9.2   RELATIONSHIP BETWEEN RISK AND HAZARD

The relationship between the risk of death and the hazard rate is based on

$$F(t) = 1 - \exp\left[-\int_0^t \lambda(u)\,du \right] = 1 - \exp[-\Lambda(t)]$$

Therefore the RR of the two groups of patients with covariates vectors $\mathbf{x}_1$ and $\mathbf{x}_2$ may be written

$$\mathrm{RR} = \frac{1 - \exp[-\Lambda(t, \mathbf{x}_2)]}{1 - \exp[-\Lambda(t, \mathbf{x}_1)]} \tag{A9.1}$$

If the cumulative hazards $\Lambda(t, x_1)$ and $\Lambda(t, \mathbf{x}_2)$, are "sufficiently small", the two

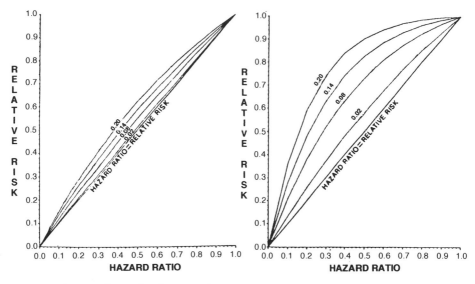

Figure A9.1   Relationship between hazard ratio (θ) and relative risk (RR) for different values of failure rate in the reference category. The curves on the left refer to the example of table 9.6 and those on the right refer to the example of table 9.7 (from Boracchi *et al.*, 1990, reproduced by permission).

terms of the ratio can be expanded in Taylor series

$$RR = \frac{1 - (1 - \Lambda(t, \mathbf{x}_2) + o(\Lambda(t; \mathbf{x}_2)))}{1 - (1 - \Lambda(t, \mathbf{x}_1) + o(\Lambda(t; \mathbf{x}_1)))} \tag{A9.2}$$

where o represents the symbol of infinitesimal ratio.

By omitting the o terms (A9.2) becomes

$$RR \simeq \frac{\Lambda(t, \mathbf{x}_2)}{\Lambda(t, \mathbf{x}_1)} = \frac{\lambda(t, \mathbf{x}_2)}{\lambda(t, \mathbf{x}_1)} = \theta \tag{A9.3}$$

To clarify what "sufficiently small" means, figure A9.1 has to be carefully considered; the graphs on the left refer to the first example (see table 9.6) and those on the right to the second example (see table 9.7) in section 9.5.2

In the abscissa different values of hazard ratio (θ) are reported and in the ordinate the corresponding relative risk (RR); each curve is drawn in relation to a given value of the force of mortality of the reference category specified on each of them.

The hazard ratio is equal to the risk ratio if the points on the graph lay on the straight line with null intercept and the slope equal to one. As far as one moves from this line upwards, figure A9.1 shows that RR always overestimates θ and, symmetrically, θ underestimates RR. Furthermore, for a given θ, the bias increases as far as the force of mortality of the reference group increases.

In oncological research, most multiple regression analyses on survival data are performed by means of proportional hazards models, thus θ is the parameter with a key role in interpreting the results. In the following comments, attention focuses on the size of bias which arises in using the relative rate (θ) to assess the relative risk (RR).

In the first example, the reference category has a death rate of $\hat{\lambda} = 0.08$ per one woman–year at risk and a period of follow-up of five years. Thus, the second curve from the straight line of equivalence informs us that θ slightly underestimates RR, reaching its maximum bias of 5% at $\theta = 0.5$.

In the second example, let us consider the follow-up period of 22 weeks. Since the death rate of the reference category $\hat{\lambda}(22) = 0.1384$ per one person–week, the third curve from the equivalence line shows that θ heavily underestimates RR and the maximum bias is now 34% at $\theta = 0.4$.

The different degree of underestimation in these two examples is easily understood if one considers the two cumulative hazards. At the end of the follow-up period considered in each example, namely after five years in the first one and 22 weeks in the second one, the cumulative hazards are $0.08 \times 5 = 0.40$ and $0.1384 \times 22 = 3.04$, respectively. The difference between these two values accounts for the different degree of underestimation.

10

Competing Risks

10.1 INTRODUCTION

Statistical methods commented upon in previous chapters are suitable for facing problems in which a single, possibly censored, failure time is recorded on each observational or experimental unit. Furthermore it was stated that the key factor in making such methods work right is the assumption of independence of the failure time and censoring time distributions (section 1.2.2). However, think of an epidemiologist aiming at assessing the effect of reducing exposure to an environmental carcinogen; he cannot focus only on the decrease of mortality for cancer, but he should study also effects on other causes of death. Analogously, in the clinical setting, several measures of therapeutical efficacy rather than overall survival are frequently recorded. Consider, for example, early breast cancer: in trials of radiation therapy, local and regional recurrences, distant metastases and non-cancer death are events of clinical concern. It is likely that the time to local/regional recurrence and the time to distant metastases be correlated and, furthermore, an early death may prevent the investigator from evaluating the effect of the treatment on the local recurrence.

Analysis of data generated in situations like those here outlined in which an individual is exposed to two or more causes of failure can be tackled by methods which make allowance for the "competing risks" of the causes to be studied.

Competing risks literature deals with observable and non-observable probabilities. The former are quantities expressed as a function of the hazard rate specific to a given cause, corresponding to the original observations with all causes of risk acting. The latter, non-observable, are needed when an epidemiologist, for instance, is interested in predicting what would happen if one or more causes of risk were removed. An example is the *net* survival probability from a given cause, defined as the probability that an individual's failure time for that cause exceeds a preassigned time t in the hypothetical condition that the only possible risk of failure is due to such a cause. Computations of non-observable probabilities are strictly dependent on the distributional models of failure times one assumes. The concept of "latent" or "potential" failure time for each cause has been found mathematically convenient for the theoretical modelling of competing risks. For example, by assuming latent independent failure times, David and Moeschberger (1978) suggested their classical multiple decrement model. However, as noted by Prentice *et al.* (1978): "...except in circumstances of complete biological or physical independence among system components giving rise to the failure types, it is unrealistic to suppose that general statistical methods can be put forward that will encompass all possible mechanisms for cause removal". Thus in the following sections we will focus only on methods suitable for analysing *observable* probabilities; some considerations on the latent failure time approach and on the non-identifiability problem will be outlined in section 10.6.

10.2 THE PRODUCT LIMIT ESTIMATOR IN THE PRESENCE OF COMPETING EVENTS

Let us consider the hypothetical data reported in table 10.1; they regard two treatment groups A and B and two events, say local relapse and distant metastases, each of them as the *first* observed event. Information concerns 35 individuals in each group; at a glance it appears that times to occurrence of event 1 and censored times are quite similar in the two groups, while times to occurrence of event 2 are notably different.

If one considers the two events jointly and uses the K–M technique, the overall survival curves obtained are shown in figure 10.1. Moreover the M–H test gives $Q_{M-H} = 5.3609$ ($p = 0.0206$), leading to the rejection of the null hypothesis of hazard rate ratio equal to 1 in the two groups for the first failure event. The estimate of the curves and the test result are statistically correct as, dealing with the two events jointly, the "competing" effect for the same subject is overcome. They, however, may appear of limited usefulness for the clinician aiming at studying the possibly different effect of treatments on the occurrence of one or the other of the two events.

Table 10.1 Example of two competing events (1 and 2) and two treatment groups (A and B); time in weeks (hypothetical data).

	Group A	Group B
Event of type 1	1, 13, 17, 30, 34, 41, 78, 100, 119, 169	7, 16, 16, 20, 39, 49, 56, 73, 93, 113
Event of type 2	1, 6, 8, 13, 13, 15, 33, 37, 44, 45, 63, 80, 89, 89, 91, 132, 144, 171, 183, 240	1, 2, 4, 6, 8, 9, 10, 13, 17, 17, 17, 18, 18, 27, 29, 39, 50, 69, 76, 110
Censored data	34, 60, 63 149, 207	34, 60, 63, 78, 149

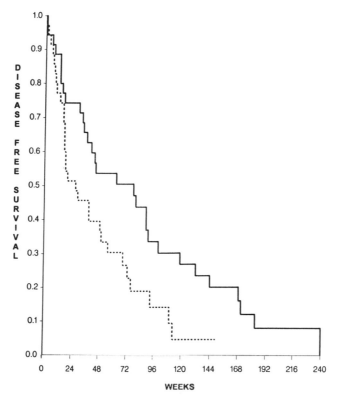

Figure 10.1   Disease free survival curves; hypothetical data from table 10.1. Group A———; group B------.

In trying to describe the pattern of occurrence of event 1, the investigator could think of regarding event 2 failures as censored at the subject's failure time and, similarly, in studying event 2, of regarding event 1 failures as censored. Even in this case, application of the K–M technique is straightforward and the resulting cause-specific survival curves are reported in figure 10.2. It clearly appears that the two curves for event 1 are notably divergent, more than those drawn for event 2. This pattern is in contrast with what one might expect from inspecting data in table 10.1 because an unexpected difference between the two curves for event 1 tends to emerge; it appears to be more an artifact of the method adopted to estimate the event free survival curve than the possibly differential effect of treatments on the two events and suggests the need of alternative statistical techniques.

Some further definitions are needed before introducing methods suitable for analysing data in the presence of competing effect of different failure types.

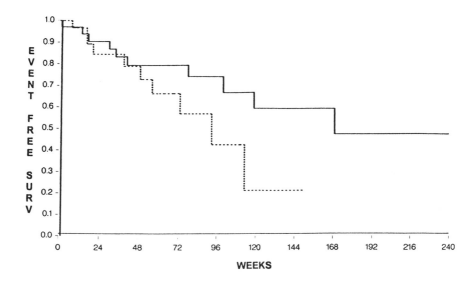

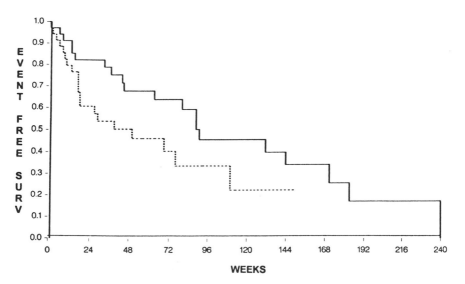

Figure 10.2 Event free survival curves; upper part: event of type 1; lower part: event of type 2. Data from table 10.1. Group A———; group B--------.

10.3 CAUSE-SPECIFIC HAZARD FUNCTION, RELATED FUNCTIONS AND THEIR ESTIMATORS

We suppose that an individual be exposed to C mutually exclusive causes of failure denoted by $L \in \{1, 2, \ldots C\}$ and, when a failure occurs, we observe the time T and the cause of failure L.

In section 5.2 the hazard function for a single cause of failure was defined; according to Prentice *et al.* (1978) it can be extended in a natural way in the presence of several causes of failure. In a homogeneous population, the cause-specific hazard function is defined as

$$h_l(t) = \lim_{\Delta t \to 0^+} \frac{\Pr\{t \leqslant T < t + \Delta t, L = l | T \geqslant t\}}{\Delta t} \tag{10.1}$$

for $l = 1, 2, \ldots, C$

We are dealing with the hazard function describing the instantaneous rate of failure for the cause l at time t in the presence of the remaining $(C - 1)$ causes. The *conditional* probability of failure for the cause l in the interval $(t, t + \mathrm{d}t)$, given the individual is free from failure of any type at time t and all the other causes are acting simultaneously, is approximately $h_l(t)\,\mathrm{d}t$. Analogously to (5.5), we define the cause-specific cumulative hazard as

$$H_l(t) = \int_0^t h_l(u)\,\mathrm{d}u \tag{10.2}$$

Furthermore, according to (5.8), we can define the functions

$$S_{cs \cdot l}(t) = \exp\left[-H_l(t)\right] \tag{10.3}$$

Note that these functions, in general, cannot be interpreted in terms of survival functions for $C > 1$. In fact, when more than one cause of failure is acting, not failing for the type l cause at time t does not imply remaining free from failure of any type beyond t.

Since the set of C causes is exhaustive of all the possible causes concurring to failure, the *overall* hazard to failure is

$$\lambda(t) = \sum_{l=1}^{C} h_l(t) \tag{10.4}$$

Therefore it results that the cumulative overall hazard is

$$\Lambda(t) = \int_0^t \lambda(u)\,\mathrm{d}u = \sum_{l=1}^{C} H_l(t) \tag{10.5}$$

From (10.5) and (10.3), we have that

$$S(t) = \exp\left[-\sum_{l=1}^{C} H_l(t) \right] = \prod_{l=1}^{C} S_{\text{cs}\cdot l}(t) \tag{10.6}$$

regardless of whether the risks of failure are independent or not. If the C causes may be assumed to act independently from each other, expression (10.6) enables us to associate probability statements to the functions $S_{\text{cs}\cdot l}(t)$ because in this case, each of them represents the probability that the failure of a given type occurs beyond t, even in the presence of the other risks of failure.

The *unconditional* probability of type l failure in the interval $(t, t + dt)$ is obtained by multiplying the probability of being free from any failure at time t, given by (10.6), times the conditional probability of occurrence of type l failure:

$$f_l(t)\, dt = S(t)h_l(t)\, dt \tag{10.7}$$

This is the *subdensity* function for failures of type l in the presence of the remaining $(C - 1)$ risks of failure.

The *crude* cumulative incidence or *subdistribution* function for failures of type l is defined

$$I_l(t) = \int_0^t f_l(u)\, du = \int_0^t S(u)h_l(u)\, du \tag{10.8}$$

The corresponding crude survival, $S_l(t) = 1 - I_l(t)$, may be interpreted as the survival function of a r.v. Y, where $Y = T$ or $Y = \infty$ depending on whether at time $T = t$ a failure of type l occurred or not.

Note that these are *not proper* distribution functions according to the definition given in section 5.2.

Let $\pi_l = I_l(\infty)$ denote the probability of ever failing of type l event; it enables us to define the *proper* crude cumulative incidence of type l event (Elandt-Johnson and Johnson, 1980, pp. 274–276) as

$$I_l^*(t) = \frac{1}{\pi_l} \int_0^t S(u)h_l(u)\, du = \frac{I_l(t)}{\pi_l}$$

which represents the actual distribution of times of failure of type l among the individuals who present failure of type l, in the presence of C causes of failure.

The corresponding proper crude survival function is

$$S_l^*(t) = 1 - I_l^*(t)$$

and the subdensity function is

$$f_l^*(t) = \frac{S(t)h_l(t)}{\pi_l}$$

Therefore the hazard rate of $S_l^*(t)$ is

$$\lambda_l^*(t) = \frac{f_l^*(t)}{S_l^*(t)} = \frac{S(t)h_l(t)}{\displaystyle\int_t^\infty S(u)h_l(u)\,du} = -\frac{1}{\pi_l}\frac{dS_l^*(t)}{dt}$$

and represents the hazard rate at time t *conditional* on presenting a failure of type l.

All the previous expressions define *observable* probabilities in accordance with the definition given in section (10.1).

By means of (10.1) and (10.3) we can write the cause-specific failure cumulative incidence, $1 - S_{cs\cdot l}(t)$:

$$I_{cs\cdot l}(t) = \int_0^t S_{cs\cdot l}(u)h_l(u)\,du$$

and by means of (10.4) and (10.6) we can write the overall failure cumulative incidence:

$$F(t) = \int_0^t S(u)\sum_{l=1}^c h_l(u)\,du$$

Owing to the fact that: $H_l(t) \leqslant \Lambda(t)$ (see (10.5)) and $S(t) \leqslant S_{cs\cdot l}(t)$ (see (10.6)), we can see that $I_l(t) \leqslant I_{cs\cdot l}(t) \leqslant F(t)$. Furthermore, from (10.4) and (10.8), the additivity of crude cumulative incidences derives:

$$\sum_{l=1}^c I_l(t) = F(t) \tag{10.9}$$

and thus

$$\sum_{l=1}^c \pi_l = 1$$

Let $0 < t_{(1)} < \cdots < t_{(j)} < \cdots < t_{(J)}$, $J \leqslant N$ be the distinct ordered failures (of any type) observed among the N individuals under study and, coherently with the definitions in section 5.4, consider the cause-specific hazard function of type l event as having atoms h_{lj} at points $\{t_j\}$. This enables us to write the following discrete cause-specific hazard model:

$$h_l(t) = h_{lj} \quad \text{for } t = t_j$$
$$= 0 \quad \text{otherwise}$$

where h_{lj} indicates the probability of failure of type l at $t_{(j)}$ given that the individual is free from failure of any type beyond $t_{(j-1)}$. We are dealing with a "non-parametric" model as $C \cdot J$ parameters, i.e. a saturated model, are utilized to describe l types of failure and J distinct times of failure.

Let d_{lj} denote the number of individuals failing for the l type cause at t_j, $d_j = \sum_{l=1}^{C} d_{lj}$, and n_j denote the number of individuals at risk of failing at t_j. Remembering (5.19), the likelihood can be written:

$$L = \prod_{j=1}^{J} \left[\left(\prod_{l=1}^{C} h_{lj}^{d_{lj}} \right) \left(1 - \sum_{l=1}^{C} h_{lj} \right)^{n_j - d_j} \right]$$

The ML estimate of the cause-specific hazard $h_l(t)$ is shown to be

$$\hat{h}_{lj} = \frac{d_{lj}}{n_j} \tag{10.10}$$

and of the cumulative hazard $H_l(t)$ is

$$\hat{H}_l(t) = \sum_{j|t_j \leqslant t} \frac{d_{lj}}{n_j}$$

According to Kalbfleisch and Prentice (1980, p. 168) the crude cumulative incidence for failure of type l is estimated by

$$\hat{I}_l(t) = \sum_{j|t_j \leqslant t} \hat{S}(t_{j-1}) \frac{d_{lj}}{n_j} \tag{10.11}$$

where $\hat{S}(t_{j-1})$ is the estimated probability of surviving up to $t_{(j)}$, but not beyond $t_{(j)}$, coherently with the considerations developed in section (5.4). Of course, $S(t)$ is estimated by the K–M method.

10.4   EXAMPLE OF COMPUTATION OF CRUDE CUMULATIVE INCIDENCE

Data of group B in table 10.1 are used to show, in table 10.2, the computation procedure of $\hat{I}_l(t)$ for the event of type 1.

Table 10.2 Computation of crude cumulative incidence of type 1 event on data in group B of table 10.1.

Time of occurrence of type 1 event $t_{(j)}$ (1)	No. of failures d_{1j} (2)	No. exposed at risk n_j (3)	Estimated failure rate $[\hat{h}_{1B}(t_j)]$ (4)	Estimated overall survival $[\hat{S}_B(t_{j-1})]$ (5)	Crude Cumulative incidence $\hat{I}_{1B}(t_j)$ (6)
0	0	35	0.000 00	1.000 0	0.000 00
7	1	31	0.032 26	0.885 71	0.028 57
16	2	26	0.076 92	0.742 86	0.085 71
20	1	19	0.052 63	0.542 86	0.114 29
39	1	15	0.066 67	0.457 14	0.144 76
49	1	13	0.076 92	0.396 19	0.175 24
56	1	11	0.090 91	0.335 24	0.205 71
73	1	7	0.142 86	0.266 67	0.243 81
93	1	4	0.250 00	0.190 48	0.291 43
113	1	2	0.500 00	0.095 24	0.339 05

The number of data censored between $t_{(j-1)}$ and $t_{(j)}$ is obtained by adding the number of "censored data" reported in table 10.1 to the number of type 2 failures occurring in the time interval between $t_{(j-1)}$ included and $t_{(j)}$ excluded, coherently with the approach introduced in section 3.2.2. According to (10.10), the estimate of the cause-specific hazard at $t_{(j)}$ is obtained by dividing the number in column (2) by the corresponding one reported in column (3) while $\hat{I}_{1B}(t_j) = I_{1B}(t_{j-1}) + \hat{S}(t_{j-1})\hat{h}_{1B}(t_j)$ is reported in column (6).

In the upper part of figure 10.3, the crude cumulative incidence curves for type 1 event (groups A and B) are displayed in the panel on the left side and those for type 2 event in the panel on the right side. To make easy the comparison with the curves reported in figure 10.2, in the lower part of figure 10.3 the corresponding *crude* survival curves are reported. The two curves on the left differ only slightly, and this agrees with the observation made at the beginning of section 10.2 about the similar pattern of occurrence of type 1 events in the two groups.

10.5 VARIANCE OF THE CRUDE CUMULATIVE INCIDENCE AND AN EXAMPLE

Since (10.11) shows that the estimate of the crude cumulative incidence of type *l* event is the sum of products of two statistics, $\hat{h}_{lj}\hat{S}(t_{j-1})$, the variance of $\hat{I}_l$

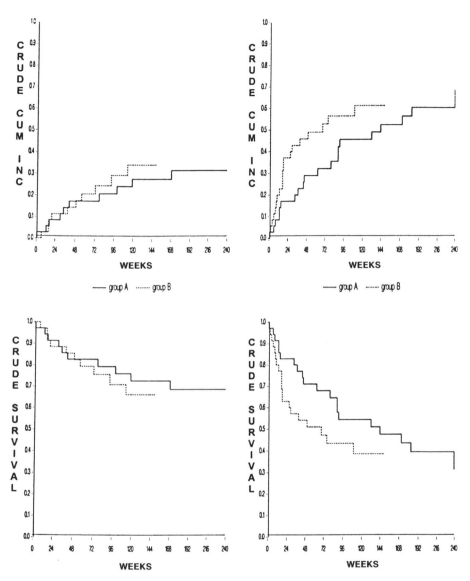

Figure 10.3 Crude cumulative incidence curves (upper panels) and crude survival curves (lower panels) of type 1 event (left side) and type 2 event (right side). Data from table 10.1. Group A———; group B---------.

at t_j is

$$var[\hat{I}_l(t_j)] = var\left[\sum_{\alpha=1}^{j} \hat{h}_{l\alpha}\hat{S}(t_{\alpha-1})\right] = \sum_{\alpha=1}^{j} var[\hat{h}_{l\alpha}\hat{S}(t_{\alpha-1})] + 2\sum_{\alpha=1}^{j-1}\sum_{\beta=\alpha+1}^{j} cov[\hat{h}_{l\alpha}\hat{S}(t_{\alpha-1}), \hat{h}_{l\beta}\hat{S}(t_{\beta-1})]$$

In Appendix A10.1 it is shown that using the delta method (Appendix A3.2) to compute variances and covariances of the above-mentioned products of two statistics, the variance of the crude cumulative incidence may be estimated by

$$var[\hat{I}_l(t_j)] = \sum_{\alpha=1}^{j} \left\{ [\hat{I}_l(t_j) - \hat{I}_l(t_\alpha)]^2 \frac{d_\alpha}{n_\alpha(n_\alpha - d_\alpha)} \right\} + \sum_{\alpha=1}^{j} [\hat{S}(t_{\alpha-1})]^2 \left(\frac{n_\alpha - d_{l\alpha}}{n_\alpha} \right) \left(\frac{d_{l\alpha}}{n_\alpha^2} \right)$$

$$- 2 \sum_{\alpha=1}^{j} [\hat{I}_l(t_j) - \hat{I}_l(t_\alpha)][\hat{S}(t_{\alpha-1})] \left(\frac{d_{l\alpha}}{n_\alpha^2} \right) \qquad (10.12)$$

where $d_j = \sum_{l=1}^{C} d_{lj}$.

Note that in the presence of one cause of risk only, the Greenwood's variance discussed in section 4.1 corresponds to the estimate given by (10.12).

With reference to table 10.2 and data on event of type 1 in table 10.1, we compute the variance of $\hat{I}_1$ (16):

$$var(\hat{I}_1(16)) = \left\{ [\hat{I}_1(16)]^2 \frac{d_1}{n_1(n_1 - d_1)} \right\} + \left\{ [\hat{I}_1(16)]^2 \frac{d_2}{n_2(n_2 - d_2)} \right\}$$

$$+ \left\{ [\hat{I}_1(16))]^2 \frac{d_4}{n_4(n_4 - d_4)} \right\} + \left\{ [\hat{I}_1(16))]^2 \frac{d_6}{n_6(n_6 - d_6)} \right\}$$

$$+ \left\{ [\hat{I}_1(16) - \hat{I}_1(7)]^2 \frac{d_7}{n_7(n_7 - d_7)} \right\} + \left\{ [\hat{I}_1(16) - \hat{I}_1(7)]^2 \frac{d_8}{n_8(n_8 - d_8)} \right\}$$

$$+ \left\{ [\hat{I}_1(16) - \hat{I}_1(7)]^2 \frac{d_9}{n_9(n_9 - d_9)} \right\} + \left\{ [\hat{I}_1(16) - \hat{I}_1(7)]^2 \frac{d_{10}}{n_{10}(n_{10} - d_{10})} \right\}$$

$$+ \left\{ [\hat{I}_1(16) - \hat{I}_1(7)]^2 \frac{d_{13}}{n_{13}(n_{13} - d_{13})} \right\} + \left\{ [\hat{S}(6)]^2 \frac{d_7(n_7 - d_7)}{(n_7)^3} \right\}$$

$$+ \left\{ [\hat{S}(15)]^2 \frac{d_{16}(n_{16} - d_{16})}{(n_{16})^3} \right\} - 2[\hat{I}_1(16) - \hat{I}_1(7)][\hat{S}(6)] \left(\frac{d_7}{n_7^2} \right)$$

which, substituting, becomes

$$var(\hat{I}_1(16)) = \left\{ (0.085\,71)^2 \frac{1}{35 \times 34} \right\} + \left\{ (0.085\,71)^2 \frac{1}{34 \times 33} \right\}$$

$$+ \left\{ (0.085\,71)^2 \frac{1}{33 \times 32} \right\} + \left\{ (0.085\,71)^2 \frac{1}{32 \times 31} \right\}$$

$$+ \left\{ (0.085\,71 - 0.028\,57)^2 \frac{1}{31 \times 30} \right\}$$

$$+ \left\{ (0.085\,71 - 0.028\,57)^2 \frac{1}{30 \times 29} \right\}$$

$$+ \left\{ (0.085\,71 - 0.028\,57)^2 \frac{1}{29 \times 28} \right\}$$

$$+ \left\{ (0.085\,71 - 0.028\,57)^2 \frac{1}{28 \times 27} \right\}$$

$$+ \left\{ (0.085\,71 - 0.028\,57)^2 \frac{1}{27 \times 26} \right\}$$

$$+ \left\{ (0.885\,71)^2 \frac{30}{(31)^3} \right\} + \left\{ (0.742\,86)^2 \frac{24 \times 2}{(26)^3} \right\}$$

$$- 2 \times \left\{ (0.085\,71 - 0.028\,57)(0.885\,71) \frac{1}{(31)^2} \right\}$$

$$= 0.000\,006\,1733 + 0.000\,006\,5474 + 0.000\,006\,9566$$

$$+ 0.000\,007\,4054 + 0.000\,003\,5107 + 0.000\,003\,7529$$

$$+ 0.000\,004\,0209 + 0.000\,004\,3188 + 0.000\,004\,6510$$

$$+ 0.000\,789\,99 + 0.001\,5071 - 0.000\,105\,33$$

$$= 2.2391 \times 10^3$$

Therefore $\mathrm{SE}[\hat{I}_1(16)] = 0.047\,32$; this enables us to compute approximate 95% confidence limits by the formula (Keiding and Andersen, 1989):

$$\exp\{\log \hat{I}_l(t) \pm 1.96\,[\mathrm{SE}(\hat{I}_l(t))]/\hat{I}_l(t)\}$$

namely

$$\exp\left\{\log 0.085\,71 \pm 1.96 \frac{0.047\,32}{0.085\,71}\right\} = (0.029\,05, 0.252\,92)$$

Some point-estimates of $\hat{I}_l(t)$ and $\hat{I}_{\mathrm{cs} \cdot l}(t)$ together with their standard errors obtained from data of table 10.1 are reported in table 10.3 and allow the following observations:

(1) in both groups, $\hat{I}_l(t)$ is lower than $\hat{I}_{\mathrm{cs} \cdot 1}(t)$ as anticipated in section 10.3;
(2) for overlapping values of the estimated $\hat{I}_l(t)$ and $\hat{I}_{\mathrm{cs} \cdot l}(t)$, $\mathrm{SE}[\hat{I}_l(t)]$ and $\mathrm{SE}[\hat{I}_{\mathrm{cs} \cdot l}(t)]$ are similar and both tend to increase over time;
(3) the maximum length of follow-up is $t = 240$ weeks, and this time corresponds to the occurrence of an event of type 2 in group A. Thus the crude cumulative incidences computed at this point in group A, $\hat{I}_1(240) = 0.314$ and $\hat{I}_2(240) = 0.686$, add to 1 and can be thought of as

Table 10.3 Crude [$\hat{I}_l(t)$] and cause-specific [$\hat{I}_{cs\cdot l}(t)$] cumulative incidence along with their standard errors for data of table 10.1.

Group	Time t (weeks)	$\hat{I}_1(t)$	SE[$\hat{I}_1(t)$]	$\hat{I}_{cs\cdot 1}(t)$	SE[$\hat{I}_{cs\cdot 1}(t)$]	$\hat{I}_2(t)$	SE[$\hat{I}_2(t)$]	$\hat{I}_{cs\cdot 2}(t)$	SE[$\hat{I}_{cs\cdot 2}(t)$]
A	15	0.057	0.039	0.059	0.041	0.171	0.064	0.177	0.066
	30	0.114	0.054	0.129	0.061	0.171	0.064	0.210	0.071
	60	0.173	0.064	0.206	0.076	0.290	0.077	0.321	0.085
	120	0.274	0.078	0.407	0.117	0.456	0.088	0.544	0.100
	240	0.314	0.083	0.525	0.141	0.686	0.083	1.000	—
B	15	0.029	0.028	0.032	0.032	0.229	0.071	0.232	0.072
	30	0.114	0.054	0.154	0.072	0.429	0.084	0.460	0.088
	60	0.206	0.069	0.337	0.110	0.490	0.085	0.538	0.091
	113	0.339	0.088	0.787	0.170	0.613	0.089	0.775	0.114

estimates of π_1 and π_2 (section 10.3), that is the proportion of individuals in whom an event of type 1 or of type 2 occurs. On the contrary, in group B the last observed time, $t = 149$, corresponds to a censored datum and the sum of the cumulative incidences computed at the last time in which an event was recorded, $t = 113$, is less than 1, namely 0.952. To obtain the estimates of π_l we must make an extrapolation by assuming that the proportion of all failures of a given type will not change over time. The estimated proportion is obtained by normalizing the crude cumulative incidence as follows:

$$\tilde{\pi}_l = \hat{I}_l(t) + [1 - \sum_{l=1}^{C} \hat{I}_l(t)] \left[\frac{\hat{I}_l(t)}{\sum_{l=1}^{C} \hat{I}_l(t)} \right]$$

In our example:

$$\tilde{\pi}_1 = 0.339 + (1 - 0.952)\left(\frac{0.339}{0.952}\right) = 0.356$$

$$\tilde{\pi} = 0.613 + 0.048\left(\frac{0.613}{0.952}\right) = 0.644$$

Comparison of $\tilde{\pi}_1$ and $\tilde{\pi}_2$ to the corresponding estimates previously obtained for group A suggests that the probabilities of ever failing for events either of type 1 or of type 2 are similar in the two groups.

Situations like that observed in group B are met in the majority of medical investigations. Note, however, that even a deep knowledge of the natural course of the disease or physiopathological arguments cannot avoid the component of arbitrariness introduced by the strong assumption underlying the extrapolation to $t = \infty$. Thus, if in the range of the follow-up period, the estimated crude cumulative incidence curve shows a long flat tail beyond a time point t', it may be sensible to limit calculation of $\hat{\pi}_l$ up to time t'. This estimate can then be used to compute $\hat{I}_l^*(t)$ in the attempt to depict the distribution of times of type l failure.

10.6   LATENT FAILURE TIMES

As mentioned in section 10.1, the concept of latent failure times has been introduced in the competing risks literature purely for mathematical convenience. For an individual exposed to $C > 1$ possible causes of death, the lth latent failure time designates a r.v. Z_l indicating the age at death of the individual in the hypothetical situation in which the possibility of failure from causes other than l is removed. The joint survivorship function of the random vector $[Z_1, Z_2, \ldots,$

$Z_1, \ldots, Z_C]$ is defined as

$$S^0(t_1^0, t_2^0, \ldots, t_l^0, \ldots, t_C^0) = \Pr\{Z_1 > t_1^0 Z_2 > t_2^0, \ldots, Z_l > t_l^0, \ldots, Z_C > t_C^0\} \quad (10.13)$$

This is a right-sided cumulative distribution satisfying $S^0(0, 0, \ldots, 0) = 1$ and $S^0(\infty, \infty, \ldots, \infty) = 0$ and each $Z_l \in (0, \infty)$; (10.13) is continuous from the right and monotonic non-increasing in each argument. This implies that also each of the C *marginal*, previously termed net, survival functions,

$$S_l^0(t_l^0) = S^0(0, 0, \ldots, t_l^0, \ldots, 0) \quad (10.14)$$

is a proper survival distribution.

The *net* hazard rate or *net* force of mortality is

$$\lambda_l(t) = -\frac{d \log S_l^0(t)}{dt} = -\frac{\partial \log S^0(t_1^0, t_2^0, \ldots, t_l^0, \ldots, t_C^0)}{\partial t_l}\bigg|_{t_\gamma^0 = 0, \forall \gamma \neq l; t_l^0 = t} \quad (10.15)$$

The latent failure times Z_l are contrasted with the *actual* survival time:

$$T = \min(Z_1, Z_2, \ldots, Z_l, \ldots, Z_C)$$

Note that from $S^0(\cdot)$ as specified in (10.13) it follows that

$$S(t) = \Pr(T > t) = \Pr[Z_l > t, \forall l] = S^0(t, t, \ldots, t) \quad (10.16)$$

and this formulation is suitable whether the failure times $Z_1, Z_2, \ldots, Z_C$ are independent or not. Furthermore, Tsiatis (1975) proved that whatever the joint distribution (10.13), the derivative of the *l*th crude survival function $S_l(t)$ (section 10.3) is equal to the partial derivative of (10.13) with respect to t_l^0 evaluated at $t_1^0 = t_2^0 = \cdots = t_C^0 = t$, i.e.:

$$h_l(t) = -\frac{\partial \log S^0(t_1^0, t_2^0, \ldots, t_l^0, \ldots, t_C^0)}{\partial t_l^0}\bigg|_{t_1^0 = t_2^0 = \cdots = t_C^0 = t} \quad (10.17)$$

This implies that any given joint survival distribution determines uniquely the corresponding crude survival function. It appears from (10.16) and (10.17) that estimable quantities like $S(t)$ and $h_l(t)$ can be written in terms of $S^0(\cdot)$.

In the same paper, Tsiatis proved that a given set of C crude survival probabilities $S_l(t)$ does not identify the corresponding net survival probabilities, unless one is able to assume that in their joint distribution the r.v.'s $Z_1, Z_2, \ldots, Z_l, \ldots, Z_C$ are mutually independent. Unfortunately, this assumption cannot be directly verified from the data. Probabilities like (10.14) and (10.15) which can be estimated only under specific untestable assumptions are termed "non-identifiable".

In the case of mutual independence among failure causes,

$$S^0(t_1^0, t_2^0, \ldots, t_l^0, t_C^0) = \prod_{l=1}^{C} S_l^0(t_l^0)$$

and it follows that

$$h_l(t) = \lambda_l(t)$$

i.e. the cause-specific hazard rate equals the net hazard rate and, consequently, the cause-specific survival equals the net survival function.

10.7 CAUSE-SPECIFIC HAZARDS AND PROGNOSTIC FACTORS

All previous considerations were based upon the assumption of a homogeneous population; however, in several epidemiological or clinical applications, the main goal of the study is the evaluation of the role of risk factors or prognostic factors. This implies modelling the cause-specific hazard as a function of a covariate vector $\mathbf{x}$ pertinent to each individual. If a semi-parametric proportional hazards model is adopted, for the type l failure we have

$$h_l(t; \mathbf{x}) = h_{0l}(t) \exp(\boldsymbol{\beta}_l' \mathbf{x}), \qquad l = 1, 2, \ldots, C \tag{10.18}$$

This is a straightforward extension of model (6.1); however, it is worth noting that both the underlying hazard $h_{0l}(t) \geqslant 0$ and the vector of regression coefficients are specific to each of the C failure types. As shown by Kalbfleisch and Prentice (1980, p. 170), the parameter estimation is based upon the method of partial likelihood. By resorting to (6.8), the $\boldsymbol{\beta}$'s are estimated separately for each failure type by considering failures of the remaining types as censored observations at the individual's failure time. Furthermore, the estimate of the underlying hazard $\hat{h}_{0l}(t_j)$ is computed by assuming it to be zero except at times at which a failure of type l occurs; in this case:

$$\hat{h}_{0l}(t_j) = d_{lj} \left[\sum_{m \in R_j} \exp(\hat{\boldsymbol{\beta}}_l' \mathbf{x}_m) \right]^{-1} \tag{10.19}$$

where R_j is the set of individuals at risk just prior to t_j and d_{lj} is the number of failures of type l at time t_j. This generalizes the single-cause technique developed in section 6.3*.

By (10.4) and (10.6) we can write

$$S(t, \mathbf{x}) = \exp\left[-\int_0^t \lambda(u, \mathbf{x})\,du \right]$$

which can be estimated as shown in section 6.3*. Therefore the crude cumulative incidence of type l failure adjusted by $\mathbf{x}$ can be estimated by

$$\hat{I}_l(t, \mathbf{x}) = \sum_{t_j \leqslant t} \hat{S}(t_j, \mathbf{x})\hat{h}_l(t_j, \mathbf{x}) \tag{10.20}$$

and the crude survival function by

$$\hat{S}_l(t, \mathbf{x}) = 1 - \hat{I}_l(t, \mathbf{x})$$

The derivation of (10.19) suggests that the whole vector $\mathbf{x}$ should be inserted into the linear predictor to model each of the C cause-specific hazards. However, on the basis of *a priori* knowledge, an investigator could argue to insert one particular covariate in the predictor of the cause-specific hazard of type l, for instance, but not in those of the remaining types. This approach should be avoided because, unless the latent failure times are independent, it is likely that all cause-specific hazards depend on that covariate and such dependence is quantified by the pertinent regression coefficient.

Here attention was focused on the Cox model, but similar extensions may be done to estimate the cause-specific hazards with regression models in Chapter 8.

As an example we will discuss some of the findings by Veronesi *et al.* (1994) on 2233 early breast cancer patients treated with conservative surgery (Quart). First failure was classified in five mutually exclusive categories: (1) local recurrence (in the same breast); (2) distant recurrence (metastases of bone, lung, bowel, etc.); (3) recurrence in the contralateral breast; (4) other cancers; (5) death with no evidence of disease (NED). The estimates of the pertinent crude cumulative hazard curves are given in figure 10.4.

The paper focused on the two events: local and distant recurrences which appear to be the most relevant in determining the first failure, with the purpose of investigating the role of a set of covariates on these two types of failure. Five external covariates were considered: age of the patient at surgery; localization of the primary tumor; dimension of the primary; number of axillary metastatic lymph nodes and histological type. All these covariates were inserted into a Cox model as dummy variables as specified in table 10.4 with the code used in the output of the regression analyses. All the information was available on 2010 patients whose data were processed; as first event, 127 local and 382 distant recurrences were observed. The median follow-up time was 8.5 years.

The results of the regression analyses are reported in table 10.5; columns (2) to (5) relate to local recurrences and columns (6) to (9) to distant

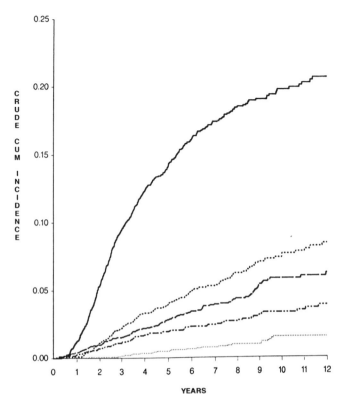

Figure 10.4 Crude cumulative hazard curves for different types of event recorded in early breast cancer (data from Veronesi *et al.*, 1994). Local failure -----; Metastases ———; Contralateral––; Other NPL ···–··–; NED Death.......

recurrences. As each of the Q_W (Wald statistics) has 1 d.f., the critical value at $p = 0.05$ is 3.84.

Let us consider firstly the findings regarding metastases. While different localizations of the primary tumor do not seem to modify the risk of developing metastases, all the other covariates do. So we can see that the risk of metastasis: (1) of young patients (35 years or less) is about twice that of old patients (66 years or more) by the highly significant coefficient of the AGE4 regressor; (2) increases some 1.5 times with the increase of the tumor size (SIZE 1 to SIZE 4); and (3) increases some 1.6 times with the increase of the number of axillary metastatic nodes (NOD1 to NOD3). As regards histology, the risk of metastasis of tumors with "other histotypes" is about half of that of infiltrating tumors (ductal or lobular). These results are in good agreement with the known "natural history" of the disease.

When we look at the findings on local failure, we note that, as previously, SITE is not significant, whilst besides AGE4, AGE3 regressor is also highly

Table 10.4 Dummy variables used to analyse the early breast cancer data, from Veronesi *et al.* 1994.

Age of the patient at surgery	AGE1	AGE2	AGE3	AGE4
⩾66 years	reference category			
56–65 years	1	0	0	0
46–55 years	0	1	0	0
36–45 years	0	0	1	0
⩽35 years	0	0	0	1

Number of axillary metastatic nodes	NOD1	NOD2	NOD3
0	reference category		
1	1	0	0
2–3	0	1	0
⩾4	0	0	1

Primary tumor

Dimension	SIZE1	SIZE2	SIZE3	SIZE4
⩽0.5 cm	reference category			
0.6–1.0 cm	1	0	0	0
1.1–1.5 cm	0	1	0	0
1.6–2.0 cm	0	0	1	0
>2.0 cm	0	0	0	1

Localization	SITE
External quadrants	reference category
Central + internal quadrants	1

Histological type	H1	H2
Invasive ductal and/or lobular carcinoma	reference category	
Other histotypes	1	0
Invasive ductal carcinoma with extensive intraductal component (EIC)	0	1

Table 10.5 Breast cancer patients: results of the regression analysis carried out by modelling cause-specific hazard rates by Cox model. Outcomes: local failure and metastasis occurred as first event (see text for coding procedure) (from Veronesi et al., 1994).

Regressor	Local failures				Metastases			
	Parameter estimate	Standard Error	Q_W	Risk ratio	Parameter estimate	Standard Error	Q_W	Risk ratio
(1)	(2)	(3)	(4)	(5)	(6)	(7)	(8)	(9)
AGE1	−0.197 09	0.469 99	0.175 84	0.821	0.151 44	0.203 73	0.552 57	1.164
AGE2	0.138 40	0.421 10	0.108 02	1.148	−0.135 91	0.198 86	0.467 09	0.873
AGE3	0.994 33	0.400 32	6.169 34	2.703	−0.163 11	0.201 88	0.652 84	0.849
AGE4	1.329 04	0.462 80	8.246 85	3.777	0.657 03	0.235 82	7.762 97	1.929
SITE	−0.130 61	0.196 23	0.443 03	0.878	0.207 56	0.109 70	3.579 72	1.231
SIZE1	0.512 19	0.440 46	1.352 23	1.669	0.546 88	0.320 98	2.902 81	1.728
SIZE2	0.640 87	0.440 00	2.121 46	1.898	0.752 50	0.317 45	5.619 23	2.122
SIZE3	0.736 19	0.453 47	2.635 67	2.088	1.015 17	0.319 32	10.107 10	2.760
SIZE4	1.150 19	0.487 61	5.563 97	3.159	1.391 73	0.330 13	17.771 79	4.022
NOD1	−0.605 34	0.288 78	4.394 05	0.546	0.266 44	0.150 77	3.123 00	1.305
NOD2	−0.699 97	0.369 80	3.582 82	0.497	0.659 16	0.153 10	18.537 89	1.933
NOD3	−0.149 64	0.353 02	0.179 67	0.861	1.411 09	0.137 74	104.952 01	4.100
H1	−0.856 79	0.459 90	3.470 81	0.425	−0.805 67	0.295 60	7.428 67	0.447
H2	0.629 86	0.280 14	5.055 23	1.877	0.014 01	0.222 90	0.003 95	1.014

significant, suggesting that patients 45 years old or less at surgery have a risk of local failure higher than that of patients 66 years old or more. As tumor size is concerned, only the SIZE4 coefficient is highly significant. The findings pertinent to lymph nodes deserve particular consideration: firstly, note that the signs of the regression coefficients are all negative, suggesting, if any, a decreased risk of developing local failures for patients with axillary metastatic nodes in comparison with those without axillary metastatic nodes; and secondly, the risk ratio tends to one as the number of metastatic nodes tends to increase. The order of the regressors appears to be, in some way, reversed with respect to that found for metastases as first event. Figure 10.5 may aid in understanding these findings; for the four classes of nodal involvement it reports the cumulative hazard curves of metastases and of local failures in the upper and lower panel respectively. The slopes of the four curves in the upper graph appear to differ clearly in the first 7–8 years of follow-up and tend to become roughly parallel afterwards, suggesting that the nodal involvement effect declines over time. Let us consider now the N^- group; the significantly lowest hazard for metastases as first event translates into a larger proportion of these patients being at risk of failure at any given time. Therefore, in the N^- group, a proportion of patients greater than that in the three other groups may fail for local failure as first event; this may be seen in the pertinent curve in the lower panel of figure 10.5. Moreover, from about 1.5 to 12 years from surgery, the local failure cumulative hazard of the N^- group increases linearly, namely with an estimated hazard rate of 10.209 per 1000 women-years. To interpret the apparent "protective" effect of metastatic nodes on local failure one might argue in terms either of competing risk of distant metastases or of effect of adjuvant treatment. Since metastases tend to develop in a relatively short period of time after surgery, and the risk becomes greater as the number of metastatic nodes increases, a proportional smaller and smaller number of patients may show local recurrence as first event. Alternatively N^+ patients could have taken advantage of the adjuvant therapy for local failures too. However, this latter hypothesis could not be reliably investigated because of the small number of N^+ women not undergoing adjuvant treatment.

The findings concerning the histological types showed that the EIC type (H1) is a highly significant prognostic factor, increasing the risk of local failure to 1.9 times that of invasive (ductal and lobular) carcinomas. On the contrary, as in the case of metastases, the category of "other histotypes" (H2) had a risk of local failure about half that of invasive carcinomas.

Note that the cause-specific hazard of a type l event for different categories of patients can differ in: (1) the proportion of patients who will ever fail for the event of interest; (2) the observed time to failure distribution of patients failing for the occurrence of that event. These two aspects could be investigated by estimating π_l and the proper cumulative distribution $I_l^*(t)$. Although on these data a long follow-up was available, we cannot exclude the possibility that the true range of failure times for any type of event is much wider than that

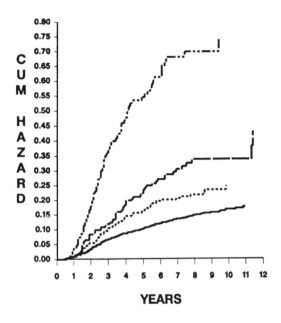

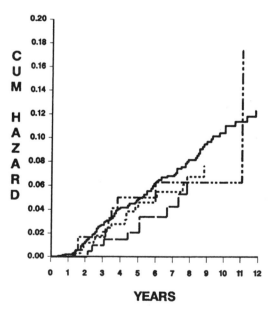

Figure 10.5 Cause–specific cumulative hazards of distant failures (upper part) and local failures (lower part) for different categories of nodal involvement (data from Veronesi *et al.*, 1994). N^-——; $1N^+$-------; $2, 3N^+$–-; $\geqslant 4N^+$··—··.

observed. Thus the device to limit estimation of π_l up to a time t' as suggested in section 10.5 cannot be reasonably applied here.

10.8   INVESTIGATION OF THE RELATIONSHIP BETWEEN DIFFERENT FAILURE TYPES

The covariates considered in the previous section are external, i.e. for each patient their value is determined at the start of follow-up. Section 6.8 commented upon the modelling of the hazard rate by means of internal time-dependent covariates which demand that the individual be alive for enabling the determination of their value. By arguments parallel to those adopted in that section, one can try to model the cause-specific hazard by internal covariates.

Consider the case in which one failure type admits a kin risk indicator systematically assessed during follow-up and suppose one inserts this indicator as time-dependent variable in the **x** vector of (10.18) pertinent to the cause-specific hazard for another failure type. According to Kalbfleisch and Prentice (1980, p. 171), a test of no association between the values assumed by such an indicator and the above-specified hazard rate would give indirect support in favour of or against a hypothesis of no association between the two failure types considered.

Think of the example of the previous section: a physician may be interested in evaluating whether distant recurrence is influenced by the occurrence of a local failure; this is the risk indicator whose value can change over time (from 0 to 1 at the occurrence of the local failure). Therefore, the patient's expected risk of developing a metastasis remains unmodified until a local failure occurs. For women developing no local failure the expected risk for metastasis is based on the patient and tumor features (external covariates) recorded at surgery. On the other hand, for patients with a local failure, the expected risk of developing a distant failure may or may not change subsequent to the occurrence of the local failure.

The number of events we must consider now is 428: 382 metastases developed as first event and 46 as second event after a local failure. The r.v. T is now time from surgery to distant recurrence as first or later event; let us call LFT the time from surgery to a local failure. The Local Failure Indicator is defined: as $\text{LFT}(t) = 1$ if $\text{LFT} < t$ and 0 otherwise. The results of the Cox regression analyses are reported in table 10.6: the results in columns (2)–(5) were obtained with model (10.18) including only external covariates, whilst those in columns (6)–(9) were obtained with model (10.18) extended to include also local failure as a time-dependent variable. Note that patients with a local failure were treated with a second surgical intervention but were not given systematic therapy, which might have influenced distant disease that may have been present.

Table 10.6 Breast cancer patients: results of the regression analysis carried out by modelling cause-specific hazard rates by Cox model. Outcomes: metastasis as first event or second event after a local failure (from Veronesi *et al.*, 1994).

Regressor (1)	External covariates only				External covariates + Internal covariate, local failure			
	Parameter estimate (2)	Standard Error (3)	Q_W (4)	Risk ratio (5)	Parameter estimate (6)	Standard Error (7)	Q_W (8)	Risk ratio (9)
LFI					1.530 30	0.165 41	85.593 43	4.620
AGE1	0.188 21	0.202 39	0.864 79	1.207	0.203 14	0.202 32	1.008 11	1.225
AGE2	−0.044 69	0.196 23	0.051 87	0.956	−0.060 82	0.196 11	0.096 20	0.941
AGE3	0.005 82	0.197 00	0.000 87	1.006	−0.073 54	0.197 54	0.138 59	0.929
AGE4	0.715 90	0.231 19	9.588 99	2.046	0.609 15	0.231 43	6.928 27	1.839
SITE	0.170 61	0.104 08	2.687 40	1.186	0.207 79	0.104 39	3.962 33	1.231
SIZE1	0.447 56	0.286 36	2.442 73	1.564	0.381 97	0.286 68	1.775 22	1.465
SIZE2	0.644 70	0.282 95	5.191 67	1.905	0.551 46	0.283 42	3.785 83	1.736
SIZE3	0.911 56	0.285 08	10.224 45	2.488	0.813 24	0.285 44	8.117 06	2.255
SIZE4	1.315 93	0.295 80	19.791 02	3.728	1.186 65	0.296 53	16.014 07	3.276
NOD1	0.113 43	0.145 64	0.606 61	1.120	0.152 21	0.145 88	1.088 68	1.164
NOD2	0.563 54	0.147 08	14.680 97	1.757	0.631 66	0.147 57	18.321 83	1.881
NOD3	1.283 29	0.131 28	95.561 50	3.608	1.267 72	0.131 57	92.833 14	3.553
H1	−0.808 28	0.273 78	8.716 00	0.446	−0.780 52	0.274 22	8.101 70	0.458
H2	0.012 90	0.208 96	0.003 81	1.013	−0.063 78	0.209 57	0.092 61	0.938

Results reported in columns (2)–(5), similar to the corresponding ones shown in table 10.5, illustrate the role of fixed covariates in predicting distant disease (first or later event). The maximum partial log-likelihood was -3019.602. When the time-dependent covariate was inserted into the model, the maximum partial log-likelihood became -2988.625. The corresponding $Q_{LR} = 61.955$ with 1 d.f. is statistically highly significant; as regards the external variables, results in columns (6)–(9) overlap those corresponding in columns (2)–(5). Thus it seems sensible to assess that, when a local failure is diagnosed to a patient who has had a conservative surgical treatment, her risk of distant disease is 4.6 times (95% C.I.: 3.34, 6.39) greater than that of a patient who belongs to the same category defined by the levels of external covariates but did not develop a local failure. To interpret this finding in clinical key, one might argue that local failure is a cause of distant metastases or, alternatively, that it indicates only that the patient was already at a greater risk of metastases when she was operated on for the primary tumor, in other words, local failure would be an "indicator" of a risk already present at surgery. Consider the results of randomized controlled trials contrasting surgical conservative treatment of early breast cancer to mastectomy; they show that the two treatments do not differ either in the crude distant disease free survival or in the mortality from all causes. Since patients randomized to mastectomy are not at risk of a recurrence in the ipsilateral breast, the second alternative seems to be plausible.

Regression analyses commented upon in this and the previous section were performed with the purpose of studying the underlying course of the disease and centred on modelling the cause-specific hazard function. This approach could be used also to test the hypothesis of equality of effect of two or more treatments by inserting pertinent dummies into the model. However, one should realize that to test this hypothesis in terms of the cause-specific hazard function for failure of type l is equivalent to considering only one of the *observable* quantities mentioned in section 10.3. A more comprehensive procedure should take into account the opportunity of testing null hypotheses regarding overall (any type) failures as well as the type l crude cumulative incidence.

10.9 COMPARISON OF TREATMENT EFFECTS

As in section 10.2, let us consider now two competing failure types (1 and 2) and two treatment groups (A and B); moreover let treatment A be the reference standard one and B a new treatment to be assessed and let us focus attention on the outcome regarding failure of type 1. We have seen in section 10.3 that the crude cumulative incidence, cause-specific incidence and overall incidence are ordered: $I_1(t) \leqslant I_{cs \cdot 1}(t) \leqslant F(t)$. Wishing to use these estimable probabilities to compare two treatment effects, the investigator should be aware of what the following relationships imply:

$$^BI_1(t) \leqslant {}^AI_1(t)$$

$$^BI_{cs\cdot1}(t) \leqslant {}^AI_{cs\cdot1}(t)$$

$$^BF(t) \leqslant {}^AF(t)$$

where A and B specify the treatment groups.

As suggested by Korn and Dorey (1992), this issue can be faced by resorting to the concept of latent failure times and specifying the null and alternative hypotheses in terms of these latter. As an example, consider the following alternatives:

(1) new treatment may prolong latent failure times of type 1 compared with the standard, but does not affect the latent times of type 2 failure;
(2) new treatment may prolong latent times of both types of failure;
(3) new treatment may prolong latent times of type 1 failure and shorten those of type 2 failure.

Coherently with (10.13) we will denote the joint survival functions of the latent times random vector $[Z_1, Z_2]$ by $^AS^0(\cdot, \cdot)$ and $^BS^0(\cdot, \cdot)$ for the treatments A and B respectively. Furthermore, if we let: $^AS^0(t_1^0, t_2^0) = {}^BS^0[\phi_1(t_1^0), \phi_2(t_2^0)]$ we can write the null as well as the previous alternative hypotheses through the $\phi(\cdot)$ functions. Namely:

Model under the null hypothesis: $\phi_1(t_1^0) = t_1^0$ $\phi_2(t_2^0) = t_2^0$
Model under the (1) alternative hypothesis: $\phi_1(t_1^0) \geqslant t_1^0$ $\phi_2(t_2^0) = t_2^0$
Model under the (2) alternative hypothesis: $\phi_1(t_1^0) \geqslant t_1^0$ $\phi_2(t_2^0) \geqslant t_2^0$
Model under the (3) alternative hypothesis: $\phi_1(t_1^0) \geqslant t_1^0$ $\phi_2(t_2^0) \leqslant t_2^0$

The null hypothesis model implies equality of all three incidence curves (crude, cause-specific and overall). Let us consider now the model under the (1) alternative hypothesis: the increase of the latent time of type 1 failure due to the new treatment is expected to reduce the overall incidence in group B compared with that is group A, $^BF(t) \leqslant {}^AF(t)$. As regards the crude incidence, it is helpful to recall the suggestion given in section 10.3 to interpret $S_1(t)$ and thus $I_1(t)$: this latter may be thought of as the cumulative incidence of a r.v. Y, where $Y = T$ or ∞ depending on whether at time $T = t$ a failure of type 1 or type 2 occurred. The increase of the latent time of type 1 failure implies a decrease of the frequency with which the r.v. Y assumes finite values in group B compared with that in group A and: $^BI_1(t) \leqslant {}^AI_1(t)$. However, there may exist t^* for which: $^BI_{cs\cdot1}(t^*) > {}^AI_{cs\cdot1}(t^*)$. Thus we can say that overall and crude incidence functions are ordered whilst cause-specific incidence function is not. By arguments parallel to these we can see that the model under the (2) alternative hypothesis implies that the overall incidence functions are ordered but not the crude and cause-specific incidence functions, while the model under the (3) alternative hypothesis implies that only the crude incidence functions are ordered.

Note that previous considerations on the three alternative models were developed without making any assumption of independence of the latent r.v.'s Z_1 and Z_2. The independence of these latter implies that the cause-specific incidence functions be ordered for all three alternative hypotheses.

Estimating cause-specific hazard functions by censoring all events not of specific concern, and testing for differences using the M–H method or the Cox model, as considered in section 10.8, may be useful to answer questions about changes in the hazard of a specific type of event in the presence of competing causes of failure. However, in the light of previous discussion, it seems that overall and crude incidence probability are more relevant when the clinician aims at evaluating the effect of two (or more) treatments in a clinical trial. From our experience the following procedure appears to allow a proper comparison of treatment effects:

(1) Test the equality of overall incidence. As this implies considering occurrence of the first event, of any type, as failure of treatment, if the two treatments differ significantly the analysis might stop. As a matter of fact the treatment with longer time to failure may be considered to be the one with better prognosis, no matter what the distribution of different events concurring to failure.

(2) Test the equality of crude cumulative incidence of one or more event(s) of special clinical interest. The non-parametric method suggested by Gray (1988) enables us to perform this test.

10.10 GRAY'S METHOD FOR COMPARING CRUDE CUMULATIVE INCIDENCES

The theory of martingales is the mathematical background adopted by Gray for the formal presentation of his test; this powerful mathematical tool has been widely used by Andersen *et al.* (1993) in their book on survival analysis. However, the reader of the present book is not expected to be familiar with counting processes and martingales. Thus this section aims only at giving a heuristic introduction to the rationale of Gray's method and showing the results of its application to two examples.

Let us take type 1 failure as that of clinical interest. The test was developed for testing the null hypothesis:

$$H_0: {}^gI_1(t) = I_1^0 \quad g = 1, 2, \ldots, G \tag{10.21}$$

where $I_1^0(\cdot)$ is an unspecified crude cumulative function and g indicates different treatment groups. From (10.8) we have: ${}^gI_1(t) = \int_0^t {}^gS(u){}^gh_1(u)\,du$; therefore, it is easy to see that the hypothesis of equality of cause specific cumulative hazard functions of failure of type 1 we considered in the previous section is different from the equality of the crude cumulative incidence functions (10.21) excluding the case in which the overall survival values $^gS(t)$ are also equal in all treatment groups.

Coherently with the definition of subdensity and subdistribution functions of the r.v. Y mentioned in section 10.3, we now define the corresponding subdistribution hazard function of the type 1 failure for the gth group as the ratio:

$$^g\eta_1(t) = \frac{^gf_1(t)}{^gS_1(t)} \tag{10.22}$$

When the effects of two treatments only, A and B, are being tested, it was shown by Gray (1988) that his test corresponds to the comparison of weighted averages of the subdistribution hazards. Furthermore, it may be shown that by a proper choice of the weights, the class of test attained by Gray's method is asymptotically equivalent to that proposed by Harrington and Fleming (1982) in the absence of competing causes of failure (see section 4.4). As in the Harrington and Fleming class, the value taken by the exponent ρ of the weighting function defines different tests. So $\rho = -1, 0, 1$ specifies that the cumulative subdistribution ratio $[^AI_1(t)/^BI_1(t)]$, the subdistribution hazard ratio $[^A\eta_1(t)/^B\eta_1(t)]$ and the subdistribution odds ratio in the two treatment groups, respectively, are constant over time. Also the class of tests suggested by Gray is shown to be asymptotically distributed as a χ^2 with 1 d.f. (in general, $G-1$ d.f. for G groups).

By considering two treatment groups and $\rho = 0$, we will focus our attention on the test we can think of as the counterpart of the M–H test in the competing risk context; under the null hypothesis it assumes that the subdistribution hazard ratio of the two groups is constant and equal to 1 over time. This test, hereafter denoted Q_{Gray}, has a numerator which may be obtained by slightly modifying the arguments which enabled us to calculate the numerator of the M–H test in section 4.2. Suppose that at time $t_{(j)}$ we observe one failure of type 1 and that just before $t_{(j)}$, ^An_j and ^Bn_j are the numbers of individuals at risk, that is free from failure of any type, in groups A and B respectively. Clearly, if $N = {}^AN + {}^BN$ is the total number of individuals randomized to the two treatments, ^An_j is obtained by subtracting from ^AN the number of failures of *any* type occurring before $t_{(j)}$ as well as the number of censored times; similarly for ^Bn_j. However, the number exposed at risk of failing for the type 1 event is expected to be greater than ^An_j or ^Bn_j as we need the numbers of individuals *free from failure of type 1*, under the assumption that the "crude effect" of this latter is acting. Thus it seems sensible to look for correction factors of ^An_j and ^Bn_j. The ratio $^AS_1(t_{j-1})/^AS(t_{j-1}) \geqslant 1$ appears to be appropriate; in fact it tends to be as much greater as the contribution of type 1 failure to the total failures observed is smaller, which agrees with the expectation that the number of subjects free of type 1 failure is as much greater than $^A n_j$ as the crude effect of cause 1 is more and more negligible. A similar factor may be computed for group B. As a consequence, the number of individuals free of type 1 failure expected in the two groups at $t_{(j)}$ is

$$\hat{R}_1(t_j) = {}^A n_j \frac{{}^A \hat{S}_1(t_{j-1})}{{}^A \hat{S}(t_{j-1})} + {}^B n_j \frac{{}^B \hat{S}_1(t_{j-1})}{{}^B \hat{S}(t_{j-1})} \tag{10.23}$$

Therefore, under the null hypothesis, the subdistribution hazard is estimated as $\hat{\eta}_1(j) = 1/\hat{R}_1(t_j)$ and the expected number of failures of type 1 in group A is

$$^A n_j \frac{{}^A S_1(t_{j-1})}{{}^A S(t_{j-1})} [\hat{R}_1(t_j)]^{-1}$$

The pertinent contribution at $t_{(j)}$ to the numerator of Q_{Gray} is in terms of observed minus expected events:

$$1 - {}^A n_j \frac{{}^A \hat{S}_1(t_{j-1})}{{}^A \hat{S}(t_{j-1})} [\hat{R}_1(t)]^{-1} \quad \text{if failure is observed in group A}$$

$$0 - {}^A n_j \frac{{}^A \hat{S}_1(t_{j-1})}{{}^A \hat{S}(t_{j-1})} [\hat{R}(t_j)]^{-1} \quad \text{if failure is observed in group B}$$

In the presence of ties of type 1 failure, the above quantities have to be adjusted accordingly. The numerator of Q_{Gray} becomes the sum of these contributions computed at each $t_{(j)}$.

As an example with reference to table 10.1, consider the event of type 1 and the time $t = 16$ The The corresponding 2×2 contingency table is:

Treatment group	No. of failures of type 1	No. of patients free from failure of any type
A	0	27
B	2	26

The ingredients for the calculation of the correction factor in group B are obtained from table 10.2. At time $t = 16$, we have from column (5) that $\hat{S}_B(t_{j-1}) = 0.742\,86$ and from column (6) we compute $\hat{S}_{1B}(t_{j-1}) = 1 - \hat{I}(t_{j-1}) = 0.971\,43$. Thus the correction factor for group B is $1.307\,68$. Analogously, for group A it is $0.942\,86/0.771\,43 = 1.222\,27$.

According to (10.23), by applying these factors to the number of individuals reported in the third column of the table we obtain the 2 × 2 table:

Treatment group	No. of failures of type 1	No. of patients expected free from type 1 failure
A	0	33
B	2	34
Total	2	67

Thus the contribution to the numerator of Gray's test is

$$0 - 2 \times \frac{33}{67} = -0.9851$$

The denominator, i.e. the variance of the weighted difference of the subdistribution hazards estimated in group A and B, cannot be obtained straightforwardly, but, as Gray (1988) remarks, one can resort to the results of Aalen (1978). The calculations, however, may be prohibitive without the aid of a computer. With reference to the data in table 10.1, computation of Gray's test for the two events gives the following results:

Type 1 failure: $Q_{Gray} = 0.222$ $p = 0.634$

Type 2 figure: $Q_{Gray} = 1.726$ $p = 0.189$

Though the log-rank test on first failure (type 1 + type 2 failure) gives $Q_{M-H} = 5.381$ ($p = 0.020$), no statistically significant difference between crude cumulative incidence is found.

Let us consider now the mastectomy vs Quart clinical trial of figure 1.3. Any of the events mentioned in section 10.7 occurring first, concur to define the disease free survival curve; the pertinent K–M estimates for the two surgical treatments are given in figure 10.6. They tend to overlap and the M–H test gives $Q_{M-H} = 0.1562$ (p = 0.693).

As previously, distant metastases and local failures are the two events whose crude incidence we are analysing. However, note that the findings relative to local failures need to be interpreted with great caution. In fact, unlike Quart-treated women, patients who underwent total mastectomy can have recurrences of tumor in the chest wall and the operative scar, but not in the ipsilateral breast. Thus a greater incidence of local failure is expected *a priori* in the Quart group.

The results of Gray's test are:

Distant failures: $Q_{Gray} = 0.115$ $p = 0.734$
Local failures: $Q_{Gray} = 6.791$ $p = 0.009$

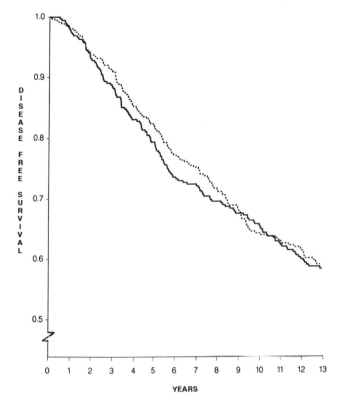

Figure 10.6   Disease free survival curves of the clinical trial presented in figure 1.3.
Mastectomy ———; Quart --------.

Resorting to the M–H test for comparing cause-specific hazards of the two
groups we obtain:

   Distant failures: $Q_{M-H} = 0.116$ $p = 0.734$
   Local failures: $Q_{M-H} = 6.258$ $p = 0.012$

These results are very similar to those obtained with Gray's test and this is to be
expected as the estimated disease free survival curves in the two groups are
overlapping. In fact, from $^gI_1(t)$ we can see that possible differences between
crude cumulative incidences in the two groups are only due to differences in the
cause-specific hazards $^gh_1(t)$ in the case that the overall disease free survival is
the same in the two groups.

APPENDIX A10.1   VARIANCE OF THE CRUDE
CUMULATIVE INCIDENCE

Let d_{lj} denote the number of subjects failing for the l type cause at t_j, $d_j = \sum_{l=1}^{C} d_{lj}$,
and n_j the number of individuals at risk of failing at t_j. From (10.11) and (10.12)

we know that the crude cumulative incidence is obtained as the sum of the products of two statistics and its variance at t_j is

$$var[\hat{I}_l(t_j)] = \sum_{\alpha=1}^{j} var(\hat{h}_{l\alpha}\hat{S}(t_{(\alpha-1)})) + 2\sum_{\alpha=1}^{j-1}\sum_{\beta=\alpha+1}^{j} cov(\hat{h}_{l\alpha}\hat{S}(t_{(\alpha-1)}), \hat{h}_{l\beta}\hat{S}(t_{(\beta-1)}))$$

In order to compute the variances and covariances to be added in this formula one can resort to the delta method, namely

$$var(\hat{h}_{l\alpha}\hat{S}(t_{\alpha-1})) = var\left(\frac{d_{l\alpha}}{n_\alpha}\prod_{a=1}^{\alpha-1}\left(1 - \frac{d_a}{n_a}\right)\right)$$

As the estimated covariances of $\hat{h}_{la}$ and $\hat{h}_{l'a'}$ are zero if $a \neq a'$ (Dinse and Larson, 1986),

$$var(g(d_{l\alpha}, d_1, \ldots, d_{(\alpha-1)})) = \left(\frac{\partial g}{\partial d_{l\alpha}}\right)^2 var(d_{l\alpha}) + \sum_{u=1}^{\alpha-1}\left(\frac{\partial g}{\partial d_u}\right)^2 var(d_u)$$

$$= \left(\frac{1}{n_\alpha}\prod_{a=1}^{\alpha-1}\left(1 - \frac{d_a}{n_a}\right)\right)^2 \left(\frac{d_{l\alpha}(n_\alpha - d_{l\alpha})}{n_\alpha}\right)$$

$$+ \sum_{u=1}^{\alpha-1}\left(\frac{d_{l\alpha}}{n_\alpha}\prod_{\substack{a=1\\(a\neq u)}}^{\alpha-1}\left(1 - \frac{d_a}{n_a}\right)\left(-\frac{1}{n_u}\right)\right)^2 \left(\frac{d_u(n_u - d_u)}{n_u}\right)$$

$$var\left(\hat{h}_{l\alpha}\hat{S}(t_{\alpha-1})\right) = (\hat{h}_{l\alpha}\hat{S}(t_{\alpha-1}))^2 \left(\frac{(n_\alpha - d_{l\alpha})}{n_\alpha d_{l\alpha}} + \sum_{u=1}^{\alpha-1}\left(\frac{d_u}{(n_u - d_u)n_u}\right)\right) \quad \text{(A10.1)}$$

Analogously one can write

$$cov(\hat{h}_{l\alpha}\hat{S}(t_{\alpha-1}), \hat{h}_{l\beta}\hat{S}(t_{\beta-1})) = cov((g(d_{l\alpha}, d_1, \ldots, d_{(\alpha-1)}), f(d_{l\beta}, d_1, \ldots, d_{(\beta-1)}))$$

$$= cov\left(\frac{d_{l\alpha}}{n_\alpha}\prod_{a=1}^{\alpha-1}\left(1 - \frac{d_a}{n_a}\right), \frac{d_{l\beta}}{n_\beta}\prod_{b=1}^{\beta-1}\left(1 - \frac{d_b}{n_b}\right)\right) = \left(\frac{\partial g}{\partial d_{l\alpha}}\right)\left(\frac{\partial f}{\partial d_\alpha}\right)cov(d_{l\alpha}, d_\alpha)$$

$$+ \sum_{u=1}^{\alpha-1}\left(\frac{\partial g}{\partial d_u}\right)\left(\frac{\partial f}{\partial d_u}\right)var(d_u) = \left(\frac{1}{n_\alpha}\prod_{a=1}^{\alpha-1}\left(1 - \frac{d_a}{n_a}\right)\right)\left(\frac{d_{l\beta}}{n_\beta}\prod_{\substack{b=1\\(b\neq\alpha)}}^{\beta-1}\left(1 - \frac{d_b}{n_b}\right)\right.$$

$$\times \left(-\frac{1}{n_\alpha}\right)\right)cov(d_{l\alpha}, d_\alpha) + \sum_{u=1}^{\alpha-1}\left(\frac{d_{l\alpha}}{n_\alpha}\prod_{\substack{a=1\\(u\neq a)}}^{\alpha-1}\left(1 - \frac{d_a}{n_a}\right)\left(\frac{-1}{n_u}\right)\right)$$

$$\times \left(\frac{d_{l\beta}}{n_\beta}\prod_{\substack{b=1\\(u\neq b)}}^{\beta-1}\left(1 - \frac{d_b}{n_b}\right)\left(\frac{-1}{n_u}\right)\right)var(d_u)$$

Since

$$cov(d_{l\alpha}, d_{\alpha}) = cov\left(d_{l\alpha}, \sum_{l=1}^{C} d_{l\alpha}\right) = \sum_{\gamma \neq l} cov(d_{l\alpha}, d_{\gamma\alpha}) + var(d_{l\alpha}) = \frac{d_{l\alpha}(n_{\alpha} - d_{\alpha})}{n_{\alpha}}$$

and

$$var(d_u) = \left(\frac{d_u(n_u - d_u)}{n_u}\right)$$

we obtain

$$cov(\hat{h}_{l\alpha}\hat{S}(t_{\alpha-1}), \hat{h}_{l\beta}\hat{S}(t_{\beta-1}))$$

$$= (\hat{h}_{l\alpha}\hat{S}(t_{(\alpha-1)}))(\hat{h}_{l\beta}\hat{S}(t_{(\beta-1)}))\left(\frac{-1}{n_{\alpha}} + \sum_{a=1}^{\alpha-1}\frac{d_a}{n_a(n_a - d_a)}\right) \qquad \text{(A10.2)}$$

Finally, summing up terms given by (A10.1) and (A10.2) we have the estimate

$$var[\hat{I}_l(t_j)] = \sum_{\alpha=1}^{j}(\hat{h}_{l\alpha}\hat{S}(t_{(\alpha-1)}))^2\left(\frac{(n_{\alpha} - d_{l\alpha})}{n_{\alpha}d_{l\alpha}} + \sum_{a=1}^{\alpha-1}\left(\frac{d_a}{(n_a - d_a)n_a}\right)\right)$$

$$+ 2\sum_{\alpha=1}^{j-1}\sum_{\beta=\alpha+1}^{j}(\hat{h}_{l\alpha}\hat{S}(t_{(\alpha-1)}))(\hat{h}_{l\beta}\hat{S}(t_{(\beta-1)}))\left(\frac{-1}{n_{\alpha}} + \sum_{u=1}^{\alpha-1}\left(\frac{d_u}{n_u(n_u - d_u)}\right)\right)$$

$$\text{(A10.3)}$$

which coincides with the formula given by Gaynor *et al.* (1993).

11
Meta-analysis

11.1 INTRODUCTION

Scientific methodology allowing investigators to combine findings from several studies, addressing the same question, has history extending over several centuries, one of the earliest examples concerning the combination of replications of astronomical measures made by Galileo.

In the statistical literature the problem of combining results of experiments planned and performed to optimize different agricultural productions was faced more than half a century ago by Cochran (1937) and Yates and Cochran (1938). Some fifteen years later Cochran (1954) published a paper which, even nowadays, may sensibly be considered the reference paper to combine outcomes collected in terms of quantitative, normally distributed variables, by means of the Analysis of Variance.

In the medical field the continuous critical review of epidemiological hypotheses and assessment of surgical or pharmacological treatments on the basis of results gathered from different studies, either planned observation or controlled clinical trials, is a crucial point in the process of accumulation of medical knowledge. Actually several prestigious medical journals customarily offer to their readers review articles on epidemiological and clinical topics. It may be guessed that these reviews are exposed to the risk of bias if the authors do not take advantage of a proper procedure of quantitative evaluation of results. In fact, even an open-minded expert tends to promote results which are more favourable to the hypotheses coherent with his own experience. The quantitative approach underlying meta-analysis appears to be a helpful device to overcome these drawbacks.

11.2 DEFINITION AND GOALS OF META-ANALYSIS

The term was coined by Glass (1976) who defined meta-analysis as "the statistical analysis of a large collection of analysis results from individual studies for the purpose of integrating the findings". As a matter of fact, "meta-analysis has a qualitative component, i.e. application of predetermined criteria of quality (e.g., completeness of data, absence of biases), and a quantitative component, i.e. integration of the numerical information. Meta-analysis includes aspects of an overview, and of pooling of data, but implies more than either of these processes" (Last, 1988).

The great majority of meta-analyses one can find in perusing medical literature concern results of controlled clinical trials; Bulpitt (1988) even upholds that only these latter should be meta-analysed. However meta-analyses of epidemiological data arising from case-control and cohort studies have recently been performed both to investigate deeply the role of a risk factor and to assess reliably its true effect (Longnecker *et al.*, 1988; Howe *et al.*, 1990, 1991).

In trying to evaluate the therapeutic effect of a given treatment it must be remembered that, nowadays, moderate effects are all that one can reasonably expect in treating chronic diseases; nevertheless, even small effects on mortality may be humanly worthwhile. Furthermore, although one cannot assume that different trials are comparable (for instance, they differ for the experimental design), nor that patients are comparable for their features (for instance for age), it is sensible to assume that if a set of trials addresses related questions then one should expect some tendency to show a similar relationship between the treatments and the course of the disease, i.e. to indicate whether or not a given treatment produces any effect of clinical relevance on the outcome. These considerations are the prerequisite of a meta-analysis of clinical trials originally planned and carried out to answer the same biologically sensible and clinically relevant question.

In this context the goals of a meta-analysis appear to be:

(1) to test the null hypothesis relative to the treatment effect;
(2) to obtain an accurate and precise estimate of the treatment effect;
(3) to offer the methodological support for a reliable generalization of findings;
(4) to enable the investigators to perform subgroup analyses whose power is very low if carried out for each of the clinical trials separately;
(5) to guide the researchers in planning new clinical trials. A recent example is given by the meta-analysis of chemotherapy in advanced ovarian cancer (1991); this study [†] , carried out on updated data from more than 8000 women, not only gave information suitable for clarifying the role of chemotherapy, but also pointed out the need for a multicentre randomized trial (Icon-2) which was activated subsequently;
(6) to balance the "overflow of enthusiasm" which often follows the introduction of a new therapy in the clinical practice.

In performing meta-analyses of clinical trials, many methodological problems of different types have to be faced. They concern selection of trials, methodology

[†] Advanced Cancer Trialist Group. Chemotherapy in advanced overian cancer: an overview of randomized clinical trials. *British Medical Journal*, **303**, 884–893, 1991.

to obtain the data, techniques of statistical analysis and proper interpretation of results. These issues have been thoroughly discussed in a workshop, the proceedings of which were reported in a special issue of *Statistics in Medicine* (1987) and by L'Abbé *et al.* (1987), Chalmers *et al.* (1987a, 1987b) and Chalmers (1991). Coherently with the objectives of this book, in the following sections our attention shall focus on the statistical methods of current use in meta-analysis of therapeutical trials; a general approach to hypothesis testing and estimation and to the identification of heterogeneity among trial results will be presented. Firstly, a model in which treatment effects are fixed will be dealt with, and later a random effect model, allowing treatment effects to vary among trials, will be presented. Before going into the details of the statistical techniques, we wish to emphasize here that simply pooling the outcomes from all trials and then comparing the totals is likely to give a misleading result. In meta-analysis instead we consider each trial as a stratum; a treatment effect is estimated within each trial, and then single effects are suitably combined to estimate the overall effect. In this way patients randomly allocated to the treatment to be assessed are compared with those randomly allocated to the control treatment of that *same* trial.

11.3 FIXED EFFECT APPROACH

A parameter ξ of clinical interest for assessing the effect of two treatments, independent estimates of this parameter computed "within" each of M trials, and the weighted means of these statistics "across" trials are the basic ingredients of every meta-analysis.

Let $\hat{\xi}_m$ $(m = 1, 2, \ldots, M)$ be the estimate of ξ calculated in the mth trial; it is assumed for the moment that $\hat{\xi}_m$ is normally distributed around ξ with variance $\sigma_{\varepsilon m}^2$ based upon the internal variability of the mth trial and let v_m be its estimate. After defining the weight $w_m = v_m^{-1}$, we consider the statistic $\sqrt{w_m}\hat{\xi}_m$, which is asymptotically normally distributed with expectation $E(\sqrt{w_m}\hat{\xi}_m) = \sqrt{w_m}\xi$ and variance

$$var(\sqrt{w_m}\hat{\xi}_m) = 1 \tag{11.1}$$

By a standard least-squares analysis, the estimate of ξ pooled among the trials is found to be

$$\bar{\hat{\xi}} = \frac{\sum_{m=1}^{M} w_m \hat{\xi}_m}{\sum_{m=1}^{M} w_m} \tag{11.2}$$

with variance

$$var(\bar{\bar{\xi}}) = \left(\sum_{m=1}^{M} w_m \right)^{-1} \tag{11.3}$$

Since $\bar{\bar{\xi}} \sim N(\xi, (\sum_{m=1}^{m} w_m)^{-1})$ it is easy:

(1) to test the null hypothesis, $H_0: \xi = 0$, by means of the standardised Gaussian deviate:

$$z = \frac{\sum_{m=1}^{M} w_m \hat{\xi}_m}{\sqrt{\sum_{m=1}^{M} w_m}} \tag{11.4}$$

(2) to calculate the $100(1 - \alpha)\%$ confidence interval of ξ by

$$\bar{\bar{\xi}} \pm \frac{z(1 - \alpha/2)}{\sqrt{\sum_{m=1}^{M} w_m}} \tag{11.5}$$

In this approach we assumed the "fixed effects" model:

$$\hat{\xi}_m = \xi + \varepsilon_m \tag{11.6}$$

where ε_m is a random component with average equal zero and variance $\sigma_{\varepsilon m}^2$. However, before merging information from the M trials the null hypothesis of homogeneity of treatment effect across all trials, $H_0: \xi_1 = \xi_2 = \cdots = \xi_m = \cdots = \xi_M = \xi$, should be tested.

Since in a meta-analysis each trial is thought of as a stratum, the pertinent test statistic is equal to (4.16):

$$Q_{\text{hom}} = \sum_{m=1}^{M} w_m (\hat{\xi}_m - \bar{\bar{\xi}})^2 \tag{11.7}$$

which is asymptotically distributed as χ^2 with $(M-1)$ degrees of freedom. This statistic coincides with the "ordinary" residual sum of squares; as it appears from (11.1), if the model (11.6) is correct, residuals have a true variance of 1. However, as the power of the statistical tests for homogeneity is often low (Breslow and Day, 1980, p. 142), we think it can serve as a warning light with high specificity but low sensitivity. Since large p values do not indicate that heterogeneity can be ignored, we suggest resorting also to graphical devices as further tools to face the problem of heterogeneity. Galbraith (1988) proposed a bivariate scatter plot of the standardized estimate of treatment effect ($\sqrt{w_m}\,\hat{\xi}_m$) against its precision $\sqrt{w_m}$; alternatively L'Abbé *et al.* (1987) proposed to investigate variations in observed risk reductions by plotting the relative frequencies of event in the treatment groups on the vertical axis and in the control groups on the horizontal axis. This latter graph appears to be particularly helpful to the clinician concerned in revealing heterogeneity by the response of the control group. Response in the control group, varying markedly from trial to trial, gives a hint to how much patients differ in baseline features among trials and suggests a legitimate objection to combining them.

The clinically relevant parameter ξ we will consider in the next subsections and in examples are suitable for meta-analysing trials whose results are expressed by binary data or by survival data.

11.3.1 Binary data

Two measures of treatment effect are usually adopted in clinical trials: the difference in the outcome probability and the log-odds-ratio. Let results of the mth trial be arranged in a 2×2 contingency table:

Treatment	Failure	Success	Total	
A	f_{Am}	s_{Am}	n_{Am}	$p_{Am} = f_{Am}/n_{Am}$
B	f_{Bm}	s_{Bm}	n_{Bm}	$p_{Bm} = f_{Bm}/n_{Bm}$
Total	$f_{\cdot m}$	$s_{\cdot m}$	$n_{\cdot m}$	

Let π_{Am} = true failure probability under treatment A, trial m, and π_{Bm} = true failure probability under treatment B, trial m.

Consider firstly the probability difference: $\xi_m = \pi_{Am} - \pi_{Bm}$. By the maximum likelihood method we obtain the estimates (see DerSimonian and Laird, 1986):

$$\hat{\xi}_m = \frac{f_{Am}}{n_{Am}} - \frac{f_{Bm}}{n_{Bm}} \tag{11.8}$$

and

$$v_m = \left(\frac{f_{Am} \cdot s_{Am}}{n_{Am}^3} + \frac{f_{Bm} \cdot s_{Bm}}{n_{Bm}^3} \right) \tag{11.9}$$

The logarithm of the odds ratio is

$$\xi_m = \log \theta_m = \log \frac{\pi_{Am}(1 - \pi_{Bm})}{\pi_{Bm}(1 - \pi_{Am})}$$

and by the maximum likelihood method we obtain the estimates

$$\hat{\xi}_m = \log \frac{f_{Am} \cdot s_{Am}}{s_{Am} \cdot f_{Bm}} \qquad (11.10)$$

and

$$v_m = [f_{Am}^{-1} + s_{Am}^{-1} + f_{Bm}^{-1} + s_{Bm}^{-1}] \qquad (11.11)$$

Instead of using (11.10) and (11.11), we can adopt the approach mentioned in section 4.5, namely

$$\hat{\xi}_m = f_{Am} - \frac{f_{\cdot m} n_{Am}}{n_{\cdot m}} = O_m - E_m \qquad (11.12)$$

where O_m and E_m are the observed and expected number of failures on treatment A, say the experimental treatment to be tested, with variance

$$v_m = \frac{f_{\cdot m} s_{\cdot m} n_{Am} n_{Bm}}{n_{\cdot m}^2 (n_{\cdot m} - 1)} \qquad (11.13)$$

11.3.2   Survival data

In Chapters 5 and 6 it was shown that under the assumption of proportional hazard, the treatment effect is measured by the hazard ratio; thus:

$$\xi = \log \frac{\lambda_A(t)}{\lambda_B(t)}$$

If Cox's propotional hazards model was fitted in each trial and the regression coefficient estimate $(\hat{\beta})$ pertinent to the treatment is quoted, together with its standard error [SE($\hat{\beta}$)], the meta-analysis can be performed straightforwardly by means of these statistics.

More frequently, however, meta-analyses are based upon the one-step approximations to the M–L estimator of $\log \theta$ under the Cox model; therefore from section 4.3 we have

$$\hat{\xi}_m = \frac{U_m}{var(U_m)} = \frac{O_m - E_m}{var(U_m)}$$

with

$$v_m = [var(U_m)]^{-1}$$

Thus it is immediate to see that

$$\sum_{m=1}^{M} w_m \hat{\xi}_m = \sum_{m=1}^{M} (O_m - E_m)$$

and

$$(\bar{\hat{\xi}}) = \frac{\displaystyle\sum_{m=1}^{M} (O_m - E_m)}{\displaystyle\sum_{m=1}^{M} var(U_m)} \tag{11.14}$$

with variance

$$var(\bar{\hat{\xi}}) = \left[\sum_{m=1}^{M} var(U_m) \right]^{-1} \tag{11.15}$$

11.4 RANDOM EFFECT MODEL

Suppose that the analyses accomplished as mentioned in previous sections convinced the investigator that the separate estimates $\hat{\xi}_m$ differ by more than they should under the hypothesis of treatment effect homogeneity. The reasons for presence of heterogeneity could then be investigated; the following considerations may be helpful:

(1) variances used to compute (11.2) and (11.7) are based on unrealistic assumptions about the data of each trial. Think, for instance, of not making allowance for the inflation of variance due to intrapatient correlation in studying patency of distal anastomoses in patients with aorto-coronary bypass grafts (Marubini *et al.*, 1993);

(2) choice of improper scale of measuring treatment effect. By going to a new definition of ξ, one could eliminate heterogeneity;

(3) different features of the experimental design adopted in the M trials. A big issue here is whether the explanation of heterogeneity was suggested by prior considerations or derived by examining the data;

(4) baseline characteristics unbalances;

(5) variation among the ξ_m can be due to one or a few trials with anomalous data, then the inconsistent trial could be investigated separately and the remaining ones combined by (11.2);

(6) heterogeneous responses are unexplainable. This would be the situation in which the investigator would feel least comfortable in reaching conclusions and drawing generalizations.

A model, alternative to (11.6), was proposed by DerSimonian and Laird (1986) to deal with these situations. Basically, in the absence of a specific explanation of the variation in ξ, this is taken as random and the model becomes

$$\begin{cases} \hat{\xi}_m = \xi_m + \varepsilon_m \\ \xi_m = \xi + \phi_m \end{cases} \qquad (11.16)$$

where ϕ_m are assumed to be independent r.v normally distributed with mean zero and variance σ_ϕ^2. This implies assuming that the M estimates $\hat{\xi}_m$ are independently normally distributed with mean ξ and variance $\sigma_{\varepsilon m}^2 + \sigma_\phi^2$ and, consequently, the weights to be used in the random-effect model are

$$w'_m = (\sigma_{\varepsilon m}^2 + \sigma_\phi^2)^{-1} \qquad (11.17)$$

As in the previous section $\sigma_{\varepsilon m}^2$ is estimated by v_m; DerSimonian and Laird (1986) suggested estimating σ_ϕ^2 by the method of moments as shown in Appendix A11.1. Using Q_{hom} from (11.7),

$$s_\phi^2 = \max\left\{0, [Q_{\text{hom}} - (M-1)] \Bigg/ \left[\sum_{m=1}^{M} w_m - \left(\sum_{m=1}^{M} w_m^2\right)\left(\sum_{m=1}^{M} w_m\right)^{-1}\right]\right\} \qquad (11.18)$$

The estimator of ξ is now

$$\bar{\bar{\xi}} = \sum_{m=1}^{M} \frac{\hat{\xi}_m}{v_m + s_\phi^2}\left[\sum_{m=1}^{M} (v_m + s_\phi^2)^{-1}\right]^{-1} \qquad (11.19)$$

with variance

$$var(\bar{\bar{\xi}}) = \left[\sum_{m=1}^{M} (v_m + s_\phi^2)^{-1}\right]^{-1} \qquad (11.20)$$

Therefore, in the presence of real heterogeneity and aiming at summarizing the difference of treatment effects in a clinically relevant measure, the researcher can assume that the additional variability, "between trials", in a given sense be random and incorporate the heterogeneity of effects in the estimation of the overall treatment effect by (11.18) and (11.19). However, it appears sensible to interpret with great caution the results of meta-analyses carried out according to the random effect model for the following considerations:

(a) suppose that the addition of the random effect considered in model (11.16) modifies substatially inferences reached with model (11.6). The degree of heterogeneity will be so large as to tend to nullify the value of any summary statistic (including or excluding the random effect). Therefore there is a need to further investigate possible sources of variability among trials;

(b) in typical applications no empirical, biological or clinical justification can be advocated for specific distributional forms of the random-effect model;

(c) the summary statistic computed under the random-effect model "... has no population-specific interpretation, but instead represents the mean of a distribution that generates effects. Unlike a standardized rate ratio, it does not correspond to an average effect in a population. In essence, then, a random-effect model exchanges a questionable homogeneity assumption for a fictitious distribution of effects" (Greenland, 1987).

There has been a continuous debate over the relative merits of the fixed effect and random effect approach; however, the quality and timeliness of data are the key to reaching clinically reliable conclusions rather than the statistical model adopted to perform the analysis. In fact, only by adopting the simple, though difficult to realize, policy of collecting updated and individual data for patients recruited in all randomized trials can the investigator escape the most hazardous sources of bias: omission of unpublished studies, arbitrary exclusion of patients in published trial results, and trials incorrectly randomized.

11.5   EXAMPLES

In order to illustrate the above-presented statistical methods for binary outcome we utilize the data published by Collins *et al.* (1990) in a meta-analysis aiming at assessing the effectiveness of antihypertensive drugs after at least one year of treatment; the outcome here considered is the occurrence of fatal and non-fatal strokes. The data pertinent to 16 trials are reported in table 11.1; the correspondence with the generic 2×2 contingency table of section 11.3.1 is obtained by putting Treatment A = antihypertensive drugs and failure = stroke.

The L'Abbé *et al.* (1987) plot of these data is reported in figure 11.1; the solid diagonal line corresponds to equal event frequency in treated and control groups, and splits the quadrant in two portions: points lying in the lower one, delimited by the diagonal and the horizontal axis, indicate that treatment is better than control whilst the opposite is suggested by points lying in the portion included between the diagonal and the vertical axis. The upper and lower dotted lines indicate a relative difference (see Chapter 9), in favour of treatment, of 25% and 50% respectively. We realize immediately that the treatment compares favourably with the control, and the majority of trial results, being concentrated around the lower dotted line, tend to indicate a clinically relevant treatment effect; furthermore, the relative frequency of failure in the control groups is lower than 24% except for the Carter trial where a clearly higher failure percentage (43.8) was recorded.

The analysis is performed in terms of probability difference by (11.8) and (11.9) in columns (6)–(9) and of log-odds-ratio by (11.10) and (11.11) in columns (10)–(13) of table 11.1

Table 11.1 Meta-analysis of antihypertensive randomized clinical trials. Data from Collins *et al.* (1990).

Trial (1)	Drug		Control		Probability difference				(log) Odds ratio			
	f_{Am} (2)	n_{Am} (3)	f_{Bm} (4)	n_{Bm} (5)	$\hat{\xi}_m$ (6)	w_m (7)	$\hat{\xi}_m w_m$ (8)	$\hat{\xi}^2_m$ (9)	$\hat{\xi}_m$ (10)	w_m (11)	$\hat{\xi}_m w_m$ (12)	$\hat{\xi}^2_m w_m$ (13)
1. VA-NHLBI	0	508	0	504	—	—	—	—	—	—	—	—
2. HDFP stratum I	59	3903	88	3922	−0.007 32	106 302.7	−778.2	5.697	−0.402 40	34.68	−13.95	5.616
3. Oslo	0	406	5	379	−0.013 19	29 112.3	−384.1	5.067	—	—	—	—
4. ANBPS	13	1721	22	1706	−0.005 34	84 620.0	−452.0	2.415	−0.540 24	8.09	−4.37	2.362
5. MRC	60	8700	109	8654	−0.005 70	449 570.9	−2562.0	4.600	−0.608 06	38.35	−23.32	14.180
6. VA II	5	186	20	194	−0.076 21	1620.1	−123.5	9.410	−1.425 74	3.83	−5.46	7.780
7. USPHS	1	193	6	196	−0.025 43	5614.5	−142.8	3.631	−1.802 23	0.85	−1.53	2.759
8. HDFP stratum II	25	1048	36	1004	−0.012 00	17 651.5	−211.8	2.542	−0.419 91	14.33	−6.02	2.527
9 HSCCSG	43	233	52	219	−0.052 89	679.0	−35.9	1.900	−0.319 07	18.61	−5.94	1.895
10. VAI	1	68	3	63	−0.032 91	1071.9	−35.3	1.161	−1.208 96	0.73	−0.89	1.071
11. Wolff	2	45	1	42	0.020 63	667.9	13.8	0.284	0.645 52	0.65	0.42	0.269
12. Barraclough	0	58	0	58	—	—	—	—	—	—	—	—
13. Carter	10	49	21	48	−0.233 42	118.5	−27.6	6.454	−1.109 66	4.76	−5.28	5.855
14. HDFP stratum III	18	534	34	529	−0.030 56	5724.6	−175.0	5.348	−0.677 54	11.24	−7.62	5.162
15. EWPHE	32	416	48	424	−0.036 28	2454.2	−89.1	3.231	−0.426 52	17.44	−7.44	3.172
16. Coope	20	419	39	465	−0.036 14	3653.3	−132.0	4.771	−0.602 35	12.42.	−7.48	4.508
Total	289	18 487	484	18 407		708 861.4	−5135.5	66.511		165.98	−88.88	57.156

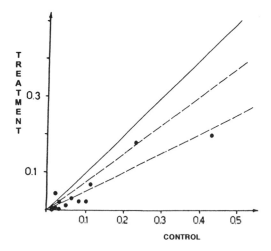

Figure 11.1   The L'Abbé *et al.* (1987) plot of data reported in table 11.1 (see text).

Probability difference:

$$\bar{\bar{\xi}} = -\frac{5135.5}{708\,861.4} = -0.007\,24 \qquad SE(\bar{\bar{\xi}}) = 1/\sqrt{708\,861.4} = 0.001\,19$$

95% confidence interval of ξ:$(-0.009\,57, -0.004\,91)$

$Q_{\text{hom}} = 66.511\,81 - (5135.5^2/708\,861.4) = 29.31$

$$\text{Odds ratio: } \bar{\bar{\xi}} = -\frac{88.88}{165.98} = -0.535\,49 \qquad SE(\bar{\bar{\xi}}) = 0.077\,62$$

95% confidence interval of ξ: $(-0.383\,35, -0.687\,62)$
$Q_{\text{hom}} = 57.156\,28 - (88.88^2/165.98) = 9.56$

$$\hat{\theta} = \exp(\bar{\bar{\xi}}) = 0.585 \qquad \text{95\% confidence interval of } \theta: (0.503, 0.682)$$

Since in two trials (VA-NHLBI and Barraclough), no failures were observed, they do not give any contribution to the total; thus Q_{hom} is asymptotically distributed as χ^2 with 13 d.f. The 99th centile, $\chi^2_{13} = 27.69$, leads us to conclude that the test for heterogeneity is statistically significant for the probability difference. The wide range of percentages of patients failing in the control group we observed in figure 11.1 goes some way to explaining this result. However, as the Q_{hom} for the log-odds-ratio is not significant, we are in the situation considered at point (2) at the beginning of the previous section and can assess that the log-odds-ratio is to be preferred to the probability difference as a measure of effect to perform this meta-analysis.

However, the result obtained in terms of probability difference will be used to complete our example with the application of the random effect model.

Firstly we estimate σ_ϕ^2 by (11.18):

$$S_\phi^2 = (29.31 - 13)/(708\,861.42 - 312\,927.58) = 4.119\,38 \times 10^{-5}$$

According to (11.17) the weight for the second trial, for example, is now

$$w_2' = (v_2 + s_\phi^2)^{-1} = (9.407\,10 \times 10^{-6} + 4.119\,38 \times 10^{-5})^{-1} = 197\,66.53$$

Calculation of the new weights for all the trials enables us to obtain the estimates:

$$\tilde{\tilde{\xi}} = -0.007\,24 \qquad \mathrm{SE}(\tilde{\tilde{\xi}}) = 0.003\,11$$

and 95% confidence interval of ξ: $(-0.013\,33, -0.001\,16)$.

As expected, the two estimates $\tilde{\xi}$ and $\tilde{\tilde{\xi}}$ are equal while, $var(\tilde{\tilde{\xi}})$ being larger than $var(\tilde{\xi})$, the confidence interval of ξ obtained by the random-effect model is wider than that obtained by the fixed-effect one. The conclusion on the effectiveness of treatment does not change, since both intervals do not include the value $\xi = 0$ defined under the null hypothesis.

However the meta-analyst takes the risk of getting into an awkward situation if the random effect model is taken formally. Think, for example, of two trials whose results are significant on their own, but where the sizes of the effect are substantially different. It may happen that, when the trials are combined by accounting for the between-trial variability, the results do not enable a claim of statistical significance of treatment effect. This is coherent with the warnings previously outlined in using the random effect model.

The example on survival data is based on table 4M of the meta-analysis performed by the Early Breast Cancer Trialists' Collaborative Group (1992) to assess the effectiveness of hormonal, cytotoxic or immune adjuvant therapy in early breast cancer. Information on 133 randomized trials for a total of 75 000 women was collected and centrally checked on recurrence and mortality for each woman accrued in any randomized trial that began before 1985. Owing to the availability of the whole set of survival data, the authors were able to compute numerator and denominator of (11.14) for 10-year survival. Table 4M, reported in table 11.2, refers results obtained in trials using hormonal therapy, namely tamoxifen. This table reports not only the ingredients needed to perform the meta-analysis, but also a graph which, originally introduced by R. Peto, is now a 'classical' component of clinical trial meta-analyses. For each trial, the odds ratio (treatment:control) is given together with the 99% confidence interval of the true value of this odds ratio. The black square indicates the actual value of the odds-ratio estimate for that trial and the horizontal line the width of

Table 11.2 Meta-analysis of adjuvant hormonal therapy in breast cancer. Outcome: overall mortality; drug: tamoxifen. Data from Early Breast Cancer Trialists' Collaborative Group (1992).

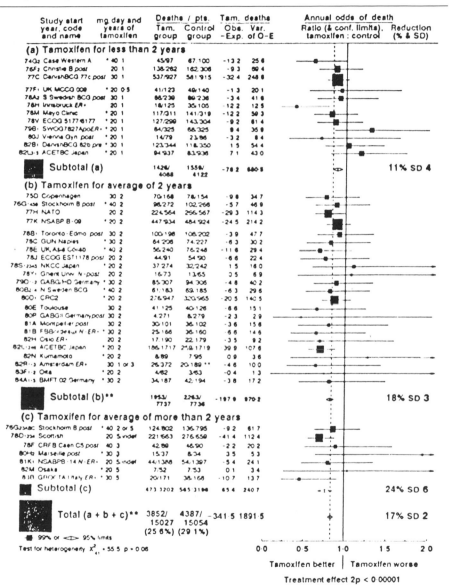

the pertinent confidence interval. (A trial yielding several failures is character-ized by a large black square and a narrow interval.) The solid vertical line indicates that the two treatment results do not differ and results to the left of this favour the treatment. The broken vertical line indicates the average relative difference of 17% and the "typical odds-ratio" (treatment:control) together with its 95% confidence interval, computed on the whole set of trials, is displayed by a diamond shape. According to (11.4) and (11.5), from the total of the two last columns it is easy to estimate: $\hat{\xi} = -341.5/1891.5 = -0.1805$ and the 95% confidence interval of ξ:

$$-0.1805 \pm \frac{1.96}{\sqrt{1891.5}} = (-0.2226, -0.1355)$$

By exponentiating we have: $\hat{\theta} = 0.83$ and 95% C.I.: $(0.80, 0.87)$.

This estimate is the overall result obtained combining all three subgroups of trials classified according to treatment duration. Subgroup estimates of treat-ment effect are indicated by the three diamonds relative to each subtotal. Note that the test of homogeneity is borderline ($\chi^2_{41} = 55.5, p = 0.06$); however, given that $z = -341.5/\sqrt{1891.5} = -7.85$, i.e. the overall difference (O − E) is nearly eight times greater than its standard error, one can be quite confident on the conclusion in favour of tamoxifen, regardless of heterogeneity.

11.6   A PROMISING APPROACH

In a stimulating paper titled: "Meta-analysis: science or religion?", Meinert (1989) faced many issues of paramount importance for the validity of meta-analysis. In discussing the timing of meta-analyses he states: "I think we need to be planning for meta-analyses in a prospective fashion in which we enroll trials for analysis, as they are started, in much the same fashion as we now enroll patients into trials. The present approach in which trials are identified after-the-fact, whether via the published literature or other means, is and continues to be open to question regarding the trials selected for analysis and the timing of the analyses. "This prospective view implies, on the one hand, the development of registers of planned and in-process randomized trials as the only tools for investigating and dealing with the problem of publication bias, and, on the other hand, the innovative approach of accomplishing a new meta-analysis as the findings of a trial of a given therapy are published." Recently, Lau *et al.* (1992) gave an example of a cumulative meta-analysis of therapeutic trials for myocar-dial infarction. Figure 11.2, corresponding to figure 2 of Lau *et al.*, gives the results of cumulative meta-analyses of 60 trials aiming at assessing the effect of intravenous thrombolytic agents on mortality in infarctuated patients. The statistics are the odds-ratios together with their 95% confidence intervals

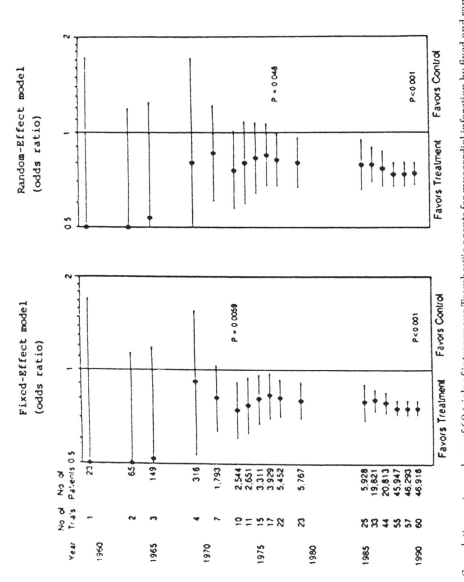

Figure 11.2 Cumulative meta-analyses of 60 trials of intravenous Tromboytic agents for myocardial infarction by fixed and random effect models. Data from Lau *et al.* (1992).

displayed on a logarithmic scale. Results obtained by the fixed-effect model are reported on the left side and by the random-effect model on the right side of the figure; they are grouped according to year of publication of the trial findings. It may be seen that, given heterogeneity in the estimates of treatment effect, statistical significance of the random-effect approach emerged in 1977, four years later than that of the fixed effect approach (1973). Nonetheless the conclusion in favour of treatment was reached on the basis of the outcomes observed in some 5 500 patients; the approximately 41 000 subsequently accrued patients did not change substantially the conclusion except in narrowing the confidence interval of θ.

When cumulative meta-analyses are performed as results of each new randomized trial are available, the problem of inflation of type I error probability arises in analogy with the multiple looks of a randomized trial considered in section 2.8. Yusuf *et al.* (1991) suggested adjusting the *p* values according to the approaches outlined in section 2.8. Alternatively, because ξ_m in (11.16) is a random variable, Bayesian arguments are applicable and might be a proper methodological support to deal with meta-analyses in a prospective fashion (Carlin, 1992; Rogatko, 1992).

APPENDIX A11.1   ESTIMATION OF σ_ϕ^2 BY THE METHOD OF MOMENTS

$$Q_{\text{hom}} = \sum_{m=1}^{M} w_m (\hat{\xi}_m - \overline{\hat{\xi}})^2$$

$$= \sum_{m=1}^{M} w_m [(\hat{\xi}_m - \xi) - (\overline{\hat{\xi}} - \xi)]^2$$

$$= \sum_{m=1}^{M} w_m (\hat{\xi}_m - \xi)^2 - \sum_{m=1}^{M} w_m (\overline{\hat{\xi}} - \xi)^2$$

Therefore

$$E(Q_{\text{hom}}) = \sum_{m=1}^{M} w_m \, var(\hat{\xi}_m) - \sum_{m=1}^{M} w_m \, var(\overline{\hat{\xi}})$$

which, given the model (11.16), can be rewritten:

$$E(Q_{\text{hom}}) = \sum_{m=1}^{M} w_m (\sigma_{\varepsilon m}^2 + \sigma_\phi^2) - \sum_{m=1}^{M} w_m \left[\left(\sum_{m=1}^{M} w_m \right)^{-1} + \frac{\sum_{m=1}^{M} w_m^2 \sigma_\phi^2}{\left(\sum_{m=1}^{M} w_m \right)^2} \right]$$

$$= (M-1) + \sigma_\phi^2 \left[\sum_{m=1}^{M} w_m - \frac{\sum_{m=1}^{M} w_m^2}{\sum_{m=1}^{M} w_m} \right] \tag{A11.1}$$

As suggested by DerSimonian and Laird (1986) a non-iterative estimate of σ_ϕ^2 can be attained by equating the sample statistic Q_{hom} with the corresponding expected value (A11.1); it is straightforward to obtain the estimator (11.18).

References

Aalen O. Nonparametric inference in connection with multiple decrement models. *Scandinavian Journal of Statistics*, **3**, 15–27, 1976.

Aalen O. Nonparametric inference for a family of counting processes. *The Annals of Statistics*, **6**, 701–726, 1978.

Aitkin M., Laird N. and Francis B. A reanalysis of the Stanford heart transplant data (with comment). *Journal of the American Statistical Association*, **78**, 264–281, 1983.

Akaike H. Use of an information theoretic quantity for statistical model identification. *Proc. Fifth Hawaii Internat. Conf. on System Sciences*. Western Periodically, North Hollywood, CA, 1972.

Altman D. G. Comparability of randomised groups. *The Statistician*, **34**, 125–136, 1985.

Altman D. G. and Andersen P. K. Bootstrap investigation of the stability of a Cox regression model. *Statistics in Medicine*, **8**, 771–783, 1989.

Altman D. G. and De Stavola, B. L. Practical problems in fitting a proportional hazards model to data with updated measurements of the covariates. *Statistics in Medicine*, 13, 301–341, 1994.

Andersen P. K. Testing goodness–of–fit of Cox's regression and life model. *Biometrics*, **38**, 67–77, 1982.

Andersen P. K. Comparing survival distributions via hazard ratio estimates. *Scandinavian Journal of Statistics*, **10**, 77–85, 1983.

Andersen P. K. Time dependent covariates and Markov processes. In: *Modern Statistical Methods in Chronic Disease Epidemiology*; Moolgavkar S.H. and Prentice R.L. (eds), John Wiley & Sons, New York, 1986.

Andersen P. K. and Gill R. D. Cox's regression model for counting processes: a large sample study. *Annals of Statistics*, **10**, 1100–1120, 1982.

Andersen P. K., Borgan O., Gill R. D. and Keiding N. Linear nonparametric tests for comparison of counting processes, with applications to censored survival data. *International Statistical Review*, **50**, 219–258, 1982.

Andersen P. K., Borgan O., Gill R. D. and Keiding N. *Statistical Models Based on Counting Processes*. Springer Verlag, New York, 1993.

Anderson J. A. and Senthilselvan A. A two-step regression model for hazard functions. *Applied Statistics*, **31**, 44–51, 1982.

Anderson J. R. and Bernstein L. Asymptotically efficient two–step estimators of the hazards ratio for follow-up studies and survival data. *Biometrics*, **41**, 733–739, 1985.

Anderson J. R., Cain K. C. and Gelber R. D. Analysis of survival by tumor response. *Journal of Clinical Oncology*, **1**, 710–719, 1983.

Arjas E. Stanford heart transplantation data revised: a real-time approach. In: *Modern*

Statistical Methods in Chronic Disease Epidemiology; Moolgavkar S.H. and Prentice R.L. (eds), John Wiley & Sons, New York, 1986.

Arjas E. A graphical method for assessing goodness of fit in Cox's proportional hazards model. *Journal of the American Statistical Association*, **83**, 204–212, 1988.

Arjas E. and Venzon D. A test for discriminating between additive and multiplicative relative risks in survival analysis. *Applied Statistics*, **37**, 1–11, 1988.

Armitage P. *Sequential Medical Trials* (2nd edition). Blackwell Scientific Publications, Oxford, 1975.

Armitage P. Interim analysis in clinical trials. *Statistics in Medicine*, **10**, 925–937, 1991.

Armitage P. and Berry G. *Statistical Methods in Medical Research* (2nd edition). Blackwell Scientific Publications, Oxford, 1987.

Armitage P., McPherson C.K. and Rowe B.C. Repeated significance tests on accumulating data. *Journal of the Royal Statistical Society, A*, **132**, 235–244, 1969.

Azzalini A. and Cox D.R. Two new tests associated with analysis of variance. *Journal of the Royal Statistical Society, B*, **46**, 335–343, 1984.

Bailey B. J. R. Confidence limits to the risk ratio. *Biometrics*, **43**, 201–205, 1987.

Barlow W. E. and Prentice R. L. Residuals for relative risk regression. *Biometrika*, **75**, 65–74, 1988.

Bennett S. Log–logistic regression models for survival data. *Applied Statistics*, **32**, 165–171, 1983.

BMDP. BMDP Statistical Software, Los Angeles, USA, 1990.

Bradford Hill A. *A Short Textbook of Medical Statistics*. Hodder and Stoughton, London 1977.

Bradley J. *Distribution Free Statistical Tests*. Prentice-Hall, Englewood Cliffs, NJ, 1968.

Breslow N. E. Covariance analysis of censored survival data. *Biometrics*, **30**, 89–99, 1974.

Breslow N. E. and Crowley J. A large sample study of the life table and product limit estimates under random censorship. *Annals of Statistics*, **2**, 437–453, 1974.

Breslow N. E. and Day N. E. *Statistical Methods in Cancer Research. Vol. 1: The Analysis of Case–control Studies*. International Agency for Research on Cancer, London, 1980.

Breslow N. E. and Day N. E. *Statistical Methods in Cancer Research. Vol. 2: The Design and Analysis of Cohort Studies*. International Agency for Research on Cancer, Lyon, 1987.

Breslow N. E. and Haug C. Sequential comparison of exponential survival curves. *Journal of the American Statistical Association*, **67**, 691–697, 1972.

Breslow N. E. and Storer B. E. General relative risk functions for case–control studies. *American Journal of Epidemiology*, **122**, 149–162, 1985.

Brieman L., Friedman J. H., Olshen R. A. and Stone C. J. *Classification and Regression Trees*. Wadsworth International Group, Belmont, CA, 1984.

Brookmeyer R. and Gail M. H. *AIDS Epidemiology: a Quantitative Approach*. Oxford University Press, New York, 1993.

Bulpitt C. J. Meta analysis. *The Lancet*, **2**, 93–94, 1988.

Byar D. P. Assessing apparent treatment–covariate interactions in randomized clinical trials. *Statistics in Medicine*, **4**, 255–263, 1985.

Byar D. P. Factorial and reciprocal control designs. *Statistics in Medicine*, **9**, 55–64, 1990.

Byar D. P. and Piantadosi S. Factorial designs for randomized clinical trials. *Cancer Treatment Reports*, **69**, 1055–1062, 1985.

Cain K. C. and Lange N. T. Approximate case influence for the proportional hazards regression model with censored data. *Biometrics*, **40**, 493–499, 1984.

Canner P. L. Monitoring treatment differences in long–term clinical trials. *Biometrics*, **33**, 603–615, 1977.

Carlin J. B. Meta–analysis for 2x2 tables: a Bayesian approach. *Statistics in Medicine*, **11**, 141–158, 1992.

Chalmers T. C. Problems induced by meta–analyses. *Statistics in Medicine*, **10**, 971–980, 1991.

Chalmers T. C., Levin H., Sacks H. S., Reitman D., Berrier J. and Nagalingam R. Meta analysis of clinical trials as a scientific discipline. I: Control of bias and comparison with large co-operative trials. *Statistics in Medicine*, **6**, 315–325, 1987a.

Chalmers T. C., Berrier J., Sacks H. S., Levin H. Reitman D. and Nagalingam R. Meta analysis of clinical trials as a scientific discipline. II: Replicate variability and comparison of studies that agree and disagree. *Statistics in Medicine*, **6**, 733–744, 1987b.

Charlson M. E., Ales K. L., Simon R. and MacKenzie C. R. Why predictive indexes perform less well in validation studies. Is it magic or methods? *Archives of Internal Medicine*, **147**, 2155–2161, 1987.

Chastang C., Byar D. and Piantadosi S. A quantitative study of the bias in estimating the treatment effect caused by omitting a balanced covariate in survival models. *Statistics in Medicine*, **7**, 1243–1255, 1988.

Chen C. and George S. L. The bootstrap and identification of prognostic factors via Cox's proportional hazards regression model. *Statistics in Medicine*, **4**, 39–46, 1985.

Chiang C. L. A stochastic study of the life table and its applications: I. Probability distributions of the biometric functions. *Biometrics*, **16**, 618–635, 1960.

Chlebowski R. T., Weiner J. M., Ryden V. M. J. and Bateman J. R. Factors influencing the interim interpretation of a breast cancer trial: danger of achieving the "expected" result. *Controlled Clinical Trials*, **2**, 123–132, 1981.

Christensen E. Multivariate survival analysis using Cox's regression model. *Hepatology*, **7**, 1346–1358, 1987.

Christensen E., Neuberger J., Crowe J., Altman D. G., Popper H., Portmann B., Doniach D., Ranek L., Tygstrup N. and Williams R. Beneficial effect of Azathioprine and prediction of prognosis in primary biliary cirrhosis. *Gastroenterology*, **89**, 1084–1091, 1985.

Ciampi A. Generalized regression trees. *Computational Statistics and Data Analysis*, **12**, 57–78, 1991.

Ciampi A., Hogg S. A. and Kates L. Regression analysis of censored survival data with the generalized F family. An alternative to the proportional hazards model. *Statistics in Medicine*, **5**, 85–96, 1986.

Clayton D. and Hills M. *Statistical Models in Epidemiology*. Oxford University Press, Oxford, 1993.

Cnaan A. and Ryan L. Survival analysis in natural history studies of disease. *Statistics in Medicine*, **8**, 1255–1268, 1989.

Cochran W. G. Problems arising in the analysis of a series of similar experiments. *Journal of the Royal Statistical Society*, **4** (suppl), 102–118, 1937.

Cochran W. G. The combination of estimates from different experiments. *Biometrics*, **10**, 101–129, 1954.

Cohn B. A., Wingard D. L., Cohen R. D., Cirillo P. M. and Kaplan G. A. Sex differences in time from self reported heart trouble to heart disease death in the Alameda County Study. *American Journal of Epidemiology*, **131**, 434–442, 1990.

Collins R., Gray R., Godwin J. and Peto R. Avoidance of large biases and large random errors in the assessment of moderate treatment effects: the need for systematic overviews. *Statistics in Medicine*, **6**, 245–250, 1987.

Collins R., Peto R., Macmahon S., Hebert P., Fiebach N. H., Eberlein K. A., Godwin J.,

Qizilbash N., Taylor J. O. and Hennekens C. H. Blood pressure, stroke and coronary heart disease. Part 2. Short term reductions in blood pressure: overview of randomised drug trials in their epidemiological context. *The Lancet*, **1**, 827–838, 1990.

Copas J. B. Regression, prediction and shrinkage. *Journal of the Royal Statistical Society, B*, **45**, 311–354, 1983.

Coronary Drug Project Research Group. Influence of adherences to treatment and response of chloresterol on mortality in the Coronary Drug Project. *New England Journal of Medicine*, **303**, 1038–1041, 1980.

Cox D. R. Regression models and life–tables (with discussion). *Journal of the Royal Statistical Society, B*, **34**, 187–220, 1972

Cox D. R. Partial likelihood. *Biometrika*, **62**, 269–276, 1975.

Cox D. R. and Hinkley D. V. *Theoretical Statistics*. Chapman and Hall, London, 1974.

Cox D. R. and Oakes D. *Analysis of Survival Data*. Chapman and Hall, London, 1984.

Cox D. R. and Snell E. J. A general definition of residuals (with discussion). *Journal of the Royal Statistical Society, B*, **30**, 248–275, 1968.

Crouchley R. and Pickles A. A specification test for univariate and multivariate proportional hazards models. *Biometrics*, **49**, 1067–1076, 1993.

Crowley J. and Hu M. Covariance analysis of heart transplant survival data. *Journal of the American Statistical Association*, **72**, 27–36, 1977.

Crowley J. and Storer B. E. Comment on paper by Aitkin M., Laird N., Francis B. *Journal of the American Statistical Association*, **78**, 277–281, 1983.

Cutler S. J. and Ederer F. Maximum utilization of the life table method in analyzing survival. *Journal of Chronic Diseases*, **8**, 699–712, 1958.

Cuzick J., De Stavola B. L., Cooper E. H., Chapman C. and MacLennan C. M. Long-term prognostic value of serum β_2 microglobulin in myelomatosis. *British Journal of Haematology*, **75**, 506–510, 1990.

Dabrowska D. M., Doksum K. A., Feduska N. J., Husing R. and Neville P. Methods for comparing cumulative hazard functions in a semi-proportional hazard model. *Statistics in Medicine*, **11**, 1465–1467, 1992.

David M. A. and Moeschberger M. L. *The theory of Competing Risks*. Griffin, London, 1978.

DeMets D. L. and Gail M. H. Use of logrank tests and group sequential methods at fixed calendar times. *Biometrics*, **41**, 1039–1044, 1985.

DeMets D. L. and Ware J. H. Group sequential methods for clinical trials with a one-sided hypothesis. *Biometrika*, **67**, 651–660, 1980.

DeMets D. L. and Ware J. H. Asymmetric group sequential boundaries for monitoring clinical trials. *Biometrika* 69, 661–663, 1982.

DerSimonian R. and Laird N. Meta–analysis in clinical trials. *Controlled Clinical Trials*, **7**, 177–188, 1986.

Dinse, G. E. and Larson, M. G. A note on semi–markov models for partially censored data. *Biometrika*, **73**, 379–386, 1986.

Dixon D. O. and Divine G. W. Multiple comparisons for relative risk regression: extension of the K-ratio method. *Statistics in Medicine*, **6**, 591–597, 1987.

Doll R. Practical steps towards the prevention of bronchial carcinoma. *Scottish Medical Journal*, **15**, 433–447, 1970.

Doll R., Morgan L. G. and Speizer F. E. Cancers of the lung and nasal sinuses in nickel workers. *British Journal of Cancer*, **24**, 623–632, 1970.

Dorey F. J. and Korn E. L. Effective sample sizes for confidence intervals for survival probabilities. *Statistics in Medicine*, **6**, 679–687, 1987.

Durrleman S. and Simon R. Flexible regression models with cubic splines. *Statistics in Medicine*, **8**, 551–561, 1989.

Early Breast Cancer Trialists' Collaborative Group. Systemic treatment of early breast cancer by hormonal, cytotoxic, or immune therapy. *The Lancet*, **339**, 1–15, 71–85, 1992.

Easton D. F., Peto J. and Babiker A. G. Floating absolute risk: an alternative to relative risk in survival and case–control analysis avoiding an arbitrary reference group. *Statistics in Medicine*, **10**, 1025–1035, 1991.

Efron B. Forcing a sequential experiment to be balanced. *Biometrika*, **58**, 403–417, 1971.

Efron B. The efficiency of Cox's likelihood function for censored data. *Journal of the American Statistical Association*, **72**, 557–565, 1977.

Efron B. and Gong G. A leisurely look at the bootstrap, the jackknife, and cross–validation. *The American Statistician*, **37**, 36–48, 1983.

Efron B. and Tibshirani R. Bootstrap methods for standard errors, confidence intervals, and other methods of statistical accuracy. *Statistical Science*, **1**, 54–77, 1986.

EGRET. Statistics and Epidemiology Research Corporation, Seattle, WA, USA, 1993.

Elandt-Johnson R. C. and Johnson N. L. *Survival Models and Data Analysis*. John Wiley & Sons, New York, 1980.

EPICURE. Hirosoft International Corporation, Seattle, WA, USA, 1993.

European Coronary Surgery Study Group. Coronary–artery bypass surgery in stable angina pectoris: survival at two years. *The Lancet*, **1**, 889–893, 1979.

Feigl P. and Zelen M. Estimation of exponential survival probabilities with concomitant information. *Biometrics*, **21**, 826–838, 1965.

Feller W. *An Introduction to Probability Theory and its Applications I* (3rd edition). John Wiley & Sons, New York, 1968.

Fielding L. P., Fenoglio-Preiser C. M. and Freedman L. S. The future of prognostic factors in outcome prediction for patients with cancer. *Cancer*, **70**, 2367–2377, 1992.

Fisher B., Redmond C., Poisson R., Margolese R., Wolmark N., Wickerman L., Fisher E., Deutsch M., Caplan R., Pilch Y., Glass A., Shibata H., Lerner H., Terz J. and Sidorovich L. Eight-year results of a randomized clinical trial comparing total mastectomy and lumpectomy with or without irradiation in the treatment of breast cancer. *New England Journal of Medicine*, **320**, 822–828, 1989.

Freedman D. A. A note on screening regression equations. *The American Statistician*, **37**, 152–155, 1983.

Freedman D. and Navidi W. Ex–smokers and the multistage model for lung cancer. *Epidemiology*, **1**, 21–19, 1990.

Freedman L. S. Tables of the number of patients required in clinical trials using the logrank test. *Statistics in Medicine*, **1**, 121–129, 1982.

Freedman L. S. and Spiegelhalter D. J. Application of Bayesian statistics to decision making during a clinical trial. *Statistics in Medicine*, **11**, 23–35, 1992.

Freiman A. J., Chalmers T. C., Smith H. and Kuebler R. R. The importance of beta, the type II error and sample size in the design and interpretation of the randomized control trial. *New England Journal of Medicine*, **299**, 690–694, 1978.

Freireich E. J., Gehan E., Frei III E., Schroeder L. R., Wolman I. J., Anbari R., Burgert E. O., Mills S. D., Pinkel D., Selawry O. S., Moon J. H., Gendel B. R., Spurr C. L., Storrs R., Haurani F., Hoogstraten B. and Lee S. The effect of 6-mercaptopurine on the duration of steroid-induced remissions in acute leukemia: a model for evaluation of other potentially useful therapy. *Blood*, **21**, 699–716, 1963.

Friedman L. M., Furberg C. D. and DeMets D. L. *Fundamentals of Clinical Trials* (2nd edition). PSG Publishing Co., Littleton, MA, USA, 1985.

Gail M. H. Evaluating serial cancer marker studies in patients at risk of recurrent disease. *Biometrics*, **37**, 67–78, 1981.

Gail M. H. "Monitoring and stopping clinical trials". In *Statistics in Medical Research*; Mike V. and Stanley, K. E. (eds), John Wiley & Sons, New York, 455–484, 1982.

Gail M. H. Adjusting for covariates that have the same distribution in exposed and unexposed cohorts. In: *Modern Statistical Methods in Chronic Disease Epidemiology*; Moolgavkar S. H. and Prentice R. L. (eds), John Wiley & Sons, New York, 1986.

Gail M. H. and Byar D. P. Variance calculation for direct adjusted survival curves with application to testing for no treatment effect. *Biometrical Journal*, **28**, 587–599, 1986.

Gail M. II. and Simon R. Testing for qualitative interactions between treatment effects and patient subsets. *Biometrics*, **41**, 361–372, 1985.

Gail M. H., Wieand S. and Piantadosi S. Biased estimates of treatment effect in randomized experiments with nonlinear regressions and omitted covariates. *Biometrics*, **71**, 431–444, 1984.

Galbraith R. F. A note on graphical presentation of estimated odds ratios from several clinical trials. *Statistics in Medicine*, **7**, 889–894, 1988.

Gardner M. J. and Altman D. G. Statistics with confidence. *British Medical Journal*, 1989.

Gart J. J. and Nam J. M. Approximate interval estimation of the ratio of binomial parameters: a review and corrections for skewness. *Biometrics*, **44**, 323–338, 1988.

Gaynor J. J., Fener E. J., Tan C. C., Wu D. H., Little C. R., Straus D. J., Clarkson B. D. and Brennan M. F. On the use of cause-specific failure and conditional failure probabilities: examples from clinical oncology data. *Journal of the American Statistical Association*, **88**, 400–409, 1993.

Gehan E. A. A generalized Wilcoxon test for comparing arbitrarily singly censored samples. *Biometrika*, **52**, 203–223, 1965.

Gehan E. A. Estimating survival functions from the life table. *Journal of Chronic Diseases*, **21**, 629–644, 1969.

Gehan E. A. and Freireich E. J. Non-randomized controls in cancer clinical trials. *New England Journal of Medicine*, **290**, 198–203, 1974.

GICOG (Gruppo Interregionale Cooperativo Oncologia Ginecologica). A randomized trial in advanced ovarian cancer comparing cisplatin, cyclophosphamide + cisplatin and cyclophosphamide + cisplatin + adriamicine. *The Lancet*, **2**, 353–359, 1987.

Gill R. D. Censoring and stochastic integrals. *Mathematical Centre Tracts*, 124. Mathematisch Centrum, Amsterdam, 1980.

Gill R. and Schumacher M. A simple test of the proportional hazards assumption. *Biometrika*, **74**, 289–300, 1987.

Glass G. V. Primary, secondary and meta-analysis of research. *Educational Research*, **5**, 3–8, 1976.

Glasser M. Exponential survival with covariance. *Journal of the American Statistical Association*, **62**, 561–568, 1967.

GLIM (release 4). Oxford University Press, Oxford, 1993.

Gore S. M., Pocock S.J. and Kerr G.R. Regression models and non–proportional hazards in the analysis of breast cancer survival. *Applied Statistics*, **33**, 176–195, 1984.

Gray R. J. A class of K–sample tests for comparing the cumulative incidence of a competing risk. *The Annals of Statistics*, **16**, 1141–1154, 1988.

Gray R. J. Some diagnostic methods for Cox regression models through hazard smoothing. *Biometrics*, **46**, 93–102, 1990.

Green S. B. and Byar D. P. Using observational data from registries to compare treatments: the fallacy of omnimetrics. *Statistics in Medicine*, **3**, 361–370, 1984.

Greenland S. Quantitative methods in the review of epidemiologic literature. *Epidemiologic Reviews*, **9**, 1–30, 1987.

Greenwood M. A report on the natural duration of cancer. *Reports on Public Health and Medical Subjects*, **33**, 1–26, H. M. Stationery Office, London, 1926.

Gruppo Italiano per lo Studio della Soppravvivenza nell'Infarto Miocardico GISSI-2. A factorial randomised trial of alteplase versus streptokinase and heparin versus no heparin among 12490 patients with acute myocardial infarction. *The Lancet*, **336**, 65–71, 1990.

Guerrero V. M. and Johnson R. S. Use of the Box–Cox transformation with binary response models. *Biometrika*, **69**, 309–314, 1982.

Hall W. J. and Wellner J. A. Confidence bands for a survival curve from censored data. *Biometrika*, **61**, 133–143, 1980.

Halperin M. Maximum likelihood estimation in truncated samples. *Annals of Mathematical Statistics*, **23**, 226–238, 1952.

Harrell F. E. Jr. and Lee K. L. Verifying assumptions of the Cox proportional hazards model. In: *Proceedings of the 11th Annual SAS Users Group International Conference*, Atlanta, GA, February 1986. Cary, NC: SAS Institute, Inc, 25–30, 1986.

Harrell F. E., Jr, Lee K. L., Califf R. M., Pryor D. B. and Rosati R. Regression modelling strategies for improved prognostic prediction. *Statistics in Medicine*, **3**, 143–152, 1984.

Harrell F. E., Jr, Lee K. L., Matchar D. B. and Reichert T. A. Regression models for prognostic prediction: advantages, problems, and suggested solutions. *Cancer Treatment Reports*, **69**, 1071–1077, 1985.

Harrell F. E., Kerry L. L. and Pollock B. G. Regression models in clinical studies: determining the relationship between predictors and response. *Journal of the National Cancer Institute*, **80**, 1198–1202, 1988.

Harrington D. P. and Fleming T. R. A class of rank test procedures for censored survival data. *Biometrika*, **69**, 553–566, 1982.

Hauck W. W. and Miike R. A proposal for examining and reporting stepwise regressions. *Statistics in Medicine*, **10**, 711–715, 1991.

Haybittle J. L. Repeated assessments of results in clinical trials of cancer treatment. *British Journal of Radiology*, **44**, 793–797, 1971.

Haynes R. B. and Dantes R. Patient compliance and the conduct and interpretation of therapeutic trials. *Controlled Clinical Trials*, **8**, 12–19, 1987.

Howe G., Hirohata T., Hislop T. G., Iscovich J. M., Yuan J. M., Katsouyanni K., Lubin F., Marubini E., Modan B., Rohan T., Toniolo P. and Shunzhang Y. Dietary factors and risk of breast cancer: combined analysis of 12 case–control studies. *Journal of the National Cancer Institute*, **82**, 561–569, 1990.

Howe G., Rohan T., Decarli A., Iscovich J., Kaldor J., Katsouyanni K., Marubini E., Miller A., Riboli E., Toniolo P. and Trichopoulos D. The association between alcohol and breast cancer risk: evidence from the combined analysis of six dietary case-control studies. *International Journal of Cancer*, **47**, 707–710. 1991.

Jennison C. and Turnbull B. W. Interim Analyses: the repeated confidence interval approach. *Journal of the Royal Statistical Society*, B, **51**, 305–361, 1989.

Johansen S. The product limit estimator as maximum likelihood estimator. *Scandinavian Journal of Statistics*, **5**, 195–199, 1978.

Johansen S. An extension of Cox's regression model. *International Statistical Review*, **51**, 165–174, 1983.

Johnson N. L. and Kotz S. *Continuous Univariate Distribution–I*. Houghton Mifflin Company, Boston, MA, 1970.

Kalbfleisch J. D. and Prentice R. L. *The Statistical Analysis of Failure Time Data*. John Wiley & Sons, New York, 1980.

Kaplan E. L. and Meier P. Nonparametric estimation from incomplete observations. *Journal of the American Statistical Association*, **53**, 457–481, 1958.

Kay R. Proportional hazards regression models and the analysis of censored survival data. *Applied Statistics*, **26**, 227–237, 1977.

Kay R. Goodness of fit methods for the proportional hazards regression model: a review. *Revue d'Epidémiologie et Santé Publique*, **32**, 185–198, 1984.

Keiding N. and Andersen P. K. Nonparametric estimation of transition intensities and transition probabilities: a case study of a two-state Markov process. *Applied Statistics*, **38**, 319–329, 1989.

Kendall M. G. and Stuart A. *The Advanced Theory of Statistics*. Charles Griffin and Co., London, 1958.

Kim K. M. and DeMets D. L. Estimation following group sequential tests in clinical trials. University of Wisconsin Biostatistics Technical Report, No. 32, 1985.

Kish L. *Survey Sampling*. John Wiley & Sons, New York, 1965.

Koepcke W., Hasford J., Messerer D. and Zwingers T. *Proceedings of the XI International Biometric Conference*, Toulouse, 1982.

Korn E. L. Censoring distributions as a measure of follow-up in survival analysis. *Statistics in Medicine*, **5**, 255–260, 1986.

Korn E. L. and Dorey F. J. Applications of crude incidence curves. *Statistics in Medicine*, **11**, 813–829, 1992.

Korn E. L. and Simon R. Measures of explained variation for survival data. *Statistics in Medicine*, **9**, 487–503, 1990.

L'Abbé K. A., Detsky A. S. and O'Rourke K. Meta-analysis in clinical research. *Annals of Internal Medicine*, **107**, 224–233, 1987.

Lagakos S. W. General right censoring and its impact on the analysis of survival data. *Biometrics*, **35**, 139–156, 1979.

Lagakos S. W. The graphical evaluation of explanatory variables in proportional hazards regression models. *Biometrika*, **68**, 93–98, 1980.

Lagakos S. W. and Schoenfeld D. Properties of proportional–hazards score tests under misspecified regression models. *Biometrics*, **40**, 1037–1048, 1984.

Lakatos E. Sample sizes based on the log-rank statistic in complex clinical trials. *Biometrics*, **44**, 229–241, 1988.

Lan K. K. G. and DeMets D. L. Discrete sequential boundaries for clinical trials. *Biometrika*, **70**, 659–663, 1983.

Lan K. K. G., Simon R. and Halperin M. Stochastically curtailed tests in long–term clinical trials. *Communications in Statistics–Sequential Analysis*, **1**, 207–219, 1982.

Lan K. K. G., DeMets D. L. and Halperin M. More flexible sequential and non-sequential designs in long–term clinical trials. *Communications in Statistics, Theory and Methods*, **13**, 2339–2353, 1984.

Last J. M. *A Dictionary of Epidemiology*. Oxford University Press, Oxford, 1988.

Lau J., Antman E. M., Jimenez-Silva J., Kupelnick B. B. A., Mosteller F. and Chalmers T.C. Cumulative meta-analysis of therapeutic trials for myocardial infarction. *New England Journal of Medicine*, **327**, 248–254, 1992.

Laupacis A., Sackett D. L. and Roberts R. S. An assessment of clinically useful measures of the consequences of treatment. *New England Journal of Medicine*, **26**, 1728–1733, 1988.

Lausen B. and Schumacher M. Maximally selected rank statistics. *Biometrics*, **48**, 73–85, 1992.

Lawless J. F. *Statistical Models and Methods for Lifetime Data*. John Wiley & Sons, New York, 1982.

Lawless J. F. A note on lifetime regression models. *Biometrika,* **73**, 509–512, 1986.

Lawless J. F. and Singhal K. ISMOD: An all-subset regression program for generalized linear models. I: Statistical and computational background. *Computer Methods and Programs in Biomedicine,* **24**, 117–124, 1987.

LeBlanc M. and Crowley J. Survival trees by goodness of split. *Journal of the American Statistical Association,* **88**, 457–467, 1993.

Lehmann E. L. *Nonparametrics: Statistical Methods Based on Ranks.* Holden-Day, San Francisco, 1975.

Lemeshow S. *et al. Adequacy of Sample Size in Health Studies.* World Health Organization, 1990.

Lin D. Y. and Wei L. J. Goodness-of-fit tests for the general Cox regression model. *Statistica Sinica,* **1**, 1–17, 1991.

Longnecker M. P., Berlin J. A., Orza M. J. and Chalmers T. C. A meta analysis on alcohol consumption in relation to risk of breast cancer. *Journal of the American Medical Association,* **260**, 652–656, 1988.

Louis P. C. A. *Researches on the Effects of Bloodletting in Some Inflammatory Diseases, and on the Influence of Tartarized Antimony and Vescication in Pneumonitis.* Translated by C.G. Putnam, with preface and appendix by James Jackson. Boston, MA, 1836.

Lund B. and Williamson P. Prognostic factors for overall survival in patients with advanced ovarian carcinoma. *Annals of Oncology,* **2**, 281–287, 1991.

Machin D. and Campbell M. *Statistical Tables for the Design of Clinical Trials.* Blackwell Scientific Publications, Oxford, 1987.

Makuch R. W. Adjusted survival curve estimation using covariates. *Journal of Chronic Disease,* **35**, 437–443, 1982.

Mantel N. Evaluation of survival data and two new rank order statistics arising in its consideration. *Cancer Chemotherapy Report,* **50**, 163–170, 1966.

Mantel N. Pre-stratification or post–stratification. Letter to the Editor. *Biometrics,* **40**, 256–258, 1984.

Mantel N. and Byar D. P. Evaluation of response-time data involving transient states: an illustration using heart transplant data. *Journal of the American Statistical Association,* **69**, 81–86, 1974.

Mantel N. and Haenszel W. Statistical aspects of the analysis of data from retrospective studies of disease. *Journal of the National Cancer Institute,* **22**, 719–748, 1959.

Marsoni S. and Valsecchi M. G. Prognostic factors analysis in clinical oncology: handle with care. *Annals of Oncology,* **2**, 245–247, 1991.

Marubini E., Braga M., Cozzea Leite M. L., Petroccione A., Pirotta N. on behalf of SINBA group. "Within patient" dependent outcomes in graft occlusion after coronary artery bypass. *Controlled Clinical Trials,* **14**, 296–307, 1993.

McCullagh P. and Nelder J. A. *Generalized Linear Models.* Chapman and Hall, London, 1989.

McPherson K. Statistics: the problem of examining accumulating data more than once. *New England Journal of Medicine,* **290**, 501–502, 1974.

Meinert C. L. Meta-analysis: science or religion? *Controlled Clinical Trials,* **10**, 257S–263S, 1989.

Meinert C. L. *Clinical Trials. Design, Conduct, and Analysis.* Oxford University Press, New York, 1986.

Mezzanotte G., Boracchi P., Marubini E. and Valagussa P. Analisi delle sopravvivenze a lungo termine nel carcinoma mammario. *Rivista di Statistica Applicata,* **20**, 251–267, 1987.

Micciolo R., Valagussa P. and Marubini E. The use of historical controls in breast cancer. *Controlled Clinical Trials*, **6**, 259–270, 1985.

Miller R. G. What price Kaplan–Meier? *Biometrics*, **39**, 1077–1081, 1983.

Moolgavkar S. H. and Venzon D. General relative risk regression models for epidemiological studies. *American Journal of Epidemiology*, **126**, 949–961, 1987.

Morabito A. and Marubini E. A computer program suitable for fitting linear models when the dependent variable is dichotomous, polichotomous or censored survival and non–linear models when the dependent variable is quantitative. *Computer Programs in Biomedicine*, **5**, 283–295, 1976.

Moreau T., O'Quigley J. and Lellouch J. On D. Schoenfeld's approach for testing the proportional hazards assumption. *Biometrika* **73**, 513–515, 1986.

Morgan T. M. Omitting covariates from the proportional hazards model. *Biometrics*, **42**, 993–995, 1986.

Nair V. N. Confidence bands for survival functions with censored data: a comparative study. *Technometrics*, **26**, 265–275, 1984.

Nagelkerke N. J. D. A note on the general definition of the coefficient of determination. *Biometrika*, **78**, 691–692, 1991.

Oakes D. Survival times: aspects of partial likelihood. *International Statistical Review*, **49**, 235–264, 1981.

O'Brien P. C. and Fleming T. R. A multiple testing procedure for clinical trials. *Biometrics*, **35**, 549–556, 1979.

O'Quigley J. and Pessione F. Score tests for homogeneity of regression effect in the proportional hazards model. *Biometrics*, **45**, 135–144, 1989.

Parker R. L., Dry T. J., Willius F. A. and Gage R. P. Life expectancy in angina pectoris. *Journal of the American Medical Association*, **131**, 95, 1946.

Peto J. The calculation and interpretation of survival curves. In: *Cancer Clinical Trials, Methods and Practice*; Buyse M. E., Staquet M. J. and Sylvester R. J. (eds), Oxford University Press, Oxford, 1984.

Peto, R. Statistical aspects of cancer trials". In: *Treatment of Cancer*; Halnan, K. E. (ed.), Chapman and Hall, London, 867–871, 1982.

Peto R., Pike M. C., Armitage P., Breslow N. E., Cox D. R., Howard V., Mantel N., McPherson K., Peto J. and Smith P. G. Design and analysis of randomised clinical trials requiring prolonged observation of each patient. I: Introduction and design. *British Journal of Cancer*, **34**, 585–612, 1976.

Peto R., Pike M. C., Armitage P., Breslow N. E., Cox D. R., Howard V., Mantel N., McPherson K., Peto J. and Smith P. G. Design and analysis of randomized clinical trials requiring prolonged observation of each patient. II: Analysis and examples. *British Journal of Cancer*, **35**, 1–39, 1977.

Pettitt A. N. and Bin Daud I. Investigating time dependence in Cox's proportional hazards model. *Applied Statistics*, **39**, 313–329, 1990.

Pocock S. J. *Clinical trials. A Practical Approach*. John Wiley & Sons, Chichester, 1983.

Pocock S. J. Group sequential methods in the design and analysis of clinical trials. *Biometrika*, **64**, 191–199, 1977.

Pocock S. J., Gore S. M. and Kerr G. R. Long term survival analysis: the curability of breast cancer. *Statistics in Medicine*, **1**, 93–104, 1982.

Prentice R. L. Linear rank tests with right-censored data. *Biometrika*, **65**, 167–179, 1978.

Prentice R. L. and Mason M. W. On the application of linear relative risk regression models. *Biometrics*, **42**, 109–120, 1986.

Prentice R. L., Kalbfleisch J. D., Peterson A. V., Jr, Flournoy N., Farewell V. T. and

Breslow N. E. The analysis of failure times in the presence of competing risks. *Biometrics*, **34**, 541–554, 1978.

Prentice R. L., Brown K. S. and Mason M.W. HLA and disease: Relative-Risk regression methods and multiple testing considerations. *Biometrics*, **40**, 653–661, 1984.

Reid N. and Crépeau H. Influence functions for proportional hazards regression model. *Biometrika*, **72**, 1–9, 1985.

Robins J. M. and Greenland S. The role of model selection in causal inference from nonexperimental data. *American Journal of Epidemiology*, **123**, 392–402, 1986.

Rogatko A. Bayesian approach for meta-analysis of controlled clinical trials. *Communication in Statistics–Theory and Methods*, **21**, 1441–1462, 1992.

Rothman K. J. Estimation of confidence limits for the cumulative probability of survival in life table analysis. *Journal of Chronic Diseases*, **31**, 557–560, 1978.

SAS. SAS Institute Inc., Cary, NC, USA, 1991.

Sasieni P. Maximum weighted partial likelihood estimators for the Cox model. *Journal of the American Statistical Association*, **88**, 144–152, 1993.

Sato T. Confidence limits for the common odds ratio based on the asymptotic distribution of the Mantel–Haenszel Estimator. *Biometrics*, **46**, 71–80, 1990.

Sauerbrei W. and Schumacher M. A bootstrap resampling procedure for model building: application to the Cox regression model. *Statistics in Medicine*, **11**, 2093–2109, 1992.

Schemper M. The explained variation in proportional hazards regression. *Biometrika*, **77**, 216–218, 1990.

Schoenfeld D. Chi-squared goodness-of-fit tests for the proportional hazards regression model. *Biometrika*, **67**, 145–153, 1980.

Schoenfeld D. The asymptotic properties of nonparametric tests for comparing survival distributions. *Biometrika*, **68**, 316–319, 1981.

Schoenfeld D. Partial residuals for the proportional hazards regression model. *Biometrika*, **69**, 239–241, 1982.

Schwartz D., Flamant R., and Lellouch J. *L'essai Thérapeutique chez l'Homme*. Flammarion Médecine-Sciences, Paris, 1970.

Schweder T. and Spjotvoll E. Plot of P-values to evaluate many tests simultaneously. *Biometrika*, **69**, 493–502, 1982.

Shuster J. and van Eys J. Interaction between prognostic factors and treatment. *Controlled Clinical Trials*, **4**, 209–214, 1983.

Silvestri D., Mariani L. and Valsecchi M. G. Una procedura SAS per il calcolo delle curve di sopravvivenza corrette per fattori prognostici con il metodo della standardizzazione diretta. *SUGItalia*, Proceedings of the IX annual meeting, 205–215, Trieste, 20–22 October 1993.

Simon R. A critical assessment of approaches to improving the efficiency of cancer trials. In: *Cancer Clinical Trials: A Critical Appraisal*. Scheuzlen H., Kay R, and Braum M. (eds.) Springer-Verlag, Berlin, 18–26, 1988.

Simon R. and Altman D. G. Statistical aspects of prognostic factor studies in oncology. *British Journal of Cancer*, **69**, 979–985, 1994.

Simon R. and Lee Y. J. Nonparametric confidence limits for survival probabilities and median survival time. *Cancer Treatment Reports*, **66**, 37–42, 1982.

Simon R. and Makuch R. W. A non-parametric graphical presentation of the relationship between survival and the occurrence of an event: application to responders versus nonresponders bias. *Statistics in Medicine*, **3**, 35–44, 1984.

Slud E. V. and Wei L. J. Two sample repeated significance tests based on the modified Wilcoxon statistic. *Journal of the American Statistical Association*, **77**, 862–868, 1982.

Slud E. V., Byar D. P. and Green S. B. A comparison of reflected versus test-based confidence intervals for the median survival time, based on censored data. *Biometrics*, **40**, 587–600, 1984.

Spiegelhalter D. J. and Freedman L. S. Bayesian approaches to clinical trials. In: *Bayesian Statistics*, **3**, 453–477; Bernardo J.M., DeGroot M.H., Lindley D.V. and Smith A.F.M. (eds), Oxford University Press, Oxford, 1988.

Spiegelhalter D. J., Freedman L. S. and Blackburn P. R. Monitoring clinical trials: Conditional or predictive power? *Controlled Clinical Trials*, **7**, 8–17, 1986.

SPSS. SPSS Inc., Chicago, USA 1991.

Stablein D. M, Carter W. H. Jr. and Novak J. W. Analysis of survival data with nonproportional hazard functions. *Controlled Clinical Trials*, **2**, 149–159, 1981.

Stampfer M., Buring J. E., Willett W., Rosner B., Emberlein K., and Hennekens C. H. The 2 × 2 factorial design: its application to a randomized trial of aspirin and carotene in U.S. physicians. *Statistics in Medicine*, **4**, 111–116, 1985.

Storer B. E. and Crowley J. A diagnostic for Cox regression and general conditional likelihoods. *Journal of the American Statistical Association*, **80**, 139–147, 1985.

Struthers C. A. and Kalbfleisch J. D. Misspecified proportional hazards models. *Biometrika*, **73**, 363–369, 1986.

Tarone R. E. and Ware J. On distribution free tests for equality of survival distributions. *Biometrika*, **64**, 156–160, 1977.

Thall P. F. and Lachin J. M. Assessment of stratum–covariate interactions in Cox's proportional hazards regression model. *Statistics in Medicine*, **5**, 73–83, 1986.

Therneau T. M., Grambsch P. M. and Fleming T. R. Martingale hazards regression models and the analysis of censored survival data. *Biometrika*, **77**, 147–160, 1990.

Thomas D. C. General relative–risk models for survival time and matched case-control analysis. *Biometrics*, **37**, 673–686, 1981.

Thomas D. G., Breslow N. E. and Gart J. J. Trend and homogeneity analyses of proportions and life table data. *Computers and Biomedical Research*, **10**, 373–381, 1977.

Thomsen B. L. *A note on the modelling of continuous covariates in Cox's regression model.* Research Report 88/5, Statistical Research Unit, University of Copenhagen, 1988.

Thomsen B. L., Keiding N. and Altman D. G. A note on the calculation of expected survival, illustrated by the survival of liver transplant patients. *Statistics in Medicine*, **10**, 733–738, 1991.

Tsiatis A. A. A non-identifiability aspect of the problem of competing risks. *Proceedings of the National Academy of Science*, **72**, 20–22, 1975.

Tsiatis A. A. The asymptotic joint distribution of the efficient score test for the proportional hazards model calculated over time. *Biometrika*, **68**, 311–315, 1981a.

Tsiatis A. A. A large sample study of the estimate for the integrated hazard function in Cox's regression model for survival data. *Annals of Statistics*, **9**, 93–108, 1981b.

Tsiatis A. A., Rosner G. L. and Metha C.R. Exact confidence intervals following a group sequential test. *Biometrics*, **40**, 797–803, 1984.

Valsecchi M. G. L'applicazione di procedure di partizione recursiva per la predizione prognostica. Società Italiana di Statistica, *Proceedings of the XXXVI Meeting*, Pescara, 187–194, April 1992a.

Valsecchi M.G. Modelling the relative risk of esophageal cancer in a case-control study. *Journal of Clinical Epidemiology*, **45**, 347–355, 1992b.

Van Houwelingen J. C. and Le Cessie S. Predictive value of statistical models. *Statistics in Medicine*, **9**, 1303–1325, 1990b.

Van Houwelingen J. C. and Thorogood J. Construction, validation and updating of a prognostic model for kidney graft survival. *Proceedings of the XVI International Biometric Conference*, New Zealand, 1992.

Veronesi U., Saccozzi R., Del Vecchio M., Banfi A., Clemente C., De Lena M., Gallus G., Greco M., Luini A., Muscolino G., Rilke F., Salvadori B., Zecchini A. and Zucali R. Comparing radical mastectomy with quadrantectomy, axillary dissection and radiotherapy in patients with small cancers of the breast. *New England Journal of Medicine*, **305**, 6–11, 1981.

Veronesi U., Marubini E., Banfi A., Del Vecchio M., Saccozzi R., Clemente C., Greco M., Luini A., Muscolino R., Rilke F., Sacchini V., Salvadori B., Zecchini A. and Zucali R. Comparison of halsted mastectomy with quadrantectomy, axillary dissection, and radiotherapy in early breast cancer: long–term results. *European Journal Cancer Clinical Oncology*, **22**, 1085–1089, 1986.

Veronesi U., Banfi A., Salvadori B., Luini A., Saccozzi R., Zucali R., Marubini E., Del Vecchio M., Boracchi P., Marchini S., Merson M., Sacchini V., Riboldi G. and Santoro G. Breast conservation is the treatment of choice in small breast cancer: Long-term results of a randomized trial. *European Journal of Cancer*, **26**, 668–670, 1990.

Veronesi U., Marubini E., Del Vecchio M., Manzari A., Greco M., Luini A., Merson M. and Saccozzi R. Local recurrences and distant metastases after conservative treatment of breast cancer patients. *Journal National Cancer Institute*, 1994 (submitted).

Verweij P. J. M. and Van Houwelingen J. C. Cross-validation in survival analysis. *Statistics in Medicine*, **12**, 2305–2314, 1993.

Walter S. D., Feinstein A. R. and Wells C. K. Coding ordinal independent variables in multiple regression analysis. *American Journal of Epidemiology*, **125**, 319–323, 1987.

Wei L. J. The adaptive biased coin design for sequential experiments. *Annals of Statistics*, **6**, 92–100, 1978.

Wei L. J. The accelerated failure time model: a useful alternative to the Cox regression model in survival analysis. *Statistics in Medicine*, **11**, 1871–1879, 1992.

Whitehead J. *Design and Analysis of Sequential Clinical Trials*. Ellis Horwood, Chichester, 1983.

Yates F. and Cochran, W. G. The analysis of groups of experiments. *Journal of Agricultural Science*, **28**, 556–580, 1938.

Yusuf S., Held P., Teo K. K. and Toretsky E.R. Selection of patients for randomized controlled trials: implications of wide or narrow eligibility criteria. *Statistics in Medicine*, **9**, 73–86, 1990.

Yusuf S. *et al.* Update of effects of calcium antagonists in myocardial infarction or angina in light of the second Danish verapamil infarction trials (DA VIT-II) and other recent studies. *American Journal of Cardiology*, **67**, 1295–1297, 1991.

Zelen M. The role of statistics in the design and evaluation of trials in cancer medicine. In: *Clinical Trials in Cancer Medicine*; Veronesi U. and Bonadonna G. (eds), Academic Press, Orlando, FL, 1985.

Author Index

Note: Figures and Tables are indicated, in the index, by *italic page numbers*.

Index compiled by Paul Nash

Subject Index

Note: Figures and Tables are indicated, in the index, by *italic page numbers*.

Index compiled by Paul Nash